Hydrology

MADAN MOHAN DAS

Formerly Professor
Civil Engineering Department
Assam Engineering College, Guwahati
An Emeritus Fellow of AICTE
Director of Technical Education
Government of Assam

MIMI DAS SAIKIA

Professor
Civil Engineering Department
Assam Down Town University
Guwahati

PHI Learning Private Limited

Delhi-110092
2023

₹ 795.00

HYDROLOGY
Madan Mohan Das and Mimi Das Saikia

ISBN-978-81-203-3707-7 (Print Book)
ISBN-978-93-5443-017-6 (e-Book)

Published by Asoke K. Ghosh, PHI Learning Private Limited, Rimjhim House, 111, Patparganj Industrial Estate, Delhi-110092 and Printed by Syndicate Binders, A-20, Hosiery Complex, Noida, Phase-II Extension, Noida-201305 (N.C.R. Delhi).

Hydrology

To

Teachers and Students

Contents

Preface

Water is one of our most important natural resources. Without it, there would be no life on earth. The supply of water available for our use is limited by nature. Hydrology, the study of water, deals with the occurrence, distribution, movement and properties of the water on the earth. It also studies the relationship between the environment and each phase of the hydrologic cycle. The central concept of science of hydrology is the hydrologic cycle or water cycle. The knowledge of hydrology is the prerequisite for the development and evaluation of water resources engineering. The advent of computer and application of probability and statistical theory have revolutionized hydrology. Therefore, this field of water science is a subject of interest for hydrologists, hydrometeorologists and scientists from the research point of view.

Hydrology is a core paper in the curriculum of civil engineering both at undergraduate and postgraduate levels. This book on hydrology is organized in sixteen chapter which covers concept of hydrologic cycle, hydrometerology, precipitation, infiltration, evapotranspiration, runoff, hydrographs, discharge measurement, estimation of flood, flood disaster mitigation, river engineering and training, hydrologic and hydraulic flood routing, groundwater hydrology and hydrology of basin management. An appendix on Introduction to statistics and probability is also included at the end of the chapters. Different methods of numerical solutions of unsteady flood routing are given, which will be useful for the postgraduate students to solve numerical solutions with the help of computer. Solved examples are included in most of the chapters to reinforce understanding of the theory. Few unsolved problems are also included to comprehend the subject with ease. References are provided in each and every chapter.

This book is the outcome of our wide experience in teaching and research. We acknowledge the support of ex-students, friends and family members while writing this book. Our sincere thanks and appreciation are due to late Professor D.I.H. Barr, University of Strathclyde, Glasgow who encouraged us to involve in academic activities.

Madan Mohan Das
Mimi Das Saikia

Chapter 1
Introduction

1.1 INTRODUCTION

Hydrology is a science of occurrence, movement and circulation of water on or near the surface of earth. This branch of water science is concerned with water in rainfall, snowfalls, streams, lakes, reservoirs, ground water, snow and ice. It is an interdisciplinary subject in the field of fluid mechanics, meteorology, geology, geography, chemistry, physics, statistics, operation research and agriculture. It is a subject of great importance for human beings and their environments. Practical importance and application of hydrology are found in design and operation of hydrologic structure, hydropower generation, flood control, control of erosion and sediment, pollution and salinity abatement, irrigation, navigation, recreational use of water, fish and wildlife protection. In view of its tremendous practical importance and utility, the United Nations proclaimed the period 1965 to 1974 as the International Hydrological Decade (IHD) during which intensive progress in hydrologic education and research, development of analytical techniques and information on hydrological data on global basis was made in research institutions, technical institutions and universities.

1.2 GLOBAL WATER BUDGET

On the surface area of earth only 29% is occupied by land and the remaining 71% is covered by seas and oceans. These seas and oceans hold 97% of earth's total water while 2% is kept frozen in ice caps. The very deep ground water accounts for 0.31%. Thus, 99.31% of water on earth is of no practical use to the people. The only remaining 0.69% represents the fresh water resource with which the man has to deal. At any instant rivers and lakes hold only 3% of this fresh water, i.e., 0.0093% of the total water. It appears quite surprising that this most important water resource of the human beings which is in the order of only 0.0093% of total earth's water, sometimes becomes the most bitter enemy of the people in the form of flood and erosion to bring about disaster and devastation that causes great destructions, catastrophies to the society if it is not properly managed and controlled. On the other hand, this gift of water has tremendous potential to transform the society to the path of progress if this resource is systematically managed and exploited.

1.3 HYDROLOGIC CYCLE

Figure 1.1 is the schematic diagram of hydrologic cycle which is the central concept of the science of hydrology, the study of earth's water. Due to solar radiation, water from ocean, river, lakes or any other body on earth surface evaporates in gaseous form. This water vapour moves towards the sky and forms the clouds. A part of water called **transpiration** from plants on the earth surface also joins the evaporation to form the clouds. These clouds when condense fall back to ocean, river, lakes, earth surface as rain, snow, hail, sleet, etc. A part of this precipitation may evaporate in the process of falling; another part is intercepted by leaves of trees and vegetation called **interception** which may also evaporate. The major portion of precipitation falls on earth's surface. A part of this portion flows as runoff or overland flow towards depressions like river, lakes, etc. A portion infiltrates, meets the ground water storage and becomes seepage towards depression below the ground. Eventually overland or runoff goes to the outlet, i.e., river which eventually flows to the seas or oceans. Again evaporation and transpiration take place to form clouds from which precipitation again occurs. Thus, this cycle of water on the earth continues which is called **hydrologic cycle**.

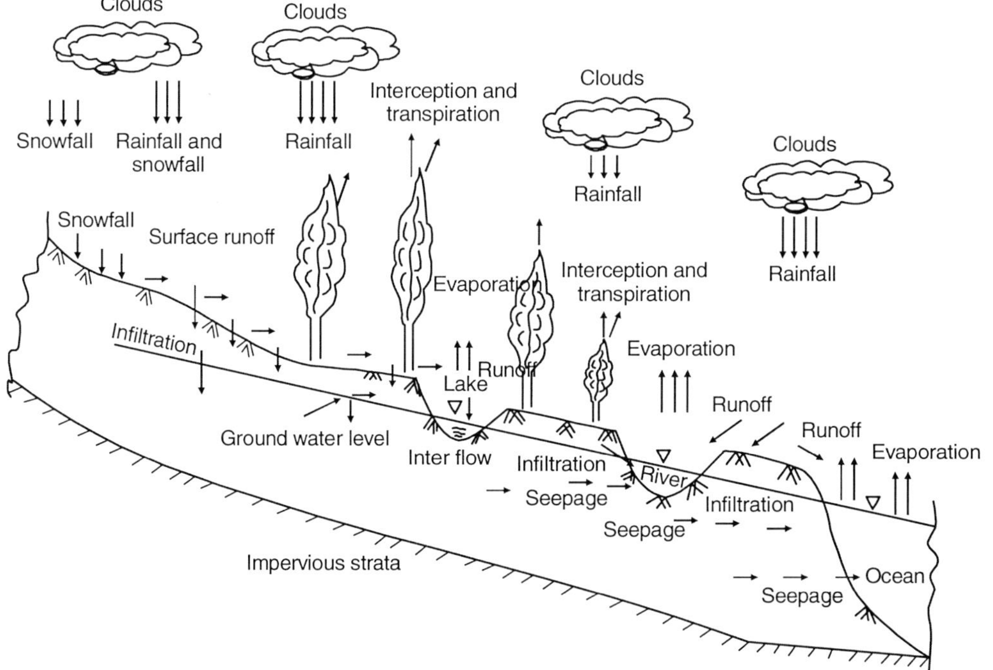

Figure 1.1 Hydrologic cycle: A schematic representation.

1.4 SYSTEM REPRESENTATION OF HYDROLOGIC CYCLE

Hydrologic cycle may also be represented as a system. A system is a set of connected parts that forms a whole. This hydrologic cycle is divided into three sub systems:

(i) The atmospheric water system, which contains the process of precipitation, evaporation, transpiration and interception. These components occur in the *atmospheric arc* of the globe.

(ii) The surface water system which contains processes of overland flow and surface runoff in *hydrosphere arc* of the globe.

(iii) The subsurface water system, which contains the processes of infiltration, seepage, ground water, recharge and ground water flow towards rivers, seas and ocean in the *lithosphere arc* of the globe.

A block diagram representing the global hydrologic cycle is shown in Figure 1.2.

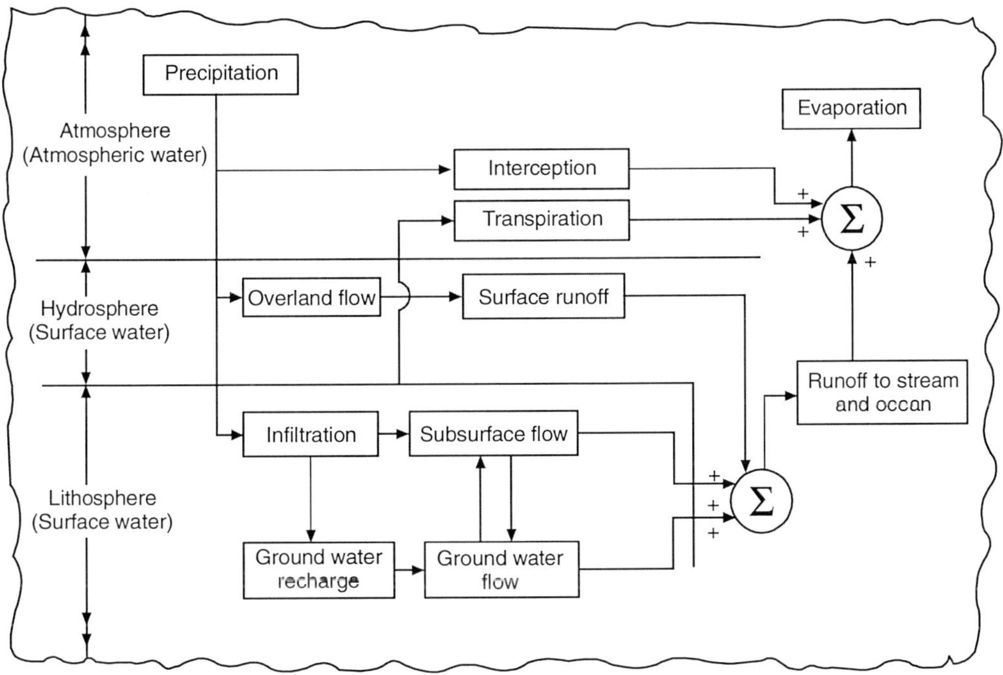

Figure 1.2 Block diagram of global hydrologic cycle.

1.5 HISTORICAL DEVELOPMENT OF HYDROLOGY

Development of hydrology is as old as human civilization. Irrigation canal, aqueducts have been found in prehistoric ruins in Egypt, Mesopotamia, Rome and India. From ancient times, the speculations regarding water circulation in hydrologic cycle by poet Homer, Philosophers like Plato, Aristotle and many others were not scientific. However, the Greek Philosopher in 400 BC believed that water is lifted to the atmosphere from sea by the sun from which it comes as rain on the earth's surface. Rain gauges were used in China, India around 1200 B.C. and 200 B.C. respectively. Leonardo da Vinci worked on velocity distribution in stream cross-section in fifteenth century. French scientist Bernard Palissy showed that rivers and streams originate from rainfall and rivers feed the oceans. Another French naturalist Pierre Perrault in the seventh century measured runoff and found it to be a fraction of rainfall. He also explained that remaining part of rainfall is lost by evaporation, transpiration etc. Hydraulic measurement and period of experimentation began in eighteenth century. Rainfall measurement instruments like tipping bucket rain gauge, velocity measurement instruments like Price current meter and Pitot tube were developed. In nineteenth century, development was rather quick. Principle of evaporation by Dalton[1], Hagen Poiseulle's capillary flow, Mulvaney's rational method of determination of peak flood flows, Darcy's[2] law in porous media, Ripple's[3] mass curve to determine reservoir capacity, ground water flow by Dupuit,[4] Thiem[5] are worth to be mentioned.

The contributions of the 20th century are unit hydrograph theory by Sherman,[6] Hazen's[7] introduction of flood of frequency analysis, infiltration model by Horton,[8] Green and Ampt,[9] Kostiakov,[10] Philips,[11] Overton, Akan, Fok and Hansen, Gumbel[12] extreme value distribution of flood estimation and by others. With the advent of computer, complex hydrologic analysis becomes possible. Details of history have been presented by Biswas.[13]

1.6 CONCLUSION

This chapter is presented to define hydrology, its importance in different fields, global water budget, hydrologic cycle and a brief historical development. In the subsequent chapters, details of hydrometeorology, precipitation, runoff, different hydrograph analysis, flood routing and flood disaster management, ground water hydrology, reservation sedimentation and control, watershed management and soil erosion will be discussed.

[1] Dalton, J., "Experimental essays on the constitution of mixed gases; on the force of steam or vapour from water and other liquids both in Torricellian vacuum and in air; on evaporation; and on expansion of gases by Heat," *Mem. Proc. Manch. Lit. Phil. Soc.,* Vol. 5, pp. 535–602, 1802.

[2] Darcy, H., *Les Fontaines Publiques de La Ville de Dijon.*, Dalmont, Paris, 1856.

[3] Ripple, W., "The capacity of storage–reservoir for water supply," *Minutes of Proceedings*, Institution of Civil Engineers, London, Vol. 71, pp. 270–278, 1883.

[4] Dupuit, A.J.E.J., *Eludes theoriques sur la mouvement des eahx decouverts et a travers lesterrains permeables,* 2nd ed., Dunod, Paris, 1863.

[5] Thiem, G., *Hydrologische Methoden*, Gebhardt Leipzig, 1906.

[6] Sherman, L.K., "Streamflow from rainfall by unit hydrograph method", *Engg. News*, Vol. 108, pp. 501–505, 1932.

[7] Hazen, A., *Flood Flows*, Wiley, New York, 1930.

[8] Horton, R.E., "The role of infiltration in hydrologic cycle", *Trans. Am. Geophysicas Union*, Vol. 14, pp. 446–460, 1933.

[9] Green, W.H. and Ampt, G., "Surface of soil physics, Part I, the flow of air and water through soil", *Jour. Agri. Science*, Vol. 4, pp. 1–24, 1911.

[10] Kostiakov, A.N., "On the dynamics of the coefficient of water percolation in soil and on the necessity for studying it from dynamic point of view for purposes of amelioration, *Trans. 6th Comm. Inter'l Soil Science*, Russion, Part A, pp. 17–21, 1982.

[11] Philips, J.R., "The theory of infiltration: The infiltration equation and its solution", *Soil Science*, Vol. 83, pp. 435–448, 1957.

[12] Gumbel, E.J., *Statistics of Extremes*, Columbia University Press, New York, 1958.

[13] Biswas, A.K., *History of Hydrology*, North Holland Publishing Co., Amsterdam, 1992.

EXERCISES

1.1 Explain with the help of sketch the hydrologic cycle in nature indicating its various phases. Also give an account of global water budget.

1.2 What is the system representation of a hydrologic cycle? Draw a block diagram of a global hydrologic cycle showing different parts. Also explain each part.

1.3 Present a brief introduction and historical developments of hydrology.

1.4 Choose the correct statement in the following:
The central concept of hydrology is:
(i) Occurrence of floods and droughts
(ii) Consumptive use of crops and water requirement of crops
(iii) Occurrence of precipitation and evaporation only
(iv) The hydrologic cycle [*Answer:* (iv)]

SUGGESTED FURTHER READINGS

Chow, V.T. (Ed.), *Handbook of Applied Hydrology*, McGraw-Hill, New York, 1964.

Dingman, S.L., Physical Hydrology, 2nd ed., Prentice Hall, Upper Saddle River, N.J. 2002.

Eagleson, P.S., *Dynamic Hydrology*, McGraw-Hill, New York, 1971.

Gray, D.M. (Ed.), *Principles of Hydrology*, Water Information Centre, Syosset, N.Y., 1970.

Linsley, K.L., Kohler, M.A. and Paulhus, J.L.H., *Hydrology for Engineers*, 3rd ed., McGraw-Hill International, New York, 1982.

Maidment, D.R. (Ed.), *Handbook of Hydrology*, McGraw-Hill, New York, 1993.

Nemec, J., *Engineering Hydrology*, Tata McGraw-Hill, New Delhi, 1973.

Raudkivi, A.D., *Hydrology*, Pergamon Press, Oxford, 1979.

Sing, V.P., *Elementary Hydrology*, Prentice Hall Inc., Upper Saddle River, N.J. 1992.

Viessman, W. and Lewis, G.L., *Introduction to Hydrology*, 4th ed., Harper Collins, New York, 1996.

Ward, A.D. and Elliot, W.J., *Environmental Hydrology*, Lewis Publishers, New York, 1995.

Wilson, E.M., *Engineering Hydrology*, Macmillan, ELBS, London, 1969.

Chapter 2
Hydrometeorology

2.1 INTRODUCTION

Hydrometeorology may be defined as the science of atmospheric phenomena. It includes the study of moisture in the atmosphere including its forms and precipitation, and thus overlaps a portion of field of hydrology. Thus, hydrometeorology is a branch of hydrology, which deals with water in the atmosphere. The recent and broad definition is: *It is the part of hydrology which concerns with atmospheric and surface water.* Breakthroughs in hydrometeorology were achieved in the latter half of the twentieth century. The works of Shaw[1], Brunt[2], Bruce and Clark[3], Chow[4] and Hoes[5] are worth to be mentioned.

2.2 HYDROMETEOROLOGICAL INSTRUMENTS

Hydrometeorological instruments are commonly used to measure hydrometeorological data: Their uses are discussed as follows:

Maximum and minimum thermometer: It indicates the lowest and highest temperature between two observations. Both the thermometers are placed in horizontal position so that the index does not move by its own weight.

Soil thermometer: It is used to measure soil temperature. The mercury bulb is placed at the end of a long bent capillary tube which is placed in the ground.

Psychrometer: A stationary psychrometer is used to measure humidity (i.e., air moisture) at hydrometeorological station. It consists of two thermometers with wet and dry bulbs. The **wet bulb thermometer**, has its mercury bulb wrapped in a wick. The other end of the wick is submerged in a container of distilled water which ensures continuous moisture supply to the wet thermometer through surface tension effect. The thermometers are ventilated by the use of a fan. Because of cooling effect of evaporation the wet bulb thermometer reads lower than the dry

[1] Shaw, W., *Manual of Meteorology*, 2nd ed., Cambridge University Press, London, 1945.
[2] Brunt, D., *Physical and Dynamical Meteorology*, Cambridge University Press, London, 1952.
[3] Bruce, J.P. and Clark, R.H., *Introduction to Hydrometeorology,* Pergamon, New York, 1966.
[4] Chow, V.T. (Ed)., *Handbook of Applied Hydrology*, McGraw-Hill, New York, 1904.
[5] Hess, S.L., *Introduction to Theoretical Meteorology*, McGraw-Hill Inc. New York, 1959.

thermometer, and the difference of these two is called **wet bulb depression**. From the difference of temperature of the two thermometers, air-moisture or humidity is calculated using psychometric table. Hair hygrometer, thermo hygrograph, dew gauge or condensation hygrometer are also used to measure humidity.

Wind vanes and anemometer: Wind velocity is a component required to study hydrometeorology. This velocity is measured by wind vanes and cup anemometer. Most common and accurate method of measurement is done by cup anemometer. It consists of 3 or 4 Robinson cups fixed in a horizontal plane to a sleeve which rotate freely about a vertical axis. The rotation of the axis is transmitted to the gauge under the cups indicating the number of rotation per minute and hence the wind's speed.

Evaporimeters: They are the instruments to measure the evaporation, i.e., water that escapes from land and water bodies into the atmosphere. The commonly used evaporimeters are Piche evaporimeters, Colorado sunken pan, US geological survey floating pan, ISI standard pan and the US weather bureau class A pan. The most commonly used evaporimeter in India is the US weather bureau class A pan evaporimeter.

Rain gauge: There are different types of rain gauges used to measure rainfall. They are Symon's non-recording rain gauge, natural siphon rain gauge, tipping bucket rain gauge and weighing bucket rain gauge. The most commonly used rain gauge in India is Symon's gauge, recommended By Indian Meteorological Department (IMD) conforming to IS 5225, 1992.

Fortin's barometer: Pressure exerted by atmosphere is a very important parameter of hydrometeorology. It is measured by Fortin's barometer. It consists of an inverted U-tube filled with mercury kept in a cistern. It is also measured by an aneroid barometer, which consists of a diaphragm deflected by the atmospheric pressure. The instruments which records the variation of atmospheric pressure continuously with time is called **barograph**. When points of equal atmospheric pressure are joined, the curve obtained is called **isobar**.

Pyranometer and radiometer: These are instruments to measure the radiation. Pyranometer (previously called **pyrhelimeter**) is used to measure the short wave solar radiation, both direct and diffused, reaching the earth's surface. It normally consists of flat circular plate mounted horizontally within a lime glass tube. The plate is divided into three portions, i.e., central white spot, middle black ring and outer white ring. The black ring absorbs radiation while the white ring reflects it. A temperature difference is setup between white and black rings which is measured by mechanism called **thermopile** that is calibrated for measuring the short wave solar radiation. The reflected short wave solar radiation from the white surface is measured by the same way of calibrated thermopile by facing a downward pyrometer. The lime glass filters out the long wave radiation.

The total incoming and outgoing radiation is also measured in the same way without filtering effect of the lime glass. This device is called **radiometer**.

Sunshine recorder: It is an instrument to measure the bright sunshine in a day. It consists of a hemispherical glass dome underneath of which a chart wrapped over a clock driven drum is placed. As the sun rays are diverted towards the centre of the dome, the intense heat produced by the sunshine makes mark by burning the chart. If the sky is clouded, such marks will be absent. The length of the chart, which is burnt, will indicate the duration of daylight sunshine. This data of sunshine hours is required to estimate potential evaporation and short-wave solar radiation in some empirical equations.

2.3 VERTICAL STRUCTURES OF ATMOSPHERE

With the aid of customary meteorological instruments, the state of atmosphere below 30.77 km has fairly been explored. The exploration of static properties of atmosphere surrounding the earth is of considerable importance to meteorologists and aeronautical engineers. In recent years, rockets and satellites have been added for direct observation. Therefore, it has become possible to outline the predominant features of atmosphere upto 1666 km (i.e., 1000 miles). On the whole, atmosphere may be distinguished by five concentric cells or spheres. Physical structure of atmosphere showing concentric cells is shown in Figure 2.1.

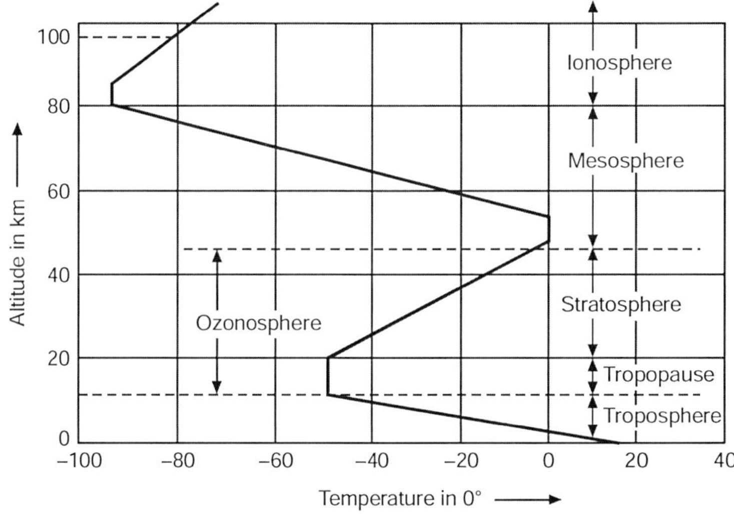

Figure 2.1 Physical stucture of atmosphere.

Troposphere: It is the lowest layer above the ground upto an elevation about 8 km. It contains above 75% of the mass and all moisture and dust of atmosphere. It has relatively steep lapse rate, i.e., change of temperature with altitude. If T is temperature, h is the altitude, then

$$\text{Lapse rate } \frac{dT}{dh} = -\text{ve}$$

The degree of static stability is low and air is often unstable.

Tropopause: It is the layer above the troposphere. Its height varies from 8 km at poles and 10 km over the equator. Temperature is almost constant in this layer.

Stratosphere: It is the third layer above tropopause varying its altitude from 10 km to approximate 45 km. Here the temperature increases, i.e., lapse rate is positive. Stratification is unstable. It contains very little moisture and dust except the dust which may be brought by major volcanic eruption.

Mesosphere: The temperature is initially high and decreases with increase of altitude. Initial high temperature is due to selective absorption of ultraviolet radiation from the sun.

Ionosphere: This major layer is normally at about above 80 km above earth. Pressure is low at this zone. Ionization processes are alive and as a result ordinary radio waves are reflected from the layer. Ionosphere is divided into number of layers with different electrical properties and varying intensities. The layer merges gradually into the outermost shell, i.e., exosphere. Ionosphere is called **thermosphere**.

The two layers *tropopause* and *stratosphere* together called **ozonosphere**.

One must be reminded that gravitational acceleration was considered to be constant. Actually, it varies with square of distance from the centre of the earth. This correction appears to be insignificant (less than 5%) within the stratosphere.

2.4 GENERAL ATMOSPHERIC CIRCULATION

The atmosphere is considered as a turbulent fluid subjected to strong thermal influences and moving over a rough rotating spherical earth. The fluid above earth surface is highly turbulent and compressible. No satisfactory theoretical and experimental technique exists for the study of such fluid in motion. Meteorologists do not have yet any accepted general circulation of the atmosphere. The following discussion on general circulation is based on Rossby's concept of general circulation.

2.4.1 Thermal Circulation on Non-rotating Earth

The thermal circulation originates from the sun as a source of origin. About 40% of radiant energy of sun is reflected back from upper surface of the clouds. The earth absorbs the remaining 60% with small losses. This in turn increases the heat of the surface. More solar energy is received at the equator and hence heated air at the equator zone rises up and flows towards the poles. To balance this, air in the lower layers must move towards the equator. Figure 2.2 shows such thermal circulation.

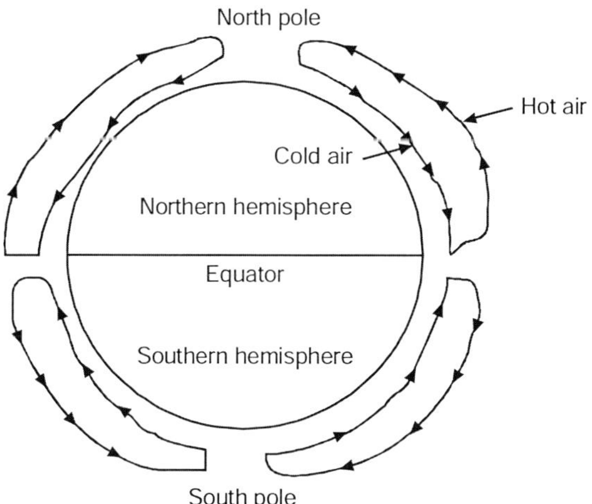

Figure 2.2 Simple thermal circulation on non-rotating earth.

But actual circulation that is shown in Figure 2.2 is different because of the following two factors:

1. The earth's rotation
2. The distribution of oceans and continents

2.4.2 The Effect of Earth's Rotation on Circulation

A parcel of air at rest relative to the earth's surface moves with a constant velocity in an orbit around the polar axis. If this parcel is forced northward in the northern hemisphere at a constant elevation, the radius of the orbit is reduced. From the principle of angular momentum eastward velocity increases. Figure 2.3 shows that air rising at the equator and moving north acquires an eastward component because of rotation of the earth. By this time, wind reaching 30° latitude losses sufficient heat so that it tends to subside. The subsiding air divides into two branches: one branch moving south develops northeast trade winds while the second part continues north and eastward in the middle latitudes. At polar cell, there is a high pressure and subsidiary air acquires a westward component as it moves southward. The details of the middle latitudes are not well established.

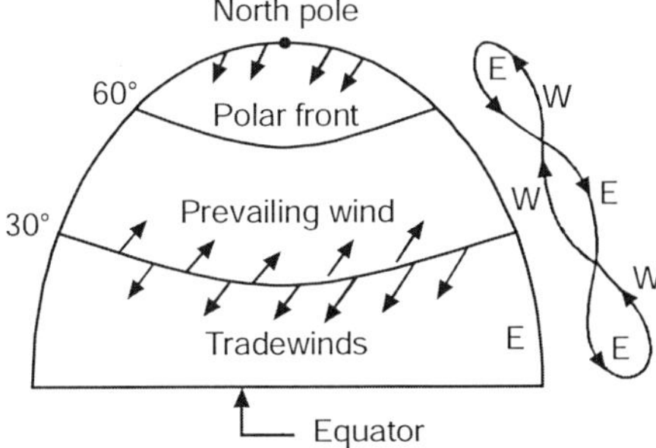

Figure 2.3 General circulation of northern hemisphere.

2.4.3 The Effect of Land and Water Distribution on Circulation

The distribution of land and sea mass causes the high and low pressure belts to break up into centres of high and low pressure. Thus, circulation is affected by

1. Mixing properties of water and land
2. Existence of barrier to air flow, i.e., resistance by earth's surface
3. Reflectivity of water surface and land
4. Lower specific heat of soil

Heat gains and losses are distributed relatively to a great depth in large water bodies by mixing while land is affected only on the surface. Consequently land surface temperature is far less equable than those of the surface of large water bodies. In winter, there is a tendency for cold dense air to accumulate over land masses and warm air over the oceans. In summer, situation is reversed.

2.5 WEATHER SYSTEM

Cold wind indicates dry weather and warm wind indicates stormy weather. Some of the terms and processes that are related to the weather system are discussed as follows:

Front: When a warm air mass meets a cold air mass, instead of their simply mixing, a definite surface discontinuity appears between them, which is called **front**. Cold air being heavier underlies warm air. If the cold air is advancing towards the warm air, the leading edge of the cold air mass is a cold front and nearly vertical in slope. If the warm air is advancing towards the cold air, the leading edge is a warm front which has a very flat slope, warm air flowing up and over the cold air.

Cyclone: Cyclone is formed in low-pressure zone of approximately circular area where wind flows towards the centre in anticlockwise direction in northern hemisphere and clockwise in southern hemisphere. It is accompanied by cloud precipitation and sometimes with violent destructive wind.

Anticyclone: Anticyclone is found in high-pressure zone with wind flowing spirally outward in clockwise direction in northern hemisphere and anticlockwise in southern hemisphere. It is accompanied by moderate wind and fair weather.

Tropical cyclone (or hurricane or typhoon): It forms at low altitude and in tropical region. It is accompanied by heavy rainfall in the area occupied by the cyclone. Tropical cyclone originates from open sea at 5° to 10° latitude and moves with a speed of 10–30 km/hr to higher latitude. It derives the energy from latent heat of condensation of ocean water vapour and increases in size as it moves on oceans. When it reaches the land, source of energy is cut off and the cyclone dissipates the energy in the form of heavy rainfall. Tropical cyclone causes heavy damage to life, properly in the form of flood on the path in which it moves on land.

Extratropical cyclone: It is the cyclone formed on location outside the tropical zone. It possesses a strong anticlockwise wind circulation in northern hemisphere. In this cycle, magnitude of velocity and precipitation is normally lower than tropical one, but duration and area covered are higher.

Tornado: Under certain conditions, activity at front may develop into strong localized convection currents. Such circulatory systems are called **tornadoes**. Tornadoes are usually of small diameter (less than 1.5 km), but their peak wind speeds may be several hundred km/hr and transitional velocity may go upto 100 km/hr. Although tornadoes are most destructive, their life span is very small that can be measured in minutes.

Convective weather: It includes showers, thunder-storms, etc. with appreciable up drift and down drift of air resulting from static instability.

Monsoon: Monsoons are wind system with an annual oscillation flowing from ocean to continent in summer and in reverse direction in winter. Monsoon is seasonal. During summer, continent is warmer than water in ocean. So wind flows from ocean to continent and this monsoon carries water to continent to cause heavy rain on land in summer. In winter time, water in ocean is warmer than on land; so cold wind blows to ocean from land. The winter monsoon wind is almost dry and hence poor rainfall occurs in winter.

2.6 CLOUDS

Cloud is formed due to the process of evaporation and condensation of water. For ordinary condition, average cloud element is a form of water droplet of size 0.01-mm diameter with an upward air movement of 0.154 m/min is sufficient to keep from falling.

Dry air coming in contact with water in oceans, rivers, lakes and damp ground vaporizes water. This air with water vapour is warmed by the contact of earth and hence it expands and rises. Water vapour rising with air causes fall in temperature and at a high altitude this water vapour condenses and they move with natural circulation. This is how clouds are formed in the sky. These clouds when condense further become heavy to remain in suspension and fall as precipitation on earth's surface.

Clouds are formed in infinite combination. These diverse formations cause the different types of clouds. They are classified as:

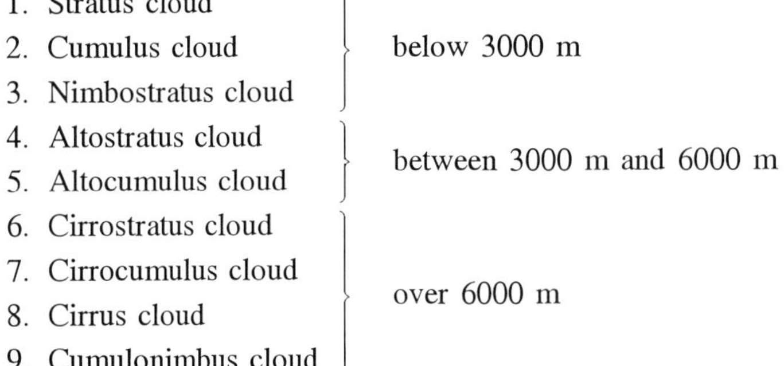

1. Stratus cloud
2. Cumulus cloud
3. Nimbostratus cloud

} below 3000 m

4. Altostratus cloud
5. Altocumulus cloud

} between 3000 m and 6000 m

6. Cirrostratus cloud
7. Cirrocumulus cloud
8. Cirrus cloud
9. Cumulonimbus cloud

} over 6000 m

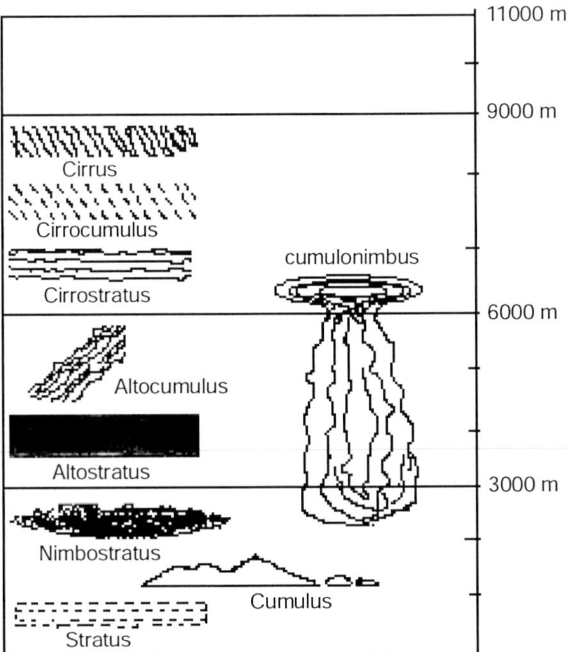

Figure 2.4 Different forms of clouds.

Stratus: It is of uniform layer. It does not touch the ground.

Cumulus: Cumulus clouds have flat bases, round tops and are usually white or grey.

Nimbostratus: It is dense, shapeless cloud from which precipitation usually occurs.

Altostratus: It is dense sheet of grey or bluish colour. They give precipitation of continuous type.

Altocumulus: Here cloud sheets are lower and flakes are larger, and often show light shadows.

Cirrostratus: It is thin white sheets of cloud.

Cirroculus: It consists of small white flakes arranged in a pattern resembling lambs' wool.

Cirrus: They consist of fine ice crystals and have a delicate silky appearance without shadow.

Cumulonimbus: They have a more rugged base and may rise up to very great height. The top of these clouds is usually bent over like a mushroom. They are very common in humid tropical region and are associated with thunderstorms.

2.7 CLOUD SEEDING

It is the process of artificial stimulation of precipitation by adding certain chemical to cloud in atmosphere.

Cloud is in colloidal stability under natural conditions and remains in suspension without falling. It can be colloidally unstable by adding dry ice, silver dioxide, frozen carbon dioxide or other chemical agents which act as freezing nuclei and produce ice crystals. Through diffusion and coalescence, they eventually grow into precipitation particles.

The seeding element such as dry ice is delivered into the cloud by aircrafts, balloons or rockets. Silver dioxide may be delivered into the cloud either by aircrafts or by ground based generator.

The success of cloud seeding with economic value is still uncertain. It is still in experimental stage. This method has not yet established whether overall gains would exceed the overall costs. Further redirecting the storm and inducing cloud to precipitate may reduce damage to property in certain areas, but it may aggravate the flood disaster in certain areas. It may also deprive certain areas from required rainfall for agriculture, industry, domestic water supply etc. Weather modifications by cloud seeding may need international regulation as its effort could extend far beyond the area to be precipitated. Thus, the cloud seeding is still in experimental and research stage. Various investigators like Dennis[6] worked on cloud seeding, Byers[7], Macon[8] and Fletcher[9] on physics of rainclouds etc.

---

[6] Dennis, A.S., *Weather Modification by Cloud Seeding*, Academic Press, New York, 1980.
[7] Byers, H.R., *Elements of Cloud Physics*, University of Chicago Press, Chicago, 1965.
[8] Mason, B.J., *Physics of Clouds*, 2nd ed., Oxford University Press, London, 1971.
[9] Fletcher, A.S., *Physics of Rain Clouds*, Cambridge University Press, London, 1962.

2.8 TERMINOLOGY OF HYDROMETEOROLOGY

Lapse rate: It is the rate of change of temperature with altitude in free atmosphere. The greatest variations in lapse rate are found in the layer of air just above the land surface. Mathematically, lapse rate = dT/dh (negative) since temperature T decreases with increase of altitude h.

Dry adiabatic lapse rate: It is the rate of change of temperature of unsaturated air resulting from expansion or contraction due to change in pressure with altitude without heat being added or removed. Expanding air will drop in temperature at the rate of 10° for each km of ascent.

Super adiabatic lapse rate: Near the surface, lapse rate is larger than the dry adiabatic lapse rate due to surface heating, and this lapse rate near the surface is called **super adiabatic lapse rate**. It is more than 17.55°C/km.

Saturation adiabatic lapse rate: When air saturated with water vapour is lifted adiabatically, it expands and cools dynamically. The cooling of the water vapour in the air causes condensation and results in the release of latent heat of vaporization. This heat serves to reduce the rate of cooling of rising parcel of air. This lapse rate is called **saturation adiabatic lapse rate** or **moist adiabatic lapse rate**. In the lower layers of atmosphere, the average value of saturated adiabatic lapse rate is 0.3°C/30.5 m.

Albedo: It is the ratio of amount of solar radiation reflected by the surface to the amount of incident solar radiation on it. It is expressed in percentage. It depends on angle of incidence and type of surface. Albedo of earth's surface is about 14%.

Solar radiation: Solar radiation is the main source of energy which determines the earth's weather and climate. Large part of solar radiation reaching the outer limit of atmosphere is scattered and absorbed in atmosphere and reflected from cloud and earth surfaces. For clear dry sky, half the radiation in blue range is scattered by air molecules, thus accounting for a blue sky cloud that reflects much incident radiation depending on the amount and type of cloud and their albedos. Only about half the incident radiation at outer limit of atmosphere reaches earth's surface. Part of it is reflected back to atmosphere and space depending on albedos of earth's surface. Other part is absorbed by earth's surface, and is used to warm adjacent air and surface substances. Short wave solar radiation on earth's surface mainly depends on number of sunshine hours in the sky in a day. Das[10] has shown that it also fully depends on elevation and latitude.

Latent heat: It is the amount of heat released or absorbed by the unit mass of substance without change in temperature when changing from one state to another.

Humidity: It denotes the moisture content in air at given temperature and expressed in percentage.

Relative humidity: It is the percentage ratio of amount of moisture in a given space to the amount of moisture the space would contain if saturated. Hence it is also the ratio of actual vapour pressure to the saturation vapour pressure, which is expressed as percentage, i.e.,

[10] Das, M.M., The Dependence of Solar and Atmospheric Radiation Formula on Elevation and Latitude, A thesis submitted to Cornell University for M.S. degree in Civil Engineering Department, Ithaca, New York, 1969.

$$f = 100\frac{e_v}{e_s}$$

where e_v is vapour pressure and e_s is the saturation vapour pressure.

Specific humidity: Specific humidity is the mass of water vapour per unit mass of moist air.

Absolute humidity: Absolute humidity is the mass of water vapour per unit volume of space.

Dew point: It is the temperature at which the air becomes saturated when cooled under constant pressure and with constant water vapour content. It is thus the temperature having a saturation vapour pressure e_s equal to the existing vapour pressure e.

Virtual temperature: It is the temperature of moist air at which dry air under pressure would have the equivalent density.

2.9 ROLE OF NUCLEI IN CONDENSATION PROCESS

Condensation of water into cloud droplets takes place on certain hygroscopic particles, which are commonly called **condensation nuclei**. With such nuclei, condensation could be initiated on certain large ions, but the relative humidity would then have to be several hundred per cent. In natural air, the degree of super saturation is usually a fraction of one per cent. The most active nuclei are particles of sea salt and such products of combustion are those which contain sulphurous and nitrous acids. The customary nuclei are generally less than 1 micron in diameter, but on occasion a few giant nuclei (5 μ or so) may be present. The number of salt nuclei varies from 10 to 1000 per cubic centimetre. The combination nuclei are generally small and their number varies greatly with industrial activity.

2.9.1 Solute and Curvature Effect

It is known from physical chemistry that the equilibrium vapour pressure is reduced when salt and similar substances are dissolved in water. The reduction is expressed by

$$\frac{e_1}{e_s} = 1 - CM$$

where e_s is the saturation vapour pressure, e_1 is the equilibrium pressure of the molar aqueous salt solution containing M moles of solute per litre and C is the temperature variable factor which depends on the particular substance.

Now, if vapour condenses on a nucleus, M decreases so thus solute effect. When the radius has increased to about 2 microns, the solute effect is negligible.

From the theory of surface tension, it is known that equilibrium vapour pressure e_2 over a curved droplet is larger than over a plain water surface. The relation is well-represented by an approximate formula:

$$\frac{e_2}{e_s} = \left(\frac{1-k}{r}\right)^{-1}$$

where r is the radius of droplet and k is constant at any given temperature. k is a very small quantity and curvature effect is generally negligible for values of r greater than 2 microns.

The solute and curvature effect only exist when cloud droplets have radii less than about 2 microns. In natural air, the process follows a path as shown in Figure 2.5, which is a compromise between effects of solute and curvature.

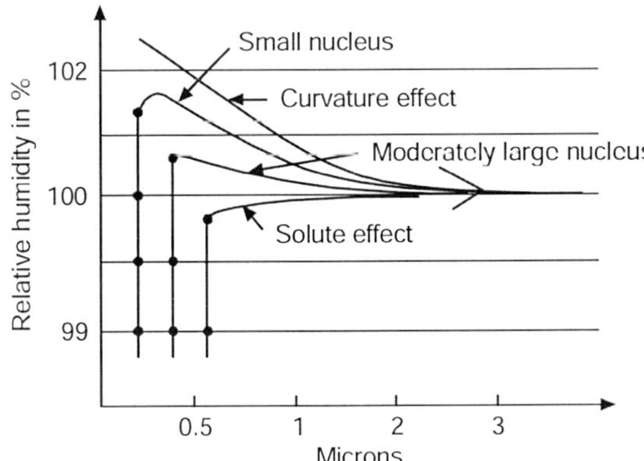

Figure 2.5 Effects of solute and curvature: large nucleus, small nucleus on relative humidity vs microns.

It is seen that larger the original nucleus, smaller is the supersaturation pressure needed for droplet to grow while the 100 per cent level of relative humidity is approached. Computation shows that under normal conditions, it takes for a nucleus a few seconds to grow droplets of 10 microns, 3 hours to grow 1000 microns and about 24 hours to become a small droplet (3 mm diameter). It follows that condensation is incapable of producing raindrops though it well produces oversize cloud droplets.

2.10 PROCESSES OF PRECIPITATION

There are two processes that cause the clouds to release the precipitation:

1. Coalescence process
2. Ice-crystal process

Coalescence process: The droplet increases in size by coalescence due to collision of cloud and precipitation element. A large size droplet moves faster and catches the smalls one in collision thereby increases in size.

Coalescence occurs through electrostatic attraction amongst the oppositely charged cloud particles. Difference of vapour pressure results from difference of temperature may cause net transfer of water from warmer to colder air leading coalescence; the motion due to difference in sizes brings together the elements of different size, temperature and electric charge.

A water drop in the air is exposed to the force of gravity and frictional drag. After a relatively short while, the two forces balance and drop has then acquired a maximum terminal velocity. This velocity is proportional to the square of the radius with the result that large drops fall faster and may collect small drop on their downward movement. Hence the large drops will grow as a result of (i) direct capture of small drops on their forward side (ii) wake capture of small drops of equal size.

As the drop grows to a diameter of 7 mm, the fall velocity increases to about 10 m/sec. At such high speed the drop flattens out and breaks into size of small rain or drizzle. Each of these falls faster than the cloud droplets with the result that the capture processes are repeated.

Ice-crystal process: Water droplets may exist in clouds at subfreezing temperatures down to –40°C. In this range, solidification occurs in connection with certain particles which are called **freezing nuclei**. The most prominent freezing nuclei are clay minerals and organic and common salts. When ice elements formed in an under cooled cloud, an imbalance is created. The equilibrium vapour pressure over the water drops is higher than over the ice element. As a result, the water tends to evaporate and condense on the ice, thus bringing about a growth of some cloud elements and shrinking of others. The uneven size distribution resulting there favours further growth through collision and coalescence. In warm clouds which do not reach upto the layers with subfreezing temperature, giant salt nuclei serve to create some unusually large drops which through fragmentation may develop the precipitation process.

2.11 NECESSARY CONDITIONS TO PRODUCE RAINFALL

There are following four conditions:

Mechanism of cooling: When air ascends to upper level, pressure reduction takes place accompanied by lowering the temperature to account for all precipitation.

Cooling lowers the capacity of a given volume to contain water vapours. In thermodynamics, this rate of containing water vapour is expressed in terms of vertical speed of the air and the mean temperature of a saturated layer of air. If the rate of precipitation is equal to the rate of production of moisture excess, heavy rainfall occurs. Alternative partial explanation of high rate of rainfall rates are:

1. Horizontal convergence of falling raindrops causes the rainfall rate to be greater than the rate of production of moisture excess.
2. Large amount of liquid water supported temporarily by a vertical current, fall rapidly when current slackens or ends.

Mechanism of condensation: Condensation in atmosphere takes place on *Hygroscopic nuclei.* Small particles of substances that have an affinity for water even air is not saturated. Products of combustion, oxides of nitrogen and salt crystals, sodium chloride, sulphur trioxide are the principal types of condensation nuclei. In ordinary air, each drop forms around a small particle of foreign substance, known as **condensation nucleus**, much smaller than dust particle. The growth of water droplets from energy consideration shows that saturation vapour pressure of surface increases as the curvature increases. Therefore, condensation occurs on a small droplet whose surface has a large curvature. As water is attracted to the particle, curvature of the droplet decreases; so less supersaturation is required for condensation. The effect of curvature and hygroscopicity becomes negligible after the droplet has attained a radius of 10^{-4} cm. Therefore, growth of droplet must be studied by consideration of the process of molecular diffusion, by which water passes from the vapour state in the surrounding air to the liquid state as a part of the drop.

Mechanism of droplet growth: Clouds are regarded as colloidal like suspensions. A tendency for droplet to remain small and therefore, not to fall, is called **colloidal stability**. On the other hand, if the droplet tends to coalesce, thereby becoming large enough to overcome the frictional resistance to falling, the cloud is said to be *colloidally unstable*. The most effective process for the coalescence of cloud droplets to form rain drops are: (i) difference in speed between large droplets and small droplets (ii) the co-existence of ice crystals and water droplet.

Mechanism of accumulation of moisture: Simple continuity considerations demand that there must be a good amount of moisture present in order that the evaporation losses between cloud and ground are over compensated if there is to be appreciable rain. The heavy rainfall amounts exceed by far the amount of water vapour in a vertical column at the beginning of rainfall over the rainfall area. For this reason, there must be a large net horizontal inflow of water vapour into the column above the rain area. This process is called **convergence** which is defined as *the net horizontal influx of air per unit area.*

2.12 CONCLUSION

There are three broad problems of hydrometeorology:

1. Measurement, recording and collection of data
2. Analysis of data to develop and expand the fundamental theory
3. Application of these theories and data to various practical problems

As meteorology is a portion of hydrology, hydrologist must have an understanding of the meteorological processes that determine the regional climate. The effect of wind, temperature, radiation, humidity, evaporation, etc. determine the hydro-meteorological features of a particular area. Hydrometeorology, a part of hydrology, is a new subject. Therefore, broad line problems between hydrology and meteorology are to be studied. Thus, the scope of this field is plenty and extensive works will be developed in the near future in this new branch of science.

EXERCISES

2.1 What is hydrometeorology? Describe the different hydrometeorological instruments used for measuring different atmospheric parameters.

2.2 Draw a sketch to describe different vertical structures of atmosphere. Give their altitude and explain how they behave with temperature.

2.3 Describe the following in brief:
- (i) General atmospheric circulation
- (ii) Thermal circulation on non-rotating earth
- (iii) The effect of earth's rotation on circulation
- (iv) The effect of land and water distribution on circulation
- (v) The effect of land and water distribution on circulation

2.4 Describe weather system and the different terms that are connected to it.

2.5 How clouds are formed? Describe with sketch the different types of clouds. What is cloud seeding?

2.6 Describe four different lapse rates, albedo, solar radiation, latent heat, specific and relative humidity, dew point and virtual humidity.

2.7 What is the role of nuclei in condensation process? What are the solute and curvature effects on nuclei with humidity?

2.8 What are the two processes that cause precipitation? Describe the necessary conditions to produce rainfall. Describe the scope of hydrometeorology.

SUGGESTED FURTHER READINGS

Ahrens, C.D., *Meteorology Today: An Introduction to Weather, Climate and the Environment*, 6th ed., Brooks and Cole Publishers, Pacific Grove, CA, 2000.

Brutsaert, W., *Evaporation into the Atmosphere: Theory, History and Applications*, D. Reidel Publishers Co., Dordrecht, Holland, 1982.

Chow, V.T., Maidment, D.R. and Mays, L.W., *Applied Hydrology*, McGraw-Hill, New York, 1988.

Gringorten, I.I., Fitting Meteorology Extremes by Various Distributions, *Q.J.R. Meteorol. Soc.*, vol. 88, pp. 170–176, 1962.

Hidore, J.J. and Oliver, J.E., *Climatology: An Atmospheric Science*, Macmillan, New York, 1993.

Linsley, R.L, Kohler, M.A. and Paulhus, J.L.H., *Hydrology for Engineers*, 3rd ed., McGraw-Hill, New York, 1982.

Maidment, D.R. (Ed.), *Handbook of Hydrology*, McGraw-Hill, New York, 1993.

Raghunath, H.M., *Hydrology: Principle-Analysis-Design*, New Age International, New Delhi, 1985.

Reddy, P.J., *A Text Book of Hydrology*, Lakshmi Publications, New Delhi, 1986.

Smithsonian Meteorological Tables Published by Smithsonian Institution 1918 & 1951.

Subramanya, K., *Engineering Hydrology*, Tata McGraw-Hill, New Delhi, 1984.

Sutton, O.G., *Micrometeorology*, McGraw-Hill Inc., New York, 1953.

Varshney, R.S., *Engineering Hydrology*, Nem Channel and Bros., Roorkee, 1979.

Weisner, C.J., *Hydrometeorology*, Chapman and Hall, London, 1979.

Williams, J., *The Weather Book*, 2nd ed., Random House, New York, 1997.

World Meteorological Organization, *Guide to Hydrological Practice*, 3rd ed., WHO, No. 168, Geneva, 1974.

World Meteorology Organization, *Guide to Hydrometerological Practices*, 3rd ed., WHO, No. 168, Tech. Paper 82, Geneva, 1974.

Chapter 3

Precipitation

3.1 INTRODUCTION

Water in the form of rainfall, snowfall, hail, frost and dew when reaches the earth from atmosphere is called **precipitation**. In India, major portion of precipitation is due to rainfall and small portion of snowfall in the Himalayan region. Some of the points relating to precipitation have been already discussed in Chapter 2. Other important points of precipitation like its forms, types and measurement by different rain gauges, adequacy of rain gauge stations, interpolation and missing precipitation data by double mass curve, intensity duration and depth area duration curves, and methods to determine average rainfall will be discussed in this chapter.

3.2 FORMS OF PRECIPITATION

Some of the common forms of precipitation are discussed as follows:

Rainfall: It is the main form of precipitation in India. When the size of water drop is larger than 0.5 mm, it is called **rainfall**. Rainfall may be light, moderate and heavy if intensities of rainfall are 2.5 mm/hr, 2.5 mm to 7.5 mm/hr and more than 7.5 mm/hr respectively.

Snow: Snow is composed of ice crystals, chiefly in complex hexagonal form and formed into a ball of mass into snow flakes which may reach several cm in diameter. New fresh snow has an initial density from 0.06 to 0.15 g/cm^3 and the average value is assumed to be 0.1 g/cm^3. In India, snow falls in the Himalayan region.

Drizzle: It is a fine sprinkle of numerous water droplets of size 0.5 mm and intensity is less than 1 mm/hr.

Glaze: It is the ice coating formed when comes in contact with cold ground at 0°C. Water drops freeze to form an ice coating.

Sleet: They are frozen raindrops of transparent grains, which form when rain falls through sub freezing temperature.

Hail: It is the precipitation in the form of irregular lumps of ice of size more than 8 mm. Hails occur in violent thunderstorm in which vertical currents are very strong. Hailstones may be spherical, conical or irregular shape.

Rime: It is white opaque deposit of ice granules more or less separated by trapped air and formed by rapid freezing of super cooled water drops impinging on exposed objects. Specific gravity may be as low as 0.2 to 0.3.

Dew: Dew forms directly by condensation on the ground mainly during night when the surface has been cooled by outgoing radiation.

3.3 TYPES OF PRECIPITATION

There are following five types of precipitation:

Frontal precipitation: This precipitation occurs due to conflict between two air masses. When two air masses with contrasting temperatures and densities clash with each other, condensation and precipitation occur at the surface of contact. The precipitation is called **frontal** as the surface of the contact from which it precipitates is called **front**. When two air masses, both cold and warm, are downwards in a low-pressure area, the front is stationary. Precipitation from this condition is called **non-frontal**.

Convective precipitation: This type results from the upward movement of air which is warmer than its surrounding. As the density of warm air is less, it has the tendency to continue its rising until it reaches a level where it has the temperature of its environment. The temperature contrast causes convection and cools adiabatically to form the shape of the cloud like cauliflower which finally bursts into thunderstorm. It is accompanied by destructive winds, called **tornadoes**.

Orographic precipitation: This precipitation caused by orographic lifting of air over mountain barrier from a large water body. This air from water body carries a lot of moisture and the high mountain barrier causes the air to lift up. At certain height it condenses and precipitates in the windward slope. Generally rainfall is low.

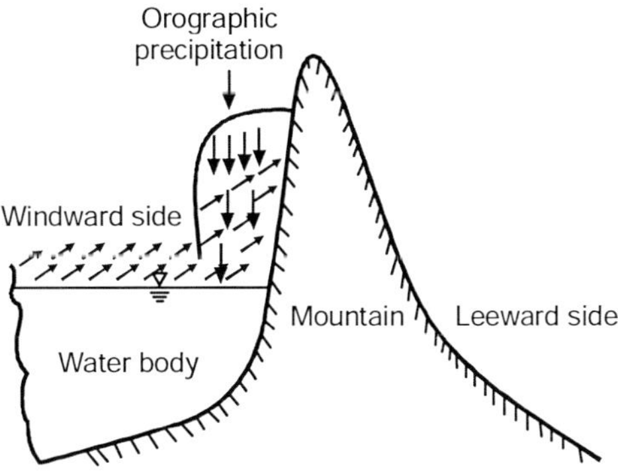

Figure 3.1 Orographic precipitation.

Precipitation due to turbulent ascent: Air mass is forced to rise up due to greater friction of earth's surface after its travel over ocean. It continues to move up because of increased turbulence and friction. When it ultimately condenses, precipitation occurs. In India, the winter rainfall in Tamil Nadu and nearby areas is mainly due to this process.

Cyclonic precipitation: This precipitation is the result of lifting of air mass, which converges into a low-pressure area. Due to difference in pressure created by unequal heating of earth's surface, air mass goes up. Here winds blow spirally inward in anticlockwise direction in northern hemisphere. There are two main types of cyclone. One is tropical cyclone called **hurricane** or **typhoon** of comparatively small diameter of 300–1500 km causing high wind velocity and heavy rainfall. The other one is extra-tropical cyclone of large diameter upto 3000 km causing wide spread frontal type precipitation.

3.4 MEASUREMENT OF PRECIPITATION

Precipitation (i.e., rainfall) which falls on earth's surface for a specified time is measured in depth of water. The instruments used for its measurement are called **rain gauges**. The following are some rain gauges used to measure the rainfall:

Symon's (non-recording) type: Figure 3.2 shows the Symon's type raingauge which is most commonly used in India. Details of the gauge with dimensions and arrangements are shown in the figure. It is an improved version of original Symon's type, recommended by Indian

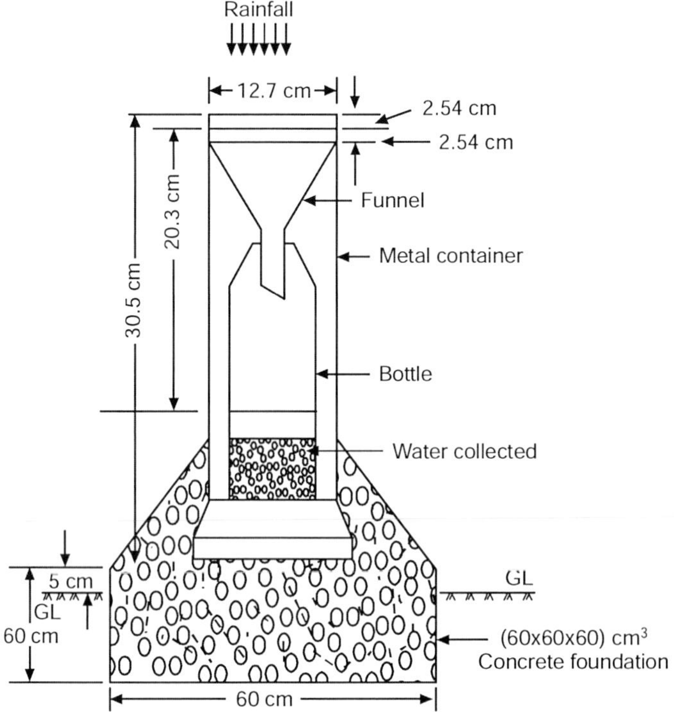

Figure 3.2 Symon's type raingauge.

Meteorological Department (IMD) as per IS 5225:1992. The foundation of the gauge is (60 × 60 × 60) cm^3 concrete block. Rainwater catch space is of 12.7 cm (5 inch) diameter, that is connected to a funnel as shown. The body is made of metal containing the funnel and the rainwater-collecting bottle. The rain collected in the bottle is measured by a graduated cylinder correct to 0.1 mm. As the rainfall collected in the bottle is not directly measured, it is called **non-recording type**.

Weighing bucket type: In Figure 3.3, the components of weighing bucket gauge have been shown. The rainfall on the receiver flows to the bucket through the funnel. The weight on the bucket is recorded by the mechanism of a pen, chart and clock-work-revolving drum. This mechanism of the instrument gives the graph of the accumulated rainfall against the elapsed time, i.e., mass curve of rainfall. This type has the advantage of measuring any type of precipitation like snow as it is based on weight.

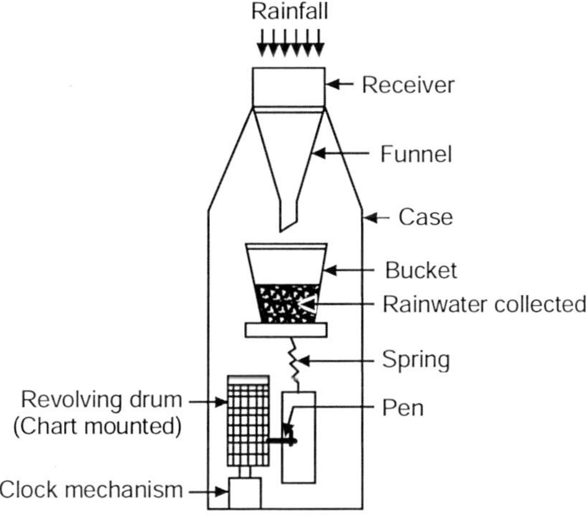

Figure 3.3 Weighing type gauge.

Tipping bucket raingauge: Figure 3.4 is the schematic view of tipping bucket raingauge. It is adopted by US Weather Bureau for use in the USA. Here rainfall received by the funnel

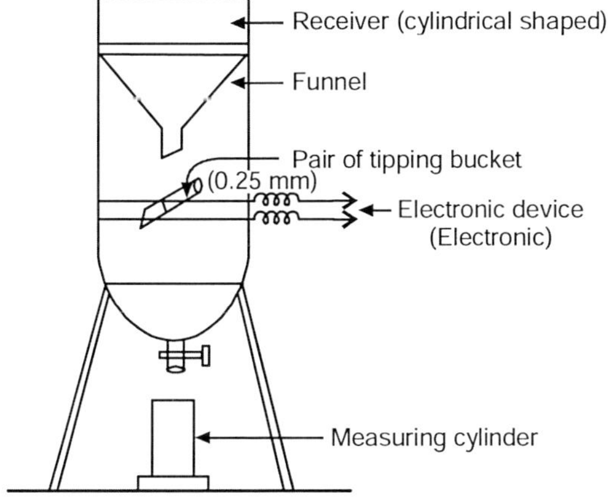

Figure 3.4 Tipping bucket raingauge.

is discharged into small pair of tipping bucket. It is designed in such a way that when water is filled in any one of the buckets, it tips and discharges into a measuring cylinder placed below. The other bucket then begins to collect the water. The tipping operation is fitted with an electronic recording device to compute the intensity of rainfall.

Siphon raingause or floating gauge: Figure 3.5 shows the siphon raingauge. Since a float is used, it is also called **float type gauge**. This is used by Indian Meteorology Department (IMD). It is called siphon type as it uses a siphon to empty the rain by siphonic action from the floating chamber. Rainwater enters through the funnel into the float chamber. The float rises. A pen attached to the float through a lever system records elevation of the float on a rotating drum by a clockwork mechanism. A siphon system is shown which empties the chamber when the float reaches a maximum level and pen is brought to zero level in the chart. As the rain continues, pen rises again from zero line of the chart. From the record on the chart made by the pen, rainfall can be calculated.

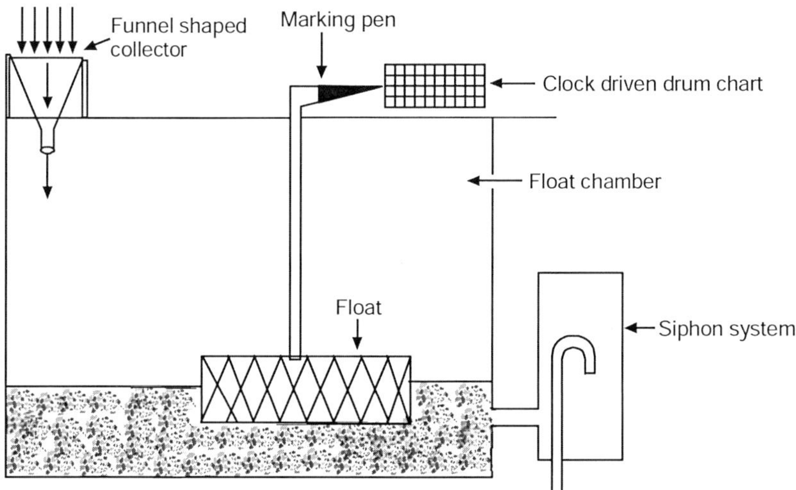

Figure 3.5 Siphon raingauge.

Storage raingauge: Storage raingauge are used in remote areas where frequent service to take note of rainfall is not possible. Weighing type storage raingauge may be used for 1 or 2 months without servicing and some non-recording type may be used as storage raingauge for entire season. Storage raingauge are charged with calcium chloride or other antifreeze solution to liquefy the snow falling on it. Losses due to evaporation are eliminated by a thin layer of oil at the top. Tower mounted storage raingauge is provided in hilly areas to eliminate obstruction such as tall trees.

Telemetering raingauge: Telemetering raingauges are of utmost important in collecting rainfall data generally from inaccessible places. It is of recording type containing electronic units to transmit the data on rainfall to a base station at regular interval. Tipping bucket type is being ideally suited.

Radar measurement of rainfall: Radar (Radio detecting and ranging) is initially devised and used to detect the aircrafts upto some km distance. But now it can be used for the measurement of rainfall also. Figure 3.6 shows the principle and working of radar. Transmitter produces electromagnetic waves, that are radiated by narrow beam antenna. The same antenna again intercepts the reflection of these waves from target. The receiver detects, amplifies and transforms them into video form on the indicator. If there is no target, the screen of the indicator is illuminated dimly and the screen becomes bright if there is a small target and a bright patch if there is an extended object such as rain.

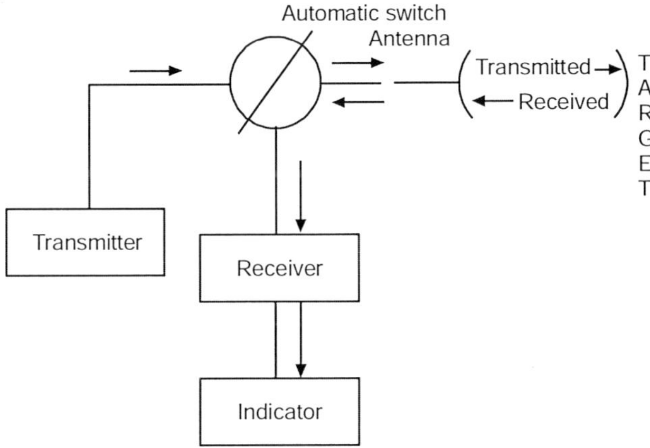

Figure 3.6 Working of radar.

Automatic radio-reporting raingauge: In mountainous areas, which are not easily accessible, this type is used to collect the rainfall data. In tipping bucket type, when buckets fill and tip, they give electric pulses equal in number to the mm of rainfall collected which are coded into messages and impressed on transmitter during broadcast. At the receiving station, these coded signals are picked by UHF receiver. This type of raingauge had been installed at Koyna Hydroelectric Project in Puna in 1966 by IMD and is working satisfactorily.

3.5 SELECTION OF RAINGAUGE SITE

In selecting a raingauge site, the following points are to be considered:

1. The site should be an open space.
2. The distance between the raingauge and the nearest object should be at least twice the height of the object.
3. The gauge should be placed on level ground, not upon a slope.
4. In hilly area, if suitable level ground is not available, it should be placed on the top of hill.
5. Again in hilly areas, site should be best shielded from high winds where wind cannot form eddies.
6. A fence, if erected to protect the gauge from cattle, should not be less than twice its height.

3.6 NETWORK OF RAINGAUGES IN A CATCHMENT AND THEIR ADEQUACY

The rain catchment areas of gauges are very small as compared to real relating to area extent of the rainfall. To get average rainfall of the whole catchment for a particular storm, number of raingauge stations should be optimum. It also depends on topography of the catchment and its accessibility. Optimum number of raingauges is always necessary to have a correct average rainfall. According to World Meteorological Organization (WMO), the recommended area densities are:

1. In flat region, 1 station for 600–3000 km^2
2. In mountainous region, 1 station for 100–1000 km^2
3. In arid and polar region, 1 station for 1500–10,000 km^2

According to Indian Standard (IS: 4987–1968),

1. In plains, 1 station/520 km^2
2. In region of average elevation of 1000 m, 1 station/260–360 km^2
3. Hilly areas with heavy rainfall, 1 station/130 km^2

Thus, to have a correct data, number of rainfall stations should be within the prescribed limit as the precipitation records are most fundamental data used in the hydrological studies. These are essential for

1. Carrying out storm analysis over a catchment
2. Depth-area-duration analysis (DAD)
3. Fixation of design flood
4. Stream flow forecasting
5. Reservoir regulation

For all such important studies, well established networks of rainfall stations are essential.

3.6.1 Adequacy of Rainfall Stations

Number of rainfall stations, i.e., raingauge stations for an area to give necessary average rainfall with certain percentage error can be obtained from statistical consideration. Following are the steps in the estimation of optimum number of raingauges:

Step 1 Calculate total rainfall.

$$P_T = P_1 + P_2 + P_3 + \cdots + P_n$$

where n is total number of existing rainfalls in the catchment or watershed.

Step 2 Calculate mean rainfall.

$$P_m = \frac{P_T}{n} \tag{3.1}$$

Step 3 Calculate the sum of the squares of all the rainfall of the gauges.

$$\Sigma P^2 = P_1^2 + P_2^2 + P_3^2 + \cdots + P_n^2 \tag{3.2}$$

Step 4 Calculate the sum.

$$\sum_{i=1}^{n} (P_i - P_m)^2 \tag{3.3}$$

Step 5 Calculate the square of standard deviation.

$$\sigma^2 = S^2 = \frac{\sum_{i=1}^{n} (P_i - P_m)^2}{n-1} \tag{3.4}$$

Step 6 Calculate the coefficient of variation.

$$C_v = \frac{100\sqrt{S^2}}{P_m} \tag{3.5}$$

Step 7 Optimum number of raingauge N that would be necessary to estimate the average rainfall within percentage error P is:

$$N = \left(\frac{C_v}{P}\right)^2 \tag{3.6}$$

For error less than equal to 10%, N is given as:

$$N = \left(\frac{C_v}{10}\right)^2 \tag{3.7}$$

Step 8 Additional number required = $(N - n)$ (3.8)

The additional numbers should be distributed in different zones in proportion to their area.

Example 3.1

The average rainfall of 5 raingauges in the base stations are 89, 54, 45, 41 and 55 cm. If the error in the estimation of rainfall should not exceed 10%, how many additional gauges may be required?

Solution: The mean rainfall is obtained as:

$$P_m = \frac{89 + 54 + 45 + 41 + 55}{5} = 56.8 \text{ cm}$$

Now, $$S^2 = \frac{(89-56.8)^2 + (54-56.8)^2 + (45-56.8)^2 + (41-56.8)^2 + (55-56.8)^2}{5-1}$$

or $S^2 = 359.2$

$\therefore$ $S = 18.95$

The coefficient of variation is calculated as:

$\therefore$ $$C_v = \frac{18.95}{56.8} \times 100 = 33.367$$

$$\therefore \qquad N = \left(\frac{C_v}{10}\right)^{10} = \left(\frac{33.367}{10}\right)^2 = 11.13$$

$$\cong 11$$

Thus Additional no. required = (11 – 5) = 6

EXAMPLE 3.2

In certain watershed (River basin), there are four raingauge stations. The normal precipitation amounts respectively to 800, 520, 440 and 420 mm. Determine the optimum number of raingauges in the watershed if it is required to limit the error in average rainfall to 6%.

Solution: Here $n = 4$, $P_T = 800 + 520 + 440 + 420 = 2180$ mm

$$\therefore \qquad P_m = \frac{2180}{4} = 545 \text{ mm}$$

Now, $\sum P^2 = 800^2 + 520^2 + 440^2 + 420^2 = 1280400 \text{ mm}^2$

Then $$S^2 = \frac{(800-545)^2 + (520-545)^2 + (440-545)^2 + (420-545)^2}{5-1} = 30766.7$$

or $S = 175.4$

Now, $$C_v = \frac{175.4}{545} \times 100 = 32.184$$

$$\therefore \qquad N = \left(\frac{32.184}{6}\right)^2 = 28.77$$

i.e. $N = 29$

Thus, Additional gauge stations required = (29 – 4)

= 25

3.7 METHODS OF COMPUTING AVERAGE RAINFALL

The average rainfall over a catchment or watershed is obtained by the following four different methods from available rainfall data of different raingauges:

Arithmetic mean method: It is the simplest method. If there are n numbers of raingauge stations in a watershed, and $P_1, P_2, P_3, \ldots, P_n$ are amount values of rainfall in those n stations in a given period (say annual precipitation), then arithmetic mean of those rainfall records give the average value of precipitation of the watershed or catchment, i.e.,

$$P_m = \frac{P_1 + P_2 + P_3 + \cdots + P_n}{n} = \frac{1}{n}\sum_{i=1}^{n} P_i \qquad (3.9)$$

Here P_m is the average rainfall in the catchment. If the catchment area is more or less flat, distribution of raingauge stations is uniform, and then average P gives good result. This method is quite rapid and easy.

Merits and demerits of this method

1. It is one of the simplest methods.
2. It is used to determine the approximate rainfall.
3. If the rainfall is uniformly distributed over the whole catchment, method gives better results.
4. Also if the number of raingauge stations are more and variation of the individual record is not far from the mean; method is taken to be accurate.
5. Even in hilly terrains, this method can yield fairly satisfactory results if the orographic influences on rainfall are considered in selection of raingauge sites.
6. The method is rapid and has got excellent adoptability for computer process.
7. To install maximum number of raingauge stations in the catchment for better results is costly and not always possible.

The Thiessen polygon method: A.M. Thiessen[1] (1911) suggested this method in which weighing effect of the area in the form of polygon closest to the station has been taken into account. Thus, it tries to eliminate the error due to non-uniform distribution of raingauges.

Consider a catchment with six measuring stations as shown in Figure 3.7. Join the station by lines. Perpendicular bisectors of those lines are drawn. These perpendicular bisectors produce the polygons A_1, A_2, A_3, A_4, A_5 and A_6 (shown in the figure). These are the Thiessen polygons.

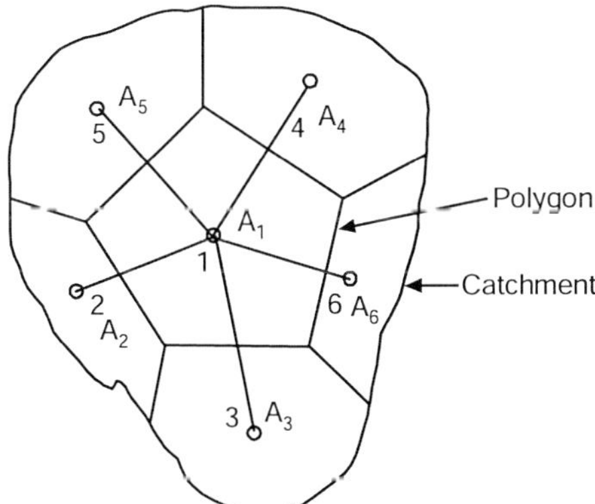

Figure 3.7 Thiessen polygon.

The polygons thus formed around each stations are the boundaries of the affected area and are assumed to be controlled by the stations. The area of the each polygon is planimetred and expressed as percentage of the whole area. If P_1, P_2, P_3, P_4, P_4, P_5 and P_6 are the rainfalls in station 1, 2, 3, 4, 5 and 6 respectively, average rainfall of the catchment is:

$$P = \frac{P_1A_1 + P_2A_2 + P_3A_3 + P_4A_4 + P_5A_5 + P_6A_6}{(A_1 + A_2 + A_3 + A_4 + A_5 + A_6)}$$

[1] Thiessen, A.M., Precipitation Average for Large Areas, *Monthly Weather Review*, pp. 1082–1084, 1911.

Thus, in general for N stations,

$$P = \frac{\sum_{i=1}^{N} P_i A_i}{A} \tag{3.10}$$

where A is the total area of the catchment.

Merits and demerits of this method.

1. Results are accurate than arithmetical mean method.
2. Thc grcatcst limitation of the method is its flexibility. A new Thiessen polygon diagram is to be drawn every time for the catchment if there is a change in the gauge network.
3. This method does not consider the orographic influences.
4. This method simply assumes linear variation of precipitation between stations and assigns each segment of area to the nearest station.
5. If gauging stations are few compared to the size of area, Thiessen polygon method should be used.
6. Station weights remain constants when the same number of stations are used.
7. If catchment area is large and raingauging stations are also quite large in number, it is then adoptable to computer computation.

EXAMPLE 3.3

Using Thiessen polygon method, compute the depth of average precipitation over a basin using the following data:

Raingauge	*Area of polygon*	*Observed depth of precipitation in* m
1	480	50
2	2860	65
3	2200	85
4	1600	90
5	710	60
6	270	70

Solution: Average rainfall is given by

$$P = \frac{P_1A_1 + P_2A_2 + P_3A_3 + P_4A_4 + P_5A_5 + P_6A_6}{A_1 + A_2 + A_3 + A_4 + A_5 + A_6}$$

$$= \frac{50 \times 480 + 65 \times 2860 + 85 \times 2200 + 90 \times 1600 + 60 \times 710 + 70 \times 270}{(480 + 2860 + 2200 + 1600 + 710 + 270)}$$

$$= \frac{602400}{8120}$$

$$= 74.1872 \text{ mm}$$

Isohyetal map method: This the most accurate method of finding average precipitation over an area provided it is used by an experienced analyst. Location of stations and amount of precipitation are plotted on a suitable map and contours of equal precipitations called **Isohyets** are drawn in Figure 3.8.

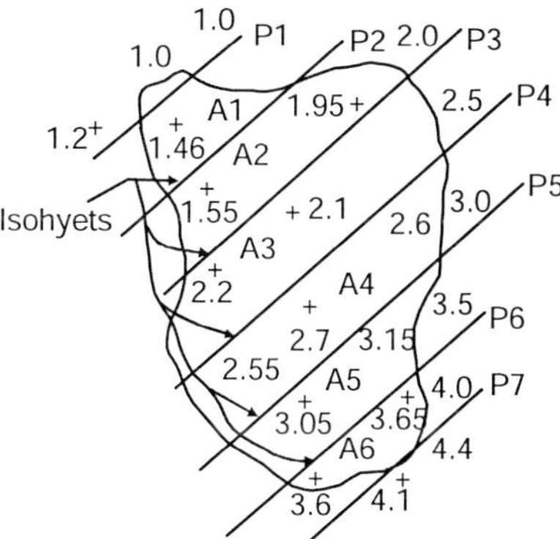

Figure 3.8 Isohyetal map method.

Let the isohyets represent the rainfall P_1, P_2, ..., P_n and areas between the successive isohyets are A_1, A_2, ..., A_{n-1}. Average precipitation is then calculated as:

$$P = \frac{A_1\left(\frac{P_1 + P_2}{2}\right) + A_2\left(\frac{P_2 + P_3}{2}\right) + \cdots + A_{n-1}\left(\frac{P_{n-1} + P_n}{2}\right)}{A_1 + A_2 + A_3 + A_4 + \cdots + A_{n-1}}$$

or

$$P = \frac{A_1\left(\frac{P_1 + P_2}{2}\right) + A_2\left(\frac{P_2 + P_3}{2}\right) + \cdots + A_{n-1}\left(\frac{P_{n-1} + P_n}{2}\right)}{A} \quad (3.11)$$

Merits and demerits

1. The isohyetal method permits the use and interpretation of all available data and is well adapted to display and discussions.
2. If the analyst has the knowledge of orographic effect, he can use it in constructing the isohyetal map. Also he must have knowledge of storm morphology. Then the final map prepared by the experienced analyst should represent a more realistic precipitation pattern than just obtained from gauged data.
3. In the hilly and rugged basin, this method is most suitable.
4. In sufficient dense network of rainfall stations within storm area, this method may give reasonable accurate indication of rainfall distribution.
5. Overall impression may be concluded that the isohyetal method is superior to the other two methods.

Weighted length method: Considering linear variation of rainfall between two raingauge stations, mean rainfall of a catchment is equal to the summation of product of the two lengths of the line joining the two stations divided by total length of the line joining the raingauge stations.

Merits and demerits of this method

1. It is a tedious and laborious process of measurement of total length of all the stations if the topography of the catchment is not even.
2. It does not purely reflect the picture of average rainfall by dividing a linc instcad of dividing by areas like Theissen polygon and isohyetal method.
3. It is rarely used.
4. In catchment with plain topography and smaller number of rainfall gauges, method may be used for approximate value.

EXAMPLE 3.4

The isohyets for annual rainfall over a catchment basin were drawn. The areas of strips between the isohyets are given. Determine the average precipitation over the basin.

Isohyets	*Area in* sq km
75–85	1600
85–95	3000
95–105	2800
105–115	1000
115–135	900
135–155	700

Solution: Average precipitation is given as:

$$P = \frac{A_1\left(\dfrac{P_1+P_2}{2}\right)+A_2\left(\dfrac{P_2+P_3}{2}\right)+\cdots+A_{n-1}\left(\dfrac{P_{n-1}+P_n}{2}\right)}{A_1+A_2+A_3+\cdots+A_n}$$

$$= \frac{1600\left(\dfrac{75+85}{2}\right)+3000\left(\dfrac{85+95}{2}\right)+2800\left(\dfrac{95+105}{2}\right)+1000\left(\dfrac{105+115}{2}\right)+700\left(\dfrac{135+155}{2}\right)}{1600+3000+2800+1000+900+700}$$

$$= \frac{(1600\times 80+3000\times 90+2800\times 100+1000\times 110+900\times 125+700+145)}{10,000}$$

$$= 100.2 \text{ cm}$$

3.8 INTERPOLATION AND ADJUSTMENT OF MISSING DATA

Sometimes rainfall data in one or two stations may be missed. Inconsistency of average rainfall may occur in catchment due to various reasons like missing precipitation data, exposure of station may be changed with growth of trees and buildings, etc. In such situations, interpolations in the estimations of average rainfall are required. These interpolations are done by the following methods depending upon causes of inconsistency as reported Paulhus and Kohler[2].

Arithmetic mean method: Sometimes, some station has a break in record due to absence of observer or failure of the instrument. It is then necessary to estimate that missing data. To estimate this data at least three stations close to this station are selected. It is also necessary that these three stations are evenly distributed around the station. If normal precipitation at each of these selected stations is within 10 percentage of that for station with missing data, then simple arithmetical mean of the precipitation of those three stations will give the value of the missing station. If P_A, P_B and P_C are precipitation of the nearby stations and P_x is the estimated precipitation of the missing station, then

$$P_x = \frac{P_A + P_B + P_C}{3} \tag{3.12}$$

Equation (3.12) gives average precipitation interpolated by simple arithmetic mean of the three nearby stations.

Normal ratio method: This method is used for interpolation when the normal annual precipitation of, say, three nearby stations A, B and C N_A, N_B and N_C, differs from that of the N_X of station X with missing data from previous record by more than 10%. If P_A, P_B, P_C are average storm precipitation of A, B and C in the year when average precipitation at X, i.e., P_X is missing, then

$$P_X = \frac{1}{3}\left[\frac{N_X}{N_A} P_A + \frac{N_X}{N_B} P_B + \frac{N_X}{N_C} P_C\right] \tag{3.13}$$

Equation (3.13) gives the average precipitation at X.

EXAMPLE 3.5

The normal annual rainfall of stations A, B, C and D in a catchment are 80.97, 67.59, 76.28, 92.01 cm respectively. In the year 2006, the station D was inoperative when station A, B, C recorded annual rainfall of 91.11, 72.23, 79.89 cm respectively. Estimate the missing rainfall at D in the year 2006 by normal ratio method.

Solution: Assuming annual precipitation at A, B and C varies more than 10% that of D, normal ratio method of interpolation is used to estimate rainfall at D.

$$\begin{aligned} P_D &= \frac{1}{3}\left[\frac{N_D}{N_A} P_A + \frac{N_D}{N_B} P_B + \frac{N_D}{N_C} P_C\right] \\ &= \frac{1}{3}\left[\frac{92.01}{80.97} \times 91.11 + \frac{92.01}{67.59} \times 72.23 + \frac{92.01}{76.28} \times 79.89\right] \\ &= 99.41 \end{aligned}$$

[2] Paulhus, J.L.H. and Kohler, M.A., Interpolation of Missing Precipitation Records, *Monthly Weather Review*, Vol. 80, No. 8, pp. 129–133, 1952.

Station year method: In this method, the records of two or more are combined into one long record provided station's record are independent and areas of stations are climatologically the same. The missing record at a certain station in a particular year may be found out by the ratio of the average or by graphical comparison. For example, in a certain year total rainfall of station A is 78 cm and for the neighbouring station B, no record is available. But if average annual precipitation at A and B in previous year record are 72 and 82 cm respectively, the missing year's rainfall at B, i.e., P_B may be obtained by simple proportion as:

$$\frac{78}{72} = \frac{P_B}{82}$$

$$\therefore \quad P_B = 88.833 \text{ cm}$$

This result must be checked by another station C for better results. If the new P_B is slightly different, average of the two will provide a bit accurate P_B.

Double mass curve: Kohler[3] presented analysis, testing and adjustment of double mass curve. A double mass curve shown in Figure 3.9 is drawn for a period of 29 years of a catchment

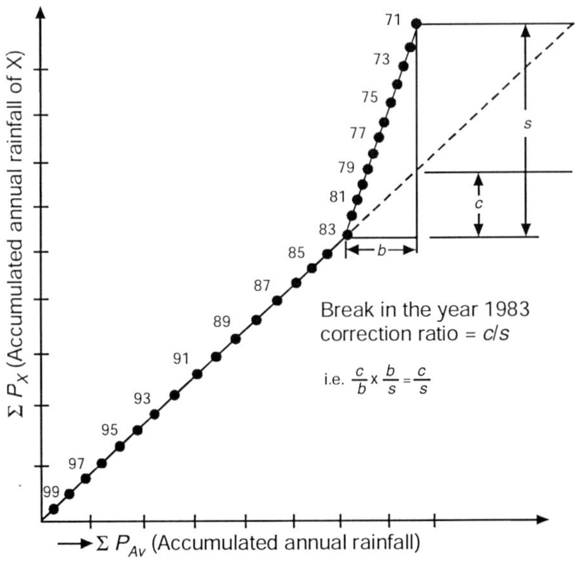

Figure 3.9 Double mass curve analysis to adjust the error.

to check the inconsistency of rainfall record of a station (say, *X*) and accordingly to adjust the incorrect results.

The cause of inconstancy may be

1. Shifting of rainfall station to a new position.
2. When ecosystem in the vicinity is changed due to land slide, forest fire, flood and earthquake.

---

[3] Kohler, M.A., Double Mass Curve Analysis for Testing consistency of Records and for Making Required Adjustments, *Bull. Am. Meteorol. Soc.*, Vol. 30, No. 5, 1949.

3. Error in observation from a certain year.
4. Change in the vicinity of the station due to growth of trees, building, fencing, cutting of forest nearby changing the wind pattern, etc.
5. Raingauge may be faulty from a certain period.
6. Site of instrument may be changed or replaced without record.

To draw this curve, a group of stations (say, 10) is taken as base station in the neighbourhood of the problem station *X*. The accumulated rainfall of station *X*, (ΣP_X) and accumulated values of average of group of base stations (ΣP_{Av}) are calculated starting from the latest record, i.e., from 1999. In Figure 3.9, the values of ΣP_X as ordinate and ΣP_{Av} as abscissa to a scale is plotted for available data of rainfall from the year 1999 to 1971. In the plot, a break in the slope has been seen from the year 1983. It indicates a change in precipitation regime of station *X*. The values at *X* beyond the period of change of regime, i.e., the point 1983 in Figure 3.9 (double mass curve) has been corrected. The original slope of the mass curve of rainfall from 1999 to 1983 is extended by a dotted line as shown in the figure. The correction to the data of rainfall at *X* from 1982 to 1971 is slope of the dotted line divided by the slope of the second mass curve of 1983 to 1971, i.e.,

$$\frac{c/b}{s/b} = \frac{c}{s}$$

The corrected data at *X* is:

$$P_{cX} = P_X \times \frac{c}{s}$$

Thus, all the inconsistent data of *X* from 1982 to 1979 are corrected by multiplying by *c*/*s*, the value of which is obtained from the plot measuring *c* and *s* as per scale.

The procedure of testing and correcting the inconsistency of data of a point by double mass curve has been presented first by Max A. Kohler in 1949.

EXAMPLE 3.6

Rainfall data for station *X* as well as the average annual rainfall measured at a group of 10 neighbouring stations located in a meteorogically homogenous region for 22 years is given as follows. Test the consistency of data at *X* by double mass curve method. Find the year in change of regime and the correction factor to be applied.

Year	*Station X* (cm)	10 *station average* (cm)	*Year*	*Station X* (cm)	10 *station average* (cm)
1999	163	161	1988	141	156
1998	130	146	1987	196	193
1997	137	130	1986	160	128
1996	130	143	1985	144	117
1995	140	135	1984	196	152
1994	142	163	1983	168	155
1993	148	135	1982	194	161
1992	95	115	1981	162	147
1991	132	145	1980	178	146
1990	145	155	1979	144	132
1989	158	164	1978	177	143

Solution: From given data rainfall at X and average rainfall of stations, ΣP_X and P_{Av} are tabulated staring from year 1999.

Year	ΣP_X (cm)	ΣP_{Av} (cm)	*Year*	ΣP_X (cm)	ΣP_{Av} (cm)
1999	163	161	1988	1661	1746
1998	293	307	1987	1857	1939
1997	430	437	1986	2017	2067
1996	560	580	1985	2161	2184
1995	700	715	1984	2357	233
1994	842	878	1983	2525	2491
1993	990	1013	1982	2719	2652
1992	1085	1128	1981	2881	2799
1991	1217	1271	1980	3059	2945
1990	1362	1426	1979	3203	3077
1989	1520	1590	1978	3380	3220

The double mass curve is plotted to a scale by taking ΣP_X as ordinate and ΣP_{Av} as abscissa. From Figure 3.10, it is found that a change of regime of station X takes place from the year 1985 and correction factor (to a scale) = s/c = 20/16 = 1.25. The precipitation data from 1985 to 1978 are to be multiplied by correction factor 1.25 to get the actual precipitation.

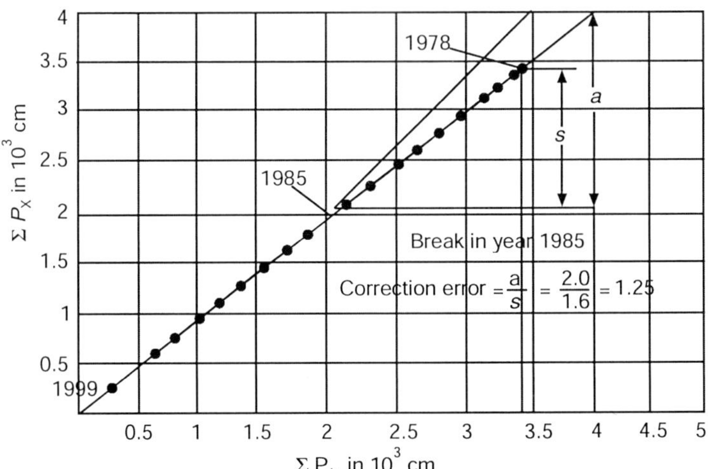

Figure 3.10 Mass curve of Example 3.6.

3.9 HYETOGRAPH AND MASS CURVE OF RAINFALL

Hyetograph is the plot of rainfall intensity in mm/hr or cm/hr of a series of rainfall against time occurring in a particular catchment. It is also called **rainfall hyetograph**. It is a discrete representation of rainfall hydrograph. It is used for hydrological analysis of the catchment for estimation of design storms for prediction of flood, estimation of runoff and derivation of unit hydrograph. The area under the hyetograph represents the total rainfall received in the period. A hyetograph of storm is shown is Figure 3.11.

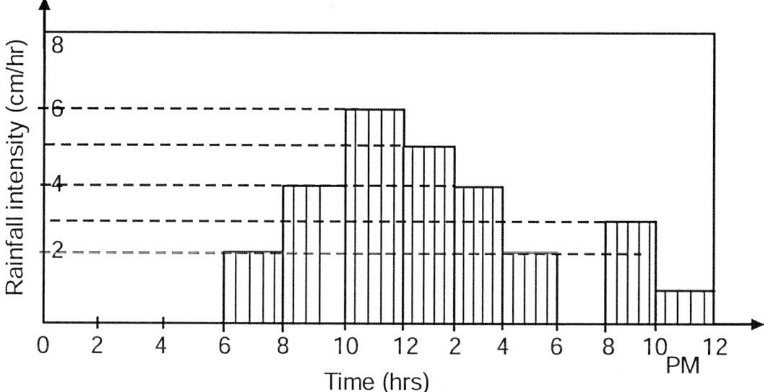

Figure 3.11 Hyetograph of a storm.

A hyetograph of a storm, shown in Figure 3.11, shows that the intensity of storm in morning 6 AM to 8 AM, 8 AM to 10 AM, 10 AM to 12 AM, 0 to 2 PM, 2 PM to 4 PM, 4 PM to 6 PM, 6 PM to 8 PM, 8 PM to 10 PM, 10 PM to 12 PM were 2 cm/hr, 4 cm/hr, 6 cm/hr, 5 cm/hr, 4 cm/hr, 2 cm/hr, zero rainfall, 3 cm/hr and 1 cm/hr respectively.

Mass curve of rainfall is a plot of accumulated rainfall against time. Records of weighing or float type gauges produce this for rainfall. A typical mass curve of rainfall is shown in Figure 3.12. It is useful to determine magnitude and duration of the storm. Slopes of the curve give the intensities at various times.

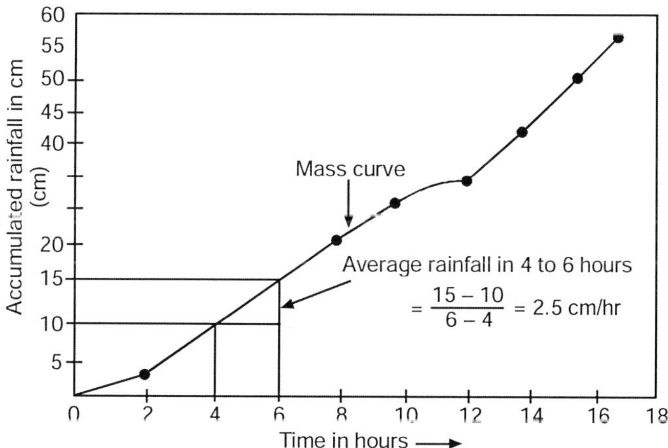

Figure 3.12 Mass curve rainfall.

3.10 INTENSITY DURATION FREQUENCY CURVE

Usually intensity of storm decreases with increase in storm duration. Further, a storm of any given duration will have a larger intensity if its return period is large. In other words, for a storm of given duration, storms of higher intensity in that duration are rarer than the storms of smaller intensity. In many design problems related to watershed management suchas runoff disposal and erosion control, it is necessary to know the rainfall intensities of different durations and different return periods. The interdependency among intensity (i cm/hr), durations (D hours) and return period (T years) is commonly expressed in general form as:

$$i = \frac{kT^x}{(D+a)^n} \tag{3.14}$$

Here k and a are constants, and x, n are exponents for a given catchment. Variations of intensity i with duration D and return period T are shown schematically in Figure 3.13(a) and (b).

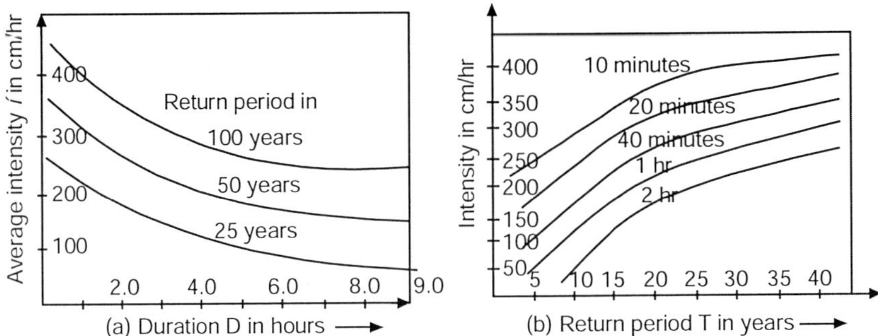

Figure 3.13 Variation of rainfall intensity *i* against D and T.

3.11 DEPTH-AREA-DURATION (DAD) ANALYSES

World Meteorological Organization (WMO)[4] made a comprehensive study on DAD analysis. The characteristics of a storm of given duration over the areas and corresponding depths of rainfall are reflected in a plot of depths of rainfall against areas on which the depths occur for given storm as shown in Figure 3.15 for storms of different durations. The curves shown in this figure is called Depth-area-duration (DAD) curves. It is seen from the curves that the storm of smaller duration has smaller depth of rainfall and as the area increases, depth decreases. DAD analysis of storm helps in determining the largest average depth of storm of different duration which is essential in determining the runoff of the catchment which in turn is required for design of culverts, irrigation projects, reservoir and dam. This analysis can be best understood with the help of following numerical example:

Example: A storm of 24 hrs duration occurring in a catchment produces the following isohyets:

Isohyets (in cm)	15	14	13	12	11	10	9	8	7
Enclosed area (in km^2)	22	78	98	115	150	195	250	300	355

Isohyets of the storm are shown in Figure 3.14. It is natural that the intensity of rainfall (15 cm) is highest in central area of 22 km^2 and gradually intensity of the storm decreases away from the centre with an increased area. Depth-area-duration (DAD) curve of this 24 hrs storm may be calculated as:

[4] World Meteorological Organisation, Manual for Depth Area-Duration Analysis of Storm Precipitation, WMO no. 237, Tech. Paper 129, pp. 1–31, Geneva, 1969.

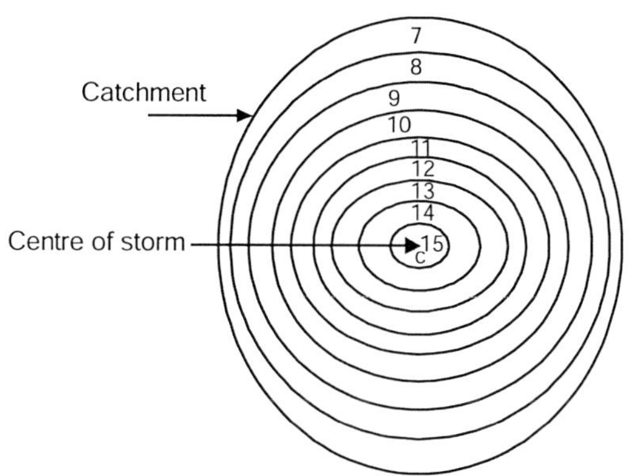

Figure 3.14 Isohyets of the storm.

Isohyets (given) cm	*Area enclosed* (*in* km^2)	*Net incremental area between isohyets* (km^2)	*Average rainfall on incremental areas* (cm)	*Rainfall volume on incremental areas* (cm-km^2). Col (3) × Col (4)	*Cumulative volume* (cm-km^2) Σ(5)	*Equivalent uniform Depths* (cm) Col (6)/Col (2)
1	2	3	4	5	6	7
15	22	22	15	330	330.00	15.00
14	78	56	14.5	812	1142.0011	14.461
13	98	20	13.5	270	1412.00	14.408
12	115	17	12.5	212.5	1624.50	14.126
11	150	35	11.5	402.5	2027.00	13.513
10	195	45	10.5	472.5	2499.50	12.818
9	250	55	9.5	522.5	3022.00	12.088
8	300	50	8.5	425.0	3447.00	11.490
7	355	55	7.5	412.5	3859.50	10.871

A curve is plotted with depths of rainfall in column 7 against corresponding areas in column 2.

Curve 1 in the plot is the depth-area-duration (DAD) curve for this 24 hours storm.

If storms of 18 hours, 12 hours and 6 hours will occur, the DAD curves can be shown in the same figures repeating the same cycle of calculation and plotting. Thus, to plot DAD curves or DAD analysis of storm, the following steps are required:

Step 1 From the data of rainfall in different places of the catchment for a particular storm, isohyetal map is prepared.

Step 2 Areas between the isohyets are planimetred.

Step 3 Thus, isohyets and area between isohyets are known.

Step 4 Net incremental area of between two isohyets is calculated.

Step 5 Average depth of rainfall between the two isohyets is the average of the two.

Step 6 Rainfall volume is calculated by multiplying net incremental area by average rainfall.

Step 7 Cumulative rainfall volume is determined.

Step 8 This cumulative volume is divided by the area enclosed between two isohyets gives average rainfall in that area.

Step 9 Same cycle of calculation is repeated for other areas.

Step 10 Plotting the obtained average rainfall depth against areas, DAD curves are obtained as shown in Figure 3.15.

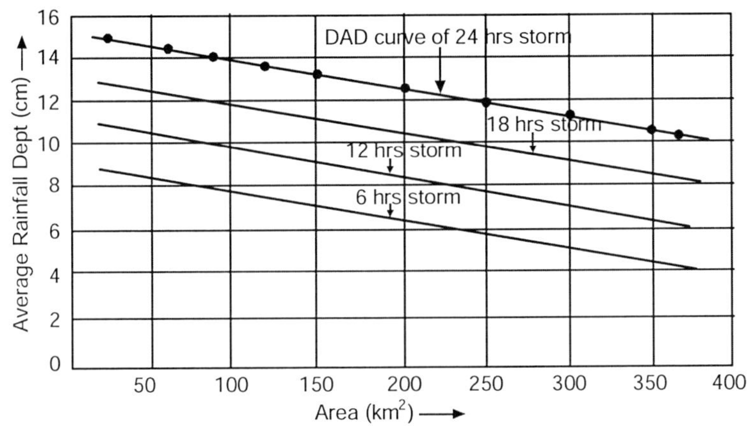

Figure 3.15 DAD curves of different storm durations.

3.11.1 Empirical Equations DAD Curve

The DAD curves, shown in Figure 3.15, can approximately be represented by empirical equation given by Horton (1940) as:

$$P = P_0 e^{-kA^n} \tag{3.15}$$

Here P is the depth of precipitation in cm over an area of A km^2, P_0 the rainfall at the centre of rainfall, k and n are constants for particular catchments. k and n also depend on duration of storm. If k and n of a particular catchment is known, Horton's equation can be used to plot DAD curve or to find P if area is known, i.e., if k and n are known, Eq. (3.15) is useful. Dhar and Bhattacharjya (1975) have computed the following values of k and n (Table 3.1) from several severe storms of Northern India.

Table 3.1 Values of k and n for storms

Duration	k	n
24 hrs	0.0008526	0.6614
48 hrs	0.0009877	0.6304
72 hrs	0.001745	0.5961

3.12 CONCLUSION

Precipitation in a catchment, its measurement by different methods, its intensity and duration is discussed in this chapter. The knowledge on precipitation is essential to determine runoff or flood in a catchment. However, before determining runoff, abstraction loss of it has to be explored. Therefore, in Chapter 4, abstraction losses will be discussed.

EXERCISES

3.1 What are the different forms of precipitation? Describe each of them.

3.2 What are the five different types of precipitations? Describe each of them.

3.3 Name the different instruments used in measuring precipitation. Describe the most common instruments used by field engineers.

3.4 What are the different points to be considered in selecting the site for raingauge station? How do you find the adequacy of raingauge stations?

3.5 The average annual rainfalls of six raingauge stations are 100, 150, 90, 70, 112 and 78 cm. It is assumed that the error of the estimation of rainfall is within 10%, compute the additional number raingauge stations that may be necessary for the basin. [*Ans:* 25]

3.6 Describe the different methods of computing average rainfall. Which method is accurate and why?

3.7 A river basin has six gauge stations having annual precipitation of 820, 1020, 1100, 980 and 1360 mm. Determine the optimum number of rainguages (i.e. additional required or not) if error in computation is limited to 10%.

[*Ans:* Optimum is 9, Additional is 3.]

3.8 A catchment has five raingauge stations and the annual precipitations are 900, 1100, 1750, 950 and 1250 mm respectively. Find extra number required or not if error in estimation is limited to 10%.

[*Ans:* Additional number = 5]

3.9 In a drainage basin of 600 km^2, isohyets drawn for a storm gave the following data:

Isohyets interval (in cm)	15–12	12–9	9–6	6–3	3–1
Inter isohyetal area (in km^2)	98	128	120	175	85

Estimate the average precipitation over the drainage basin.

[*Ans:* 7.41 cm]

3.10 In a watershed, the average annual precipitation for four sub basins was recorded as 100.84, 112.27, 84.84 and 73.406 cm. The areas of the sub basins were: 93264.3, 71243.5, 108808.2 and 168393.8 ha. Calculate the average precipitation of the total watershed using Thiessen polygon method. [*Ans:* 88.14 cm]

3.11 Raingauge station X was inoperative for a part of the month during which a storm occurred. Rainfalls recorded for this storm at A, B, C are 5, 6, 4 cm respectively. The normal annual rainfalls of X, A, B, C are 51, 54, 64, 48 cm respectively. Estimate storm precipitation of station X for the period when it was inoperative (*Hints:* Use normal ratio method.) [***Ans:*** 4.584 cm]

3.12 The annual rainfall at station X, which is assumed to be inconsistent, and the average annual precipitation of 20 base stations surrounding station X from 1999 to 1964 are given. Find the year of change of regime of X and the correction to be applied. Use double mass curve.

Year	*Rainfall in X* (cm)	*Average rainfall in 20 stations* (cm)	*Year*	*Rainfall in X* (cm)	*Average rainfall in 20 stations* (cm)
1999	70	100	1981	90	145
1998	73	90	1980	70	90
1997	120	150	1979	110	131
1996	114	116	1978	85	92
1995	80	110	1977	100	102
1994	110	115	1976	110	110
1993	70	90	1975	190	140
1992	120	145	1974	125	110
1991	90	92	1973	108	107
1990	85	115	1972	125	105
1989	85	110	1971	175	120
1988	80	95	1970	150	135
1987	110	100	1969	120	90
1986	115	130	1968	125	122
1985	80	90	1967	130	90
1984	105	90	1966	120	122
1983	96	142	1965	120	129
1982	110	120	1964	160	192

SUGGESTED FURTHER READINGS

Alvarez, F. and Henry, W.K., Raingauge Spacing and Reported Rainfall, *Bull. Int. Assoc. Sci. Hydrol.*, vol. 15, No. 3, pp. 97–107, March 1970.

American Society of Civil Engineers (ASCE), *Hydrology Handbook*, 2nd ed., ASCE Manual and Report on Engineering Practice, No. 28, New York, 1996.

Biswas, K.R. and Dennis, A.S., Formation of Rain Shower by Salt Seeding, *Jour. Appl. Meteorol.*, vol. 10, No. 4, pp. 780–784, August 1971.

Brandes, E.A., Optimizing Rainfall Estimates with the Aid of Radar, *Jour. Appl. Meteorol.*, vol. 14, No. 7, pp. 1339–1345, October 1975.

Bras, R.L., *Hydrology: An Introduction to the Hydrologic Science*, Addison-Wesley, Reading, M.A., 1990.

Chow, V.T., *Handbook of Applied Hydrology*, McGraw Hill, New York, 1964.

Das, G., *Hydrology and Soil Conservation Engineering*, Prentice-Hall of India, New Delhi, 2000.

Das, P., Role of Condensed Water in the Life Cycle of Convective Cloud, *Jour. Atmosph. Science*, vol. 21, No. 4, pp. 404–418, July 1964.

Dingman, S.L., *Physical Hydrology*, 2nd ed., Prentice-Hall, Upper Saddle River, New Jerssey, 2002.

Gum, R., Collision Characteristics of Freely Falling Water Drops, *Science*, vol. 150, No. 3697, pp. 695–791, November 1965.

Herschy, R.W. and Fiarbridge, R.W (Eds.), *Encyclopedia of Hydrology and Water Resources*, Kluwer Academic Publishers, Boston, M.A., 1998.

Huff, F.A., Sampling Errors in Measurement of Mean Precipitation, *Jour. Appl. Meteorol*, vol. 9, No. 1, pp. 35–44, February 1970.

Kessler, E. and Wilk, K.E., The Radar Measurement of Precipitation for Hydrological Purposes, Ref. WHO/IHD Project No 5, W.H.O., Geneva, 1968.

Linsley, K., Kohler, M.A. and Paulhus, J.L., *Hydrology for Engineers*, McGraw-Hill, 3rd ed., 1982.

Maidment, D.R. (Ed.), *Hand Book of Hydrology*, McGraw-Hill, New York, 1993.

Ponce, V.M., *Engineering Hydrology: Principles and Practices*, Prentice-Hall Inc., Upper Saddle River, N.J., 1989.

Raghunath, H.N., *Hydrology: Principle, Analysis Design*, New Age International, New Delhi, 1985.

Reddy, P.J., *A Textbook of Hydrology*, Lakshmi Publication, New Delhi, 1986.

Sing, V.P., (Ed.), *Environmental Hydrology*, "Climate change" by Mimikou, M.A., Kluwer Academic Publishers, Boston, MA, 1995.

Sing, V.P., *Elementary Hydrology*, Prentice-Hall Inc., Upper Saddle River, N.J., 1992.

Subramanya, K., *Engineering Hydrology*, Tata McGraw-Hill, New Delhi, 1984.

Thomas, W.A. (Ed.), *Legal and Scientific Uncertainties of Weather Modification*, Duke University Press, Durham, N.C., 1977.

Varshney, R.S., *Engineering Hydrology*, Nem Channel and Bros., Roorkee, 1979.

Viessman, W. and Lewis, G.L., *Introduction to Hydrology*, 4th ed., Harper Collins, New York, 1996.

Ward, A.D. and Elliot, W.J., *Environmental Hydrology*, Lewis Publishers, New York, 1995.

Chapter 4
Infiltration

4.1 INTRODUCTION

Infiltration is the name given to the phenomenon of liquid intake by porous media. Simple definition of infiltration is the flow of water through soil surface. It is one of the most important components of hydrological phenomena. It has great influence on rainfall, runoff, transpiration of plants and evaporation from soil surface. It is the main component of abstraction losses of rainfall. Infiltration replenishes soil moisture deficiency and excess water moves downward towards ground water storage by the force of gravity. This part of water is called **percolation**.

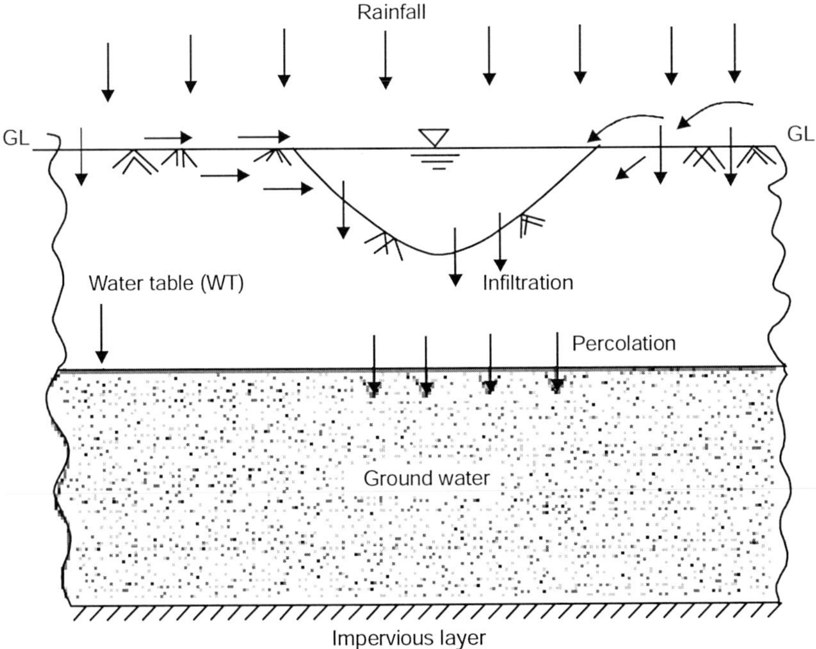

Figure 4.1 Infiltration, run off and percolation from rainfall.

The maximum rate at which the soil in any given condition is capable of absorbing water is called **infiltration capacity** f. The capacity f on dry surface or soil condition begins with a high rate f_0 initially and decreases to a fairly steady state f_c as the rain continues. Philips[1] (1969), Hillel[2] (1971), Morel-Seytoux[3] (1973) and Hadas et al[4] (1973) have presented excellent reviews of the infiltration process.

4.2 FACTORS AFFECTING INFILTRATION

Intensity of rainfall: If the intensity of rainfall is more, the impact of rain drops on the soil surface causes fines in the soils to be displaced and these in turn can clog the pore spaces in the upper layers and hence infiltration reduces. Rainfall mostly contributes runoff.

Duration of the rainfall: If rainfall duration is more, initially infiltration is more. Longer rainfall duration saturates the upper layer of soil and after that infiltration becomes steady if rain continues.

Temperature: If temperature is more, infiltration is more. Increase in temperature decreases the viscosity of water. At low viscosity, water moves faster through soil thereby increasing the infiltration rate. In winter days or season, viscosity increases at low temperature and hence infiltration decreases. Temperature variation is the main cause of seasonal variation of infiltration.

Soil characteristics: The amount of moisture present in the soil has an important effect on the infiltration. If the soil is initially dry, upper layer becomes wet at faster rate during rainfall. The lower layer of the soil remains comparatively quite dry. Therefore, a strong capillary attraction for moisture in subsurface layer develops. This capillary attraction acting in the same direction of gravity force produces high initial value of infiltration capacity. When water percolate to the lower layer, capillary attraction decreases and hence rate of infiltration decreases. If the moisture content in the soil is initially more, infiltration starts with lower rate, but with the increase of time, infiltration in both the cases becomes steady. Figure 4.2 shows variation

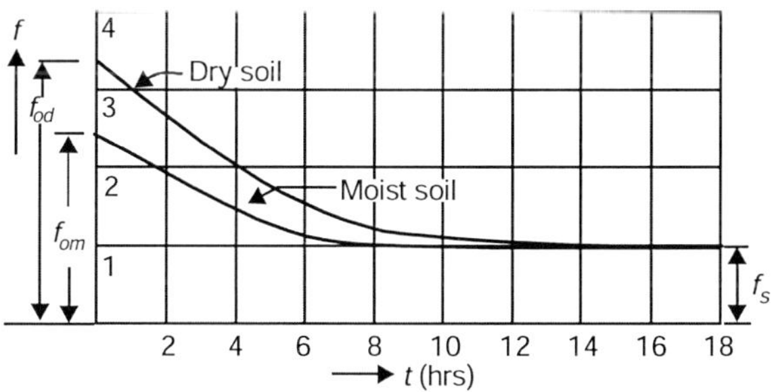

Figure 4.2 Variation of infiltration in soils with time.

[1] Philips, J.R., The Theory of Infiltration, Thc Infiltration Equation and its Solution, *Soil Sci.*, 83, pp. 435–448, 1957.
——— The Theory of Infiltration: Advances in Hydro Science, 5, pp. 215–296, 1969.

[2] Hillel, D., *Soil and Water—Physical Principles and Processes,* Academic Press, New York, 1969.

[3] Morel-Seytoux, H.J., Two Phase Flow in Porous Media: Advances in Hydro Sciences, 9, pp. 11–202, 1973.

[4] Hadas, A.D., Swartzendruber, P.E., Rijtema, P.E., Fuchs, M. and Yaron, B, *Physical Aspects of Soil Water and Salt in Ecosystems*, Springer Verlag, New York, 1973.

of infiltration f with time t. Here f_{od} is the initial infiltration in dry soil, f_{om} is the initial infiltration in moist soil and f_s is the constant steady f with increase of time.

Vegetative cover: Considerable increase of infiltration rate takes place when the soil has full dense vegetative cover. Vegetative cover decreases runoff. The soil surface cannot be compacted by raindrop. Besides vegetative cover provides a layer of decaying of organic matter which encourages the activity of burrowing animals and insects. This activity increases the permeability of soil. These two factors help in increasing infiltration capacity. Further, transpiration takes place from the vegetative cover which reduces soil moisture and, therefore, initial infiltration capacity increases. Vegetative cover by crops like potatoes and lintels is partial and values of f in such situation is relatively low.

Urban areas: The most of the urban areas are covered with pavements, building and lined drainage system. Rain falling on urban areas eventually flows as runoff through drains to a large water body. Therefore, infiltration in urban areas is very low or practically zero.

Depth of surface detention and thickness of saturated layer: Figure 4.3 shows the retention depth of water above the ground level to be d. Infiltration of water through the soil

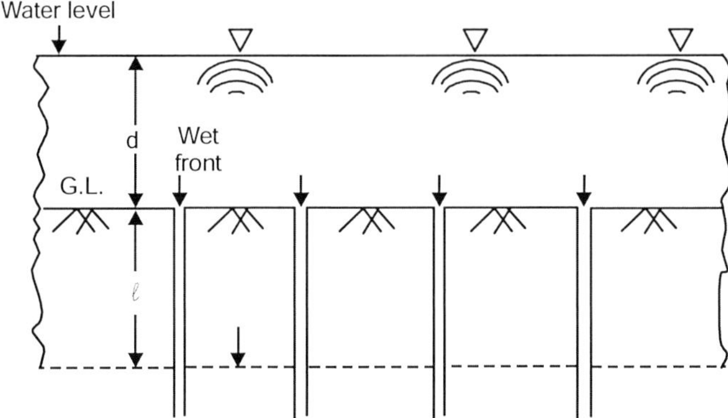

Figure 4.3 Infiltration process due to surface retention and thickness of saturated layer.

below the retention depth is assumed to take place through a series of tiny pipes of large numbers. When infiltration takes place, wet front moves downward whose length ℓ continues to grow bigger. At the beginning, ℓ is smaller than d. At that time, resistance to flow downwards is less so water enters rapidly. Thus, f is more at the beginning. As ℓ begins to grow more than d, d becomes negligible as compared to ℓ. Any further increase of ℓ, both driving force and resistance have the same effect and f begins to remain constant. It suggests, f should decrease with ℓ in a continuous rain and becomes ultimately constant f_s as shown in Figure 4.2.

Entrapped air in soil pores: The entrapped air in soil mass marginally affects the infiltration capacity because this entrapped air increases the resistance to flow.

Turbidity of water: Turbid water takes time to infiltration and hence it has the diminishing influence of infiltration capacity.

Compaction of soil: Artificial man made compaction of soil like overgrazed pastures, playground, areas subjected to vehicular traffic will have less infiltration. Compaction on clay soil takes quite easily, hence f in such compacted soil is less.

Others: Other factors like soil cracking to drying of soil, crop rotation, tillage operation presence of salts in soil, soil erosion and afforestation have affected the infiltration capacity.

4.3 MEASUREMENT OF INFILTRATION

In many situations of hydrologic problems like runoff estimation, soil moisture assessment and net irrigation water, rate of infiltration is essential. It may be determined either by plotting hyetograph and runoff hydrograph or by measuring it through infiltrometers. The difference of area of runoff hydrograph due to the same storm is the difference of losses. If evaporation and transpiration is small or neglected, then this difference gives the total infiltration volume. But this method is not accurate as the other two losses are neglected. Better results are obtained when it is determined by infiltrometers. Different methods of measurement of infiltration are discussed as follows:

Flooding type infiltrometer (Double ring infiltrometer): It is also called **double ring infiltrometer**. Figure 4.4 shows the sectional and plan views of this type of infiltrometer. Two concentric rings are driven into ground at least up to 15 cm. The diameter of the rings are

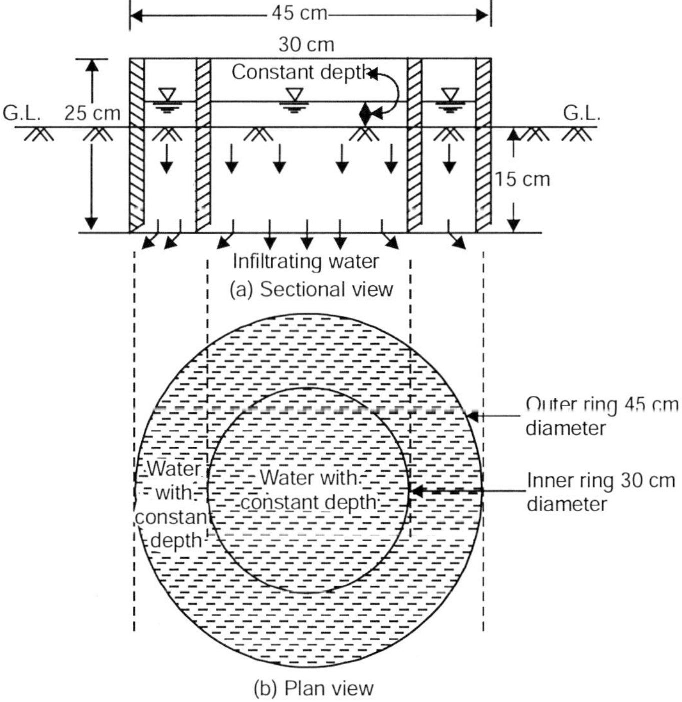

Figure 4.4 Flooding or double ring infiltrometer.

30 cm and 45 cm. Both are driven into the soil up to 15 cm without tilling or undue disturbance to the soil. Water is poured into the rings upto a constant depth, i.e., rings are flooded. Water is added at regular time intervals of 5, 10, 15, …, 60 minutes up to at least 6 hours. Points gauge is fitted at the centre of the inside ring and in the annular space between the two rings. The measurement of infiltration is taken at some interval. The function of outer ring is to prevent lateral inflow as seepage. The data obtained are plotted in Figure 4.5.

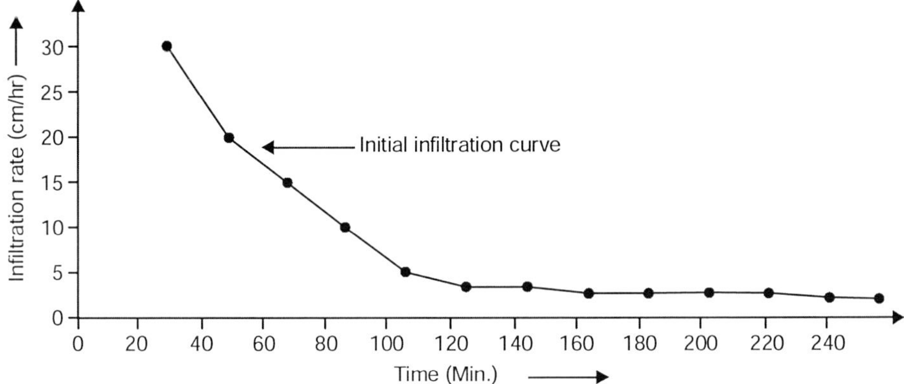

Figure 4.5 Typical infiltration curve.

Sometimes with the help of a single tube the infiltration capacity is measured. In case of single tube it is driven below the ground about 4 times the double rings or tubes.

Artificial rain simulator: A small ground area within 50 m^2 is selected. The water is allowed to fall on this area by artificial showers at uniform rate. The resulting surface runoff is measured. From the hydrographs, infiltration may be obtained.

Lysimeter under laboratory sample: A catch basin, called **Lysimeter**, under a laboratory or at some depth below the land surface is placed in order to measure infiltrating water.

Observation infiltration in pits and ponds: This method gives idea of infiltration or approximate value. If the observation of depression of the water body and evaporation from it is measured, approximate infiltration of the soil can be known.

Hydrograph analysis: From the measurement of rainfall and runoff from that rain, infiltration capacity may be determined.

4.4 INFILTRATION INDICES

When infiltration rate is assumed to be constant, it is called **infiltration index**. But in dry soil this rate is constant at the beginning. It may be more than the rainfall in the initial period of rainfall. Only after a considerable time, when the soil becomes wet, infiltration rate becomes constant or steady.

Figure 4.6 is superimposition of a rainfall hyetograph and infiltration rate curve appropriate to the soil of the catchments. The area of the hyetograph above the infiltration curve would then

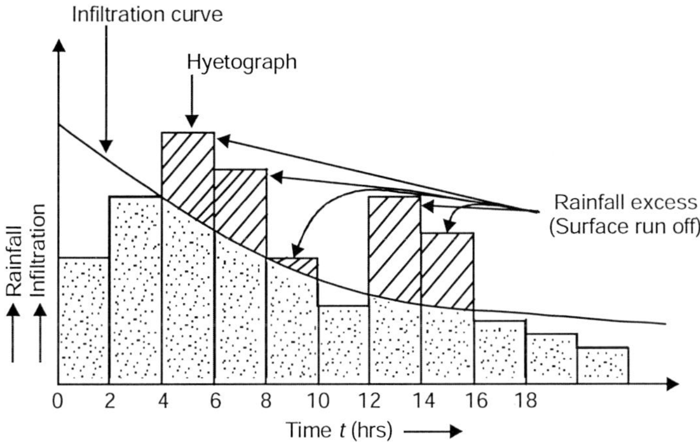

Figure 4.6 Relationship between infiltration, rainfall and runoff.

represent the runoff volume (shaded portion). The hyetograph shows that the rainfall in some hours is below the infiltration rate. The approach seems to be simple. If the rainfall intensity is always greater than the infiltration rate, results are satisfactory. But if the rainfall fluctuates above and below the infiltration rate, then problem becomes complicated. To get rid of this complicacy, infiltration indices (constant infiltration rates) are used, i.e., it is assumed to be constant (average infiltration) throughout the rainfall time. The condition of constant infiltration index is true if the soil is wet before the rainfall. Therefore, infiltration indices are best suited to use it for major flood analysis of storm occurring on wet soils. Thus, some of the commonly infiltration indices are as follows:

ϕ-Index: ϕ-Index is the average rainfall intensity above which rainfall volume is equal to runoff volume (shaded area) shown in Figure 4.7. The unshaded area below the line is also

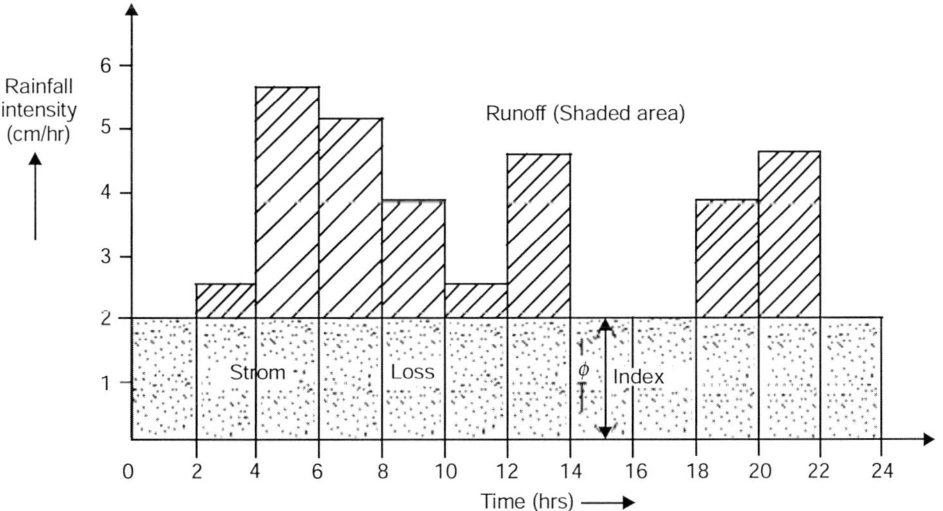

Figure 4.7 Definition sketch of ϕ-index.

measured rainfall, which is not runoff, represents loss due to infiltration. However, infiltration is the largest loss compared to other losses.

$$\phi\text{-index} = \frac{\text{All losses due to infiltration}}{\text{Duration of rainfall}}$$
$$= \frac{\text{Area under the line (Unshaded)}}{\text{Duration of rainfall}}$$

W-index: It is the infiltration obtained when (Precipitation-runoff-surface retention) is divided by duration of rainfall, i.e.,

$$\text{W-index} = \frac{P-Q-S}{t_r} \tag{4.1}$$

where P is the total rainfall, Q is the surface rainfall and S is the effective surface retention.

fav-index: In this method, an average infiltration loss is assumed throughout the storm, for rainfall $i >$ infiltration f.

Example 4.1

A seven-hour storm over a basin of 1830 km^2 produced the rainfall intensities at half an hour interval are 4, 9, 20, 18, 13, 11, 12, 2, 8, 16, 17, 13, 6 and 1 mm/hour.

If the corresponding observation runoff is 73.2×10^6 m^3, estimate the ϕ-index of the storm.

Solution: Total value runoff = 73.2×10^6 m^2
Area of the basin = 1830 km^2 = 1830×10^6 m^2

$$\therefore \quad \text{Depth runoff} = \frac{73.2\times10^6}{1830\times10^6} = 0.04 \text{ m}$$
$$= 40 \text{ mm}$$

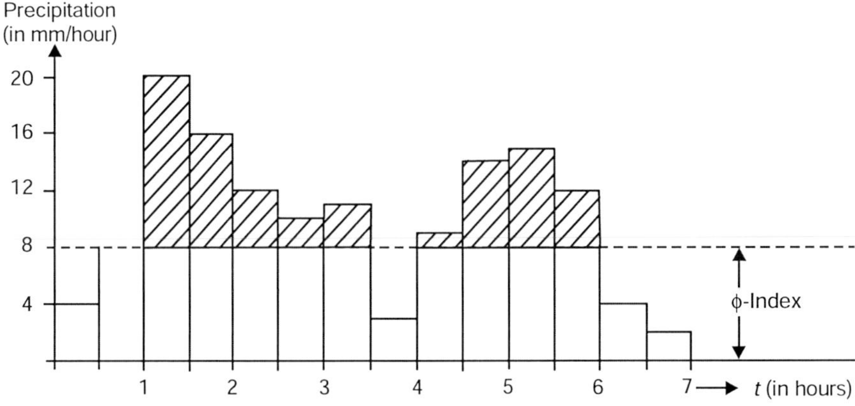

Figure 4.8 Visual of Example 4.1 to estimate ϕ-index.

Let us assume ϕ-index to be 10 mm/hr. Rainfall above the ϕ-index, i.e., runoff is:

$$0 + 0 + (20 - 10) + (18 - 10) + (13 - 10) + (11 - 10) + (12 - 10) + 0 + 0 + (16 - 10) + (17 - 10) + (13 - 10) + 0 + 0$$

$$= 10 + 8 + 3 + 1 + 2 + 7 + 6 + 3$$

$$= 40 \text{ mm}$$

This is equal to observed runoff.

$\therefore$ ϕ-index = 10 mm/hr

EXAMPLE 4.2

A storm during a dry weather flow has initial rainfall intensities of 8, 12, 40, 38, 30, 26, 28, 5, 16, 32, 36, 24, 14 and 4 mm/hr at half an hour interval. If the initial abstraction is 10 mm, what is the runoff volume of the basin which has the area of 600 km^2 and ϕ-index is 20 mm/hr.

Solution: Draw the hyetograph.

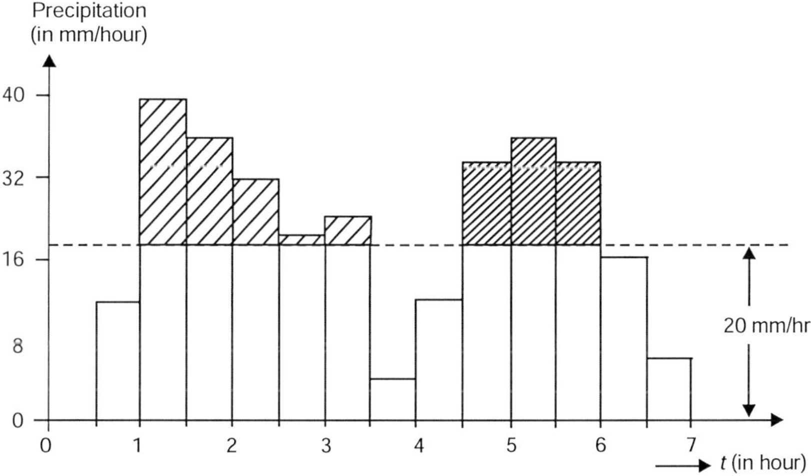

Figure 4.9 Visual of Example 4.2 to determine ϕ-index.

The rainfall excess from Figure 4.9 in half an hour basis is:

$$= 0 + 0(40 - 20) + (38 - 20) + (30 - 20) + (26 - 20) + (28 - 20) + 0 + 0 + (32 - 20) + (36 - 20) + (24 - 20) + 0 + 0$$

$$= 9 \text{ mm}$$

Hence rainfall excess in hourly basis is:

$$= 94/2$$

$$= 47 \text{ mm}$$

$\therefore$ Runoff volume = Area of the basin × Rainfall excess or runoff in depth

$$= \left(600 \times 10^6 \times \frac{47}{1000}\right)$$

$$= 28.200 \times 10^6 \text{ m}^3$$

EXAMPLE 4.3

The rates of rainfall for half an hour period of 3-hour storm are 1.6, 3.6, 5.0, 2.8, 2.2, 1.0 cm/hr. The corresponding surface runoff is estimated to 3.6 cm. Determine ϕ and W-index.

Solution:

Assuming Rainfall > ϕ-index

Runoff depth $\Sigma(i - \phi)\ \Delta\ t$

or $$(1.6 - \phi) + (3.6 - \phi) + (5.0 - \phi) + (2.8 - \phi) + (2.2 - \phi) + (1 - \phi) \times \frac{1}{2} = 3.6$$

$$16.2 - 6\phi \times \frac{1}{2} = 3.6$$

$\therefore$ $$6\phi = 9 \qquad \therefore\ \phi = 1.5 \text{ cm/hr}$$

It is seen that last rainfall 1 (cm/hr) is less than ϕ-index. Considering,

Rainfall > ϕ-index

or $$(1.6 - \phi) + (3.6 - \phi) + (5.0 - \phi) + (2.8 - \phi) \times \frac{1}{2} + (2.2 - \phi)\ \frac{1}{2} = 3.6$$

or $$15.2 - 5\phi = 3.6 \times 2 = 7.2$$

$\therefore$ $$\phi = 8 \qquad \therefore\ \phi = 1.6 \text{ cm/hr}$$

Now, W-index = (P–R–S)

$$t_r = \frac{(1.6 + 3.6 + 5.0 + 2.2 + 1) - 3.6 - 0}{6 \times \frac{1}{2}} = 1.5 \text{ cm/hr}$$

EXAMPLE 4.4

A houring rainfall of the following storm produces runoff of 6 cm. Calculate ϕ-index and W-index.

Time	1	2	3	4	5	6	7	8
Rainfall	0.1	1.0	1.4	2.6	2.0	1.5	1.0	0.4

Solution: Assuming $i > \phi$

or $(0.1 - \phi) + (1.0 - \phi) + (1.4 - \phi) + (2.6 - \phi) + (1.5 - \phi) + (1.0 - \phi) + (0.4 - \phi) \times 1 = 6$

or $$10 - 8\phi = 6$$

$\therefore$ $$8\phi = 4 \qquad \therefore\ \phi = 0.5 \text{ cm/hr}$$

Since ϕ = 0.5 cm/hr, which is greater than 1st and last rainfall, second trial is:

$$(1.0 - \phi) + (1.4 - \phi) + (2.6 - \phi) + (2.0 - \phi) + (1.5 - \phi) + (1.0 - \phi) = 6$$

$$9.5 - 6\phi = 6$$

$\therefore$ $$6\phi = 3.5 \qquad \therefore\ \phi = 0.5833 \text{ cm/hr}$$

Now, $$\text{W-index} = \frac{P-R-S}{t_r}$$

$$= \frac{10-6-0}{8} = \frac{4}{8} = 0.5 \text{ cm/hr}$$

4.5 APPROXIMATE EQUATIONS OF INFILTRATION

Literature review on infiltration indicates that quite an amount of work is done on mathematical equations or models of infiltration. Some of them are developed empirically and some on principle of flow through porous media. Starting from Darcy's law, Green and Ampt[5] (1911) derived the following equation by combining Darcy's law and unsteady continuity equation, and then integrating the differential equation.

$$F = f_c t - \frac{a}{f_c} \log_e \left[1 + \frac{f_c F}{a}\right] \tag{4.2}$$

where f_c is ultimate constant steady rate of infiltration, t is time, a is constant of proportionality and F is volume of infiltration = $\int_0^t [f(t)]dt$.

In Eq. (4.2), F is implicit even when f_c and a are known. Hence method of iteration to compute F is required.

Kostiakov[6] (1932) found it more convenient to present infiltration equation in the form of cumulative curve than a rate curve. His exponential equation for this curve is given by

$$F = at^n \tag{4.3}$$

where F is the cumulative infiltration depth from the start of rainfall to time t, a and n are two parameters of watershed to be evaluated.

4.6 HORTON'S EQUATION OF INFILTRATION

Horton[7] (1940) developed the following mathematical equation for finding rate curve of infiltration capacity $f(t)$ at any time.

$$f(t) = f_c + (f_o - f_c)e^{-kt} \tag{4.4}$$

[5] Green, W.H. and Ampt, G., "Surface of Soil Physics, Part 1, The Flow of Air and Water Through Soil, *J. Agri. Science*, 4, pp. 1–24, 1911.

[6] Kostiakov, A.N., On the Dynamics of the Coefficient of Water Percolation in Soil and on the Necessity for Studying it from a Dynamic Point of View for Purposes Amelioration, Trans. 6th Comm. Intern. Soil Soc., Russian Part A, pp. 17–21, 1932.

[7] Horton, R.E., An Approach Towards a Physical Interpretation of Infiltration Capacity, *Soil. Sci. Soc. Am. Proc.*, 5, pp. 399–417, 1940.

Here f_0 is the infiltration rate just at the beginning of rainfall, f_c is the steady state infiltration capacity, k is a constant called **Horton's infiltration constant** dependent on vegetal cover and type of soil.

Horton's curve is shown in Figure 4.10. Horton's equation is popular and mostly used

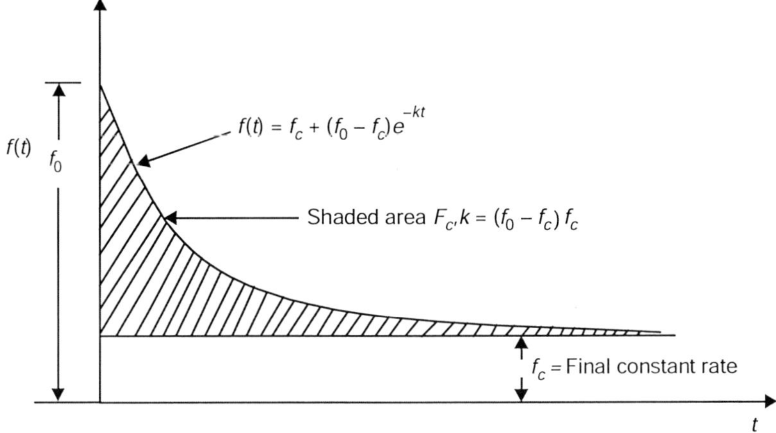

Figure 4.10 Horton infiltration curve.

4.7 SOME OTHER EQUATIONS

Philips (1957) gave the following equation:

$$f(t) = A + \frac{S}{2}t^{-1/2} \tag{4.5}$$

where A and S are two parameters to be evaluated similar to f_c and f_0 of Horton.

Overton (1964) developed another formula for infiltration, which is given as:

$$f(t) = f_c \text{see}^2\left[\tan\frac{bs_0}{\sqrt{f_c/a}}\right] \tag{4.6}$$

where s_0, a, b are constants of watershed.

Fok and Hansen (1966) presented a dimensionless infiltration equation of the following type:

$$\frac{F}{V_sH} - \log_e\left[1 + \frac{F}{V_sH}\right] = \frac{Kt_c}{V_sH} \tag{4.7}$$

where V_s is the initial moisture content in %, K is hydraulic conductivity, t_c capillary suction head and

$$H = V_s + \text{Constant water depth above the ground.}$$

Akan (1985), in his study of overland flow in pervious surface, assuming kinematic wave approximation (i.e., $s_b = s_f$) in continuity equation, gave the following partial differential equation relating infiltration rate f, rainfall i and overland flow q/unit width.

$$f = i - \left[\frac{\delta q}{\delta x} + \alpha^{1/m}\frac{\delta}{\delta t}\left(q^{1/m}\right)\right] \tag{4.8}$$

Here α and m are two constants depending upon the friction formula employed.

Smith (1972) presented the following equation after extensive experiments on the range of soil from fine clay with swelling properties to a moderately uniform soil.

$$f_p = f_a + A(t - t_0)^{-\alpha} \tag{4.9}$$

Here f_α is the saturated hydraulic conductivity or steady state infiltration rate and A, t_0, and α are some parameters unique to a soil, initial moisture and rainfall rate. For infiltration from an instantaneously ponded surface, $t_0 = 0$.

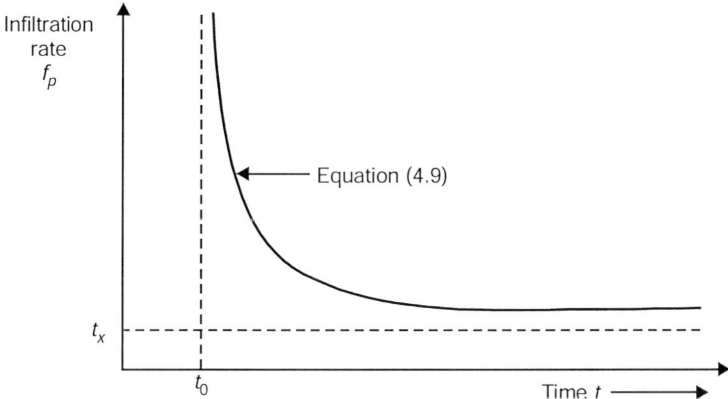

Figure 4.11 Schematic diagram of smith equation.

EXAMPLE 4.5

A 24-hour storm occurred over a catchment of 1.8 km^2 areas and the total rainfall observed is 10 cm. An infiltration capacity of 1 cm/hr initially and finally 0.3 cm/hr is obtained from Horton's curve with Horton's k = 5 hr^{-1}. An IMD pan installed in the catchment indicated a decrease of 0.6 cm in the water level (after allowing for rainfall) during 24 hours of its operation. Determine the runoff from the catchment. Assume pan coefficient = 0.7.

Solution: Total infiltration F = Area under Horton's curve

$$= ABCD + CFED$$

$$= \int_0^{15}\left[f_c + (f_0 - f_c)e^{-kt}\right]dt + (CF \times EF)$$

Here $f_c = 0.3$ cm/hr, $f_0 = 1$ cm/h, $k = 5$ hr^{-1}, $CF = (24 - 15) = 9$ hrs, $EF = f_c$

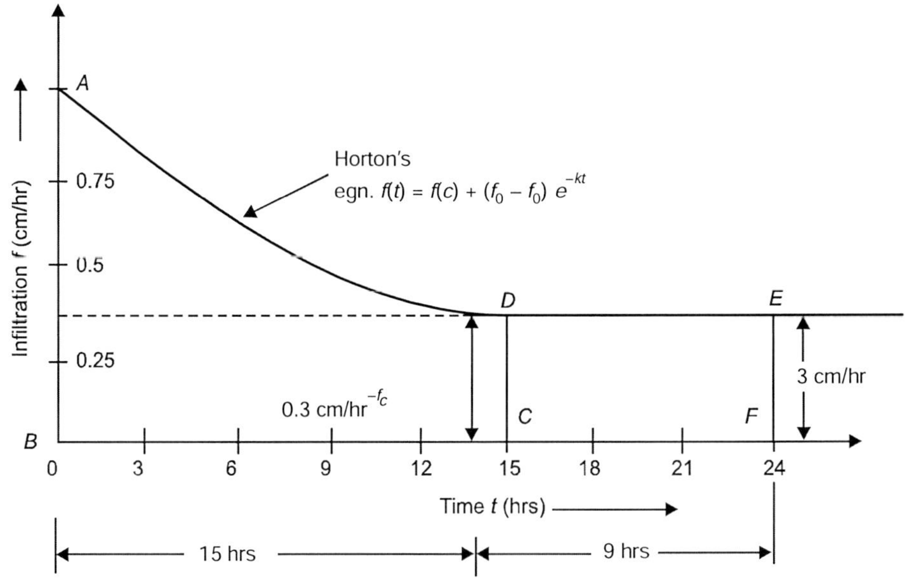

Figure 4.12 Visual of Example 4.5 to determine runoff.

After integration,

$$F = \left[0.3t + \left(\frac{0.7e^{-5t}}{-5}\right)\right]_0^{15} + (9 \times 0.3)$$

or
$$F = \left[0.3 \times 15 + \frac{0.7e^{-5\times15}}{-5} - \frac{0.7e^0}{-5}\right] + 2.7$$

or
$$F = (4.5 + 0 + 0.14) + 2.7$$

or
$$F = 7.34 \text{ cm}$$

Now,
$$\begin{aligned}\text{Runoff} &= \text{Precipitation} - \text{Infiltration} - \text{Evaporation}\\ &= 10 - 7.34 - 0.6 \times 0.7\\ &= 2.24 \text{ cm}\end{aligned}$$

$\therefore$
$$\begin{aligned}\text{Runoff volume} &= \text{Catchment area in m}^2 \times \text{Runoff in m}\\ &= \left(1.8 \times 10^6 \times \frac{2.24}{100}\right)\\ &= 40320 \text{ m}^3\end{aligned}$$

4.8 RICHARDS' PARTIAL DIFFERENTIAL EQUATION

According to Swartzendruber[8] (1969), equations in two different forms had been derived by Richards[9] (1931). These are known as **Richards' equation of infiltration**. The equations are:

$$C(h)\frac{\delta h}{\delta t}=\frac{\delta}{\delta z}\left[k(h)\frac{\delta h}{\delta z}\right]-\frac{\delta k}{\delta z} \tag{4.10}$$

and

$$\frac{\delta \theta}{\delta t}=\frac{\delta}{\delta z}\left[D(\theta)\frac{\delta \theta}{\delta z}\right]-\frac{\delta k}{\delta z} \tag{4.11}$$

Richards derived these equations starting from Darcy's law, i.e.,

$$q_s=-k\frac{\delta H}{\delta z} \tag{4.12}$$

Here q_s is the volume of water moving through soil in s direction per unit area time ($L^3/L^{-2}T^{-1}$), $\delta H/\delta S$ is the hydraulic gradient in the direction of s, k is the proportionality constant or hydraulic conductivity (L/T).

H may be considered equal to hydraulic head which is the sum of the pressure head h and distance above the datum plane. If datum plane is taken at soil surface,

$$H = h - z \tag{4.13}$$

where z is the positive distance measured downward from the surface. Again hydraulic conductivity k is a function of water content θ, and hence Eq. (4.12) may be written as:

$$q_s=-k(\theta)\frac{\delta H}{\delta S} \tag{4.14}$$

Again θ is the function of pressure h, i.e., $\theta = \theta(h)$. Hence we may write Eq. (4.14) as:

$$q_s=-k(h)\frac{\delta H}{\delta S} \tag{4.15}$$

From the principle of conservation of mass for soil water system, it may be written as:

$$\frac{\delta \theta}{\delta t}=-\nabla\cdot\overline{q} \tag{4.16}$$

For flow in vertical z direction, Eq. (4.16) may be written as:

$$\frac{\delta \theta}{\delta t}=\frac{\delta q_z}{\delta z} \tag{4.17}$$

[8] Swartzendruber, D., The Flow of Water in Unsaturated Soil in R.M. Dewiest (edition), *Flow Trough Porous Media*, pp. 215–292, Academic Press, New York, 1969.

[9] Richards, L.A., Capillary Condition Through Porous Mediums, Physics 1, pp. 312–318, 1931.

Combining Eqs. (4.15) and (4.17) and taking datum at the surface, so that $H = h - z$, yields the Richards' equation in vertical direction as:

$$C(h)\frac{\delta h}{\delta t} = \frac{\delta}{\delta z}\left[k(h)\frac{\delta h}{\delta z}\right] - \frac{\delta k}{\delta z}$$

which is Eq. (4.10), where the soil capacity $C(h)$ may be obtained from the soil water characteristic as:

$$C(h) \equiv \frac{\delta\theta}{\delta h}$$

Equation (4.10) can be also written with the water content θ as dependent variable by defining soil water diffusivity as:

$$D(\theta) = k(h)\frac{\delta h}{\delta\theta}$$

so that

$$\frac{\delta\theta}{\delta t} = \frac{\delta}{\delta t}\left[D(\theta)\frac{\delta\theta}{\delta z}\right] - \frac{\delta k}{\delta z}$$

4.9 NUMERICAL SOLUTION OF RICHARDS' EQUATION

In general case infiltration problems may involve non-uniform initial water, time dependent boundary conditions, hysteresis and heterogeneous and anisotropic porous media. These factors can be considered by using numerical methods to solve the governing equation of Richards subject to the initial and boundary conditions. Philips (1957) presented a rapidly converging numerical procedure for ponded infiltration into a deep homogenous soil with uniform initial water content. Other numerical methods to solve the equation by computer were developed by Whisler and Klute[10] (1965, 1966), Staple[11] (1966), Rubin and Steinhardt[12] (1963), Rubin[13] (1966) and Smith and Woolhiser[14] (1971).

Numerical finite difference procedures for solving Richards' equation have been developed by Rubin (1968), Amerman[15] (1969) and Freeze[16] (1971). Finite difference techniques are most commonly used. The difference schemes used are: (i) Explicit schemes (ii) Implicit schemes and

---

[10] Whisler, F.D. and Klute, A., Numerical Analysis of Infiltration Considering Hysteresis into a Vertical Soil Column at Equilibrium Under Gravity. *Soil Sci. Soc. Am. Proc.*, 29, pp. 489–494, 1965.

[11] Staple, W.J., Infiltration and Redistribution of Water in Vertical Columns of Loam Soil, *Soil, Sci. Soc. Am. Proc.*, 30, pp. 533–558, 1966.

[12] Rubin, J. and Steinhardt, R., Soil Water Relations During Rain Infiltration: Theory, *Soil Sci. Soc. Am. Proc.*, 27, pp. 246–251, 1963.

[13] Smith, R.E. and Woolhiser, D.A., Overland Flow on an Infiltrating Surface, *Water Resource. Res.*, 7(4), pp. 899–913, 1971.

[14] Rubin, J., Theoretical Analysis of 2-D Transient Flow of Water in Unsaturated and Partly Unsaturated Porous Soil, *Soil Sci. Soc. Am. Proc.*, 32, pp. 607–615, 1968.

[15] Amerman, C.R., Finite Difference Solution of Unsteady Two Dimensional Partially Saturated Porous Media Flow, Ph.D. Thesis, Purdue University, Lafayette, IN, 1969.

[16] Freeze, R.A., Three Dimensional Transient Saturated-unsaturated Flow in a Ground Water Basin, *Water Resources Res.*, 7, pp. 347–366, 1971.

(iii) Crank-Nikolson (C-N) approximation. Explicit scheme is simple, easy to handle and it provides good results with less computer runtime although sizes of Δt and Δz are small. Only explicit scheme will be discussed here.

4.9.1 Explicit Scheme

Using the grid system as shown in Figure 4.13, an explicit formulation of Eq. (4.10) yields:

$$h_i^{j+1} = h_i^j + \frac{\Delta t}{C_i^j \Delta Z_j}\left[k_{i+1/2}^j + \left(\frac{h_{i+1}^j - h_i^j}{\Delta Z} - 1\right) - k_{i-1/2}^j - \left(\frac{h_i^j - h_{i-1}^j}{\Delta Z} - 1\right)\right] \tag{4.18}$$

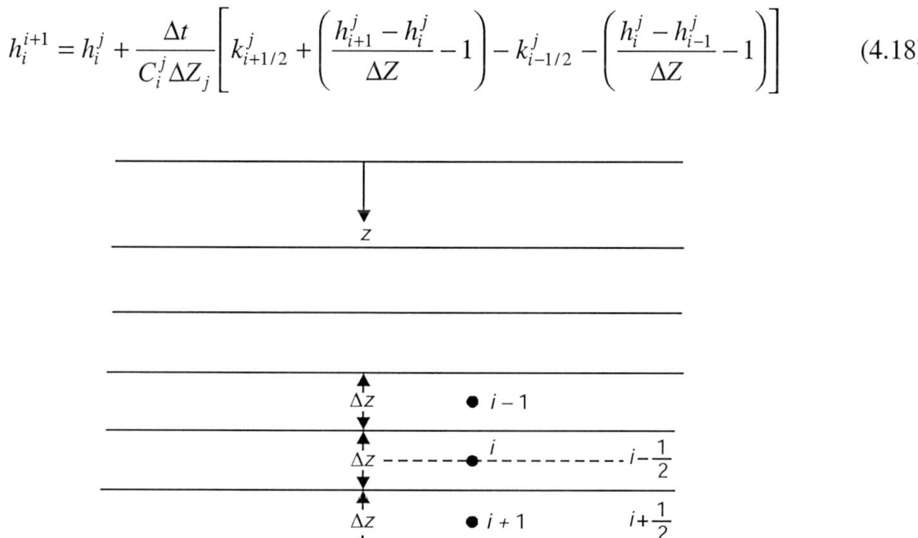

Figure 4.13 Grid for finite difference formulation of the Richards equation for flow in vertical direction.

The subscripts i refer to spatial increment and superscripts j refer to time increment. The procedures for obtaining $k_{i\pm 1/2}$ are discussed later. Equation (4.18) now can be solved from unknown h_i^{j+1} at $j + 1$ time level. The explicit scheme has been used by Ibrahim (1967) for a ponded boundary condition and by Staple (1966). This method is simple and for the method to be convergent, Richtmeyer[17] (1957) gave the following condition:

$$\Delta t < 0.25\ \Delta z^2 \frac{C}{K}$$

Procedures to obtain $k_{i\pm 1/2}$

Arithmetic mean method. This method is used quite frequently in classical finite difference method by Rubin (1968), Amerman (1968) and Whisler and Klute (1965).

$$k_{i\pm 1/2} = 0.5(k_i + k_{i\pm 1}) \tag{4.19}$$

---

[17] Richtmeyer, R.D., *Different Methods of Initial Value Problems*, Interscience Publishers, New York, 1957.

Harmonic mean method.

$$k_{i\pm1/2} = \frac{2k_i k_{i\pm1}}{k_i + k_{i\pm1}} \tag{4.20}$$

The method is used by Smith and Woolhiser (1971).

Geometric mean method.

$$k_{i\pm1/2} = \left(k_i k_{i\pm1}\right)^{1/2} \tag{4.21}$$

For details of implicit and Crank-Nicholson (C-N) schemes, readers are referred to Haverkamp[18] et al. (1977) and Hanks and Bowers[19] (1962) respectively.

4.10 CONCLUSION

Infiltration, its definition, factor affecting are discussed in the beginning. Types of different infiltrometers, infiltration indices and few numerical examples are explained. Different approximate equations developed by different investigators and actual partial differential equation of infiltration developed by Richards are presented. Numerical explicit finite difference solution is discussed for using in computer solution.

EXERCISES

4.1 Explain infiltration. Describe the factors that affect infiltration.

4.2 How infiltration in the field is measured? What are the different infiltration indices? Explain each of them by sketches where necessary.

4.3 Describe Horton's equation, Richardson's equation and some other approximate equations of infiltration.

4.4 The ordinate of mass curve of rainfall at hourly interval due to 6-hour storm over 2 km^2 basin are as follows:

Time in hours	1	2	3	4	5	6
Ordinate of mass curve (cm)	2	8	18	33	42	44

If the observed runoff is 40 ha-m of water, determine ϕ-index.

(*Hint:* Find first rainfall at different hours from the mass curve as:

Time	1	2	3	4	5	6
Rainfall	2	6	10	15	9	2

Then apply first trial assuming $i > \phi$, $\phi = 4$ cm/hour rainfall at hour 1 and hour 6 less than $\phi = 4$ cm/hr, apply second trial rainfall at 2, 3, 4 and 5.

(***Ans:*** $\phi = 5$ cm/hr)

[18] Haverkamp, R.M., Vauclin, J., Tauma, P.J.W. and Vachand, G., A Comparison of Numerical Simulation of Models for One Dimensional Infiltration, *Soil Sci. Soc. Am. Proc.,* 41, pp. 530–534, 1977.

[19] Hanks, R.J. and Bowers, S.A., Numerical Solution of Moisture Flow Equation of Infiltration in a Layered Soil, *Soil Sci. Soc. Am. Proc.,* 26, pp. 530–534, 1962.

4.5 A 7 hour storm over a catchment of 1800 sq km produced intensities at half an hour interval are 3, 6, 20, 18, 13, 11, 12, 1, 7, 16, 17, 13, 5, 2 mm/hr. If the corresponding observed runoff is 72.0×10^6 m^3, estimate the ϕ-index of the storm.

(***Ans:*** 10 mm/hr)

4.6 The ordinates of rainfall of mass curve of storm over a basin of area 850 km^2 in mm at 1 hr. interval are:

0, 12, 22, 30, 39, 45.5, 50, 55.5, 60, 64, 68

Infiltration to occur at capacity rate during this storm is given by Horton's equation with $f_0 = 6.5$ mm/hr and $f_c = 1.5$ mm/hr and $k = 0.15$ hr^{-1}. Estimate resulting runoff volume.

(***Ans:*** 23×10^6 m^3)

4.7 The rates of rainfall for the successive half an hour period of $3^1/_2$ hours storm are:

3.5, 4.0, 12.0, 8.5, 4.5, 4.5, and 3.0 cm/hr

Assuming the ϕ-index of 3.5 cm/hr find the net rainfall in cm, total rainfall and W-index.

(***Ans:*** 8 cm, 20 cm, 3.43 cm/hr)

4.8 The average rainfall over a watershed of area 45 ha for a particular storm was as follows:

Time (hr)	0	1	2	3	4	5	6	7
Rainfall (cm)	0	0.5	1.0	3.25	2.5	1.5	0.5	0

The volume of runoff from this storm was determined to be 2.25 ha-m. Determine ϕ-index.

(***Ans:*** 0.81 cm/hr)

4.9 An infiltration capacity prepared for a catchment indicated an initial infiltration capacity of 2.5 cm/hr and attains a constant rate of 0.5 cm/hr after 10 hours of rainfall with Horton's $k = 6$ day^{-1}. Determine the total infiltration loss.

SUGGESTED FURTHER READINGS

Chow, V.T. (Ed.), *Handbook of Applied Hydrology*, McGraw-Hill, New York, 1964.

Collis-George, N., Infiltration for Simple Soil System, *Water Resource Res.*, vol. 13, No. 2, pp. 395–4-5, April 1977.

Das, G., *Hydrology and Soil Conservation Engineering,* Prentice-Hall of India, New Delhi, 2000.

Dunin, F.X., Infiltration, its Simulation for Field Conditions Chapter 8 in J.C. Rodda (Ed.), *Facts of Hydrology*, pp. 199–227, Wiley, London, 1976.

Haan, C.T. (Ed.), *Hydrologic Modeling of Small Watersheds,* An ASAE Monograph No. 5, ASAE, 1982.

Horton, R.E., Analysis of Runoff Experiments with Varying Infiltration Capacity, *Trans. Am. Geophysics Union,* vol. 20, pp. 693–711, 1939.

Horton, R.E., The Role of Infiltration in Hydrologic Cycle, *Am. Geophysical Union,* vol. 14, pp. 446–460, 1933.

Mein, R.G. and Larson, C.L., Modeling Infiltration During Steady Rain, Water Resource, vol. 9, No. 9, No. 2, pp. 384–394, 1973.

Application of Infiltration Theory for the Determination of the Excess Rainfall Hyetograph, *Water Resource Bull*, vol. 17, No. 6, pp. 1012–1022, 1981.

Ponce, V.M., *Engineering Hydrology: Principles and Practices,* Prentice-Hall Inc, Upper Saddle River, N.J., 1989.

Raghunath, H.N., *Hydrology: Principles, Analysis, Design,* New Age International, New Delhi, 1995.

Rawls, W.J., Brackensiek, D.L. and Miller, N., Green Ampt Infiltration Parameters from Soil Data, *Jour, Hy. Div.*, ASCE, vol. 109, No. 1, 1983.

Ray, K.L., Kohler, M.A. and Paulhus, J.L.H., *Hydrology for Engineers,* McGraw-Hill International, New York, 1982.

Reddy, P.J., *A Textbook of Hydrology*, Lakshmi Publication, Delhi, 1989.

Sherman, L.K., Compilation of F-curves Derived by the Method of Sharp and Holton and of Sherman and Mayer, *Trans. Am. Geophysics Union*, vol. 24, Part 2, 1943.

Sing, V.P., *Elementary Hydrology,* Prentice-Hall Inc., Upper-Saddle River, N.J., 1992.

Subramanya, K., *Engineering Hydrology*, Tata McGraw-Hill, New Delhi, 1984.

Whisler, F.D. and Klute, A., Analysis of Infiltration in Stratified Soil Columns Symposium on Water in Unsaturated Zone, Wageningen, Netherlands, 1966.

Zingg, A.W., The Determination of Infiltration Rates on Small Agricultural Watersheds, *Trans. Am. Geophysics Union*, vol. 24, Part 2, 1943.

Chapter 5
Evapotranspiration

5.1 INTRODUCTION

Evaporation and transpiration jointly called **evapotranspiration (ET)** is one of the components of hydrologic cycle. Loss of water in the form of vapour from soil, snow, lake, streams, reservoir, seas and depressions to the atmosphere due to energy of sun is called **evaporation**. Transpiration is the process by which water leaves the living plant body and enters the atmosphere as vapour. The process involves collection of water in the body of the plant and finally evaporation of it to the atmosphere from stomata of the leaves. This loss of water by these two processes to the atmosphere is most commonly called **evapotranspiration** (ET). Detailed study of evapotranspiration (ET) is essential in the design of reservoir, irrigation canals, water balance on earth surface and projects relating to water.

5.2 HISTORICAL EVIDENCE AND PREVIOUS WORKS OF EVAPOTRANSPIRATION (ET)

Biswas[1] (1972) is on the opinion that transfer of water in the form of vapour by the process of ET has intrigued and fascinated the scientists since early recorded history. Aristotle wrote the first treatise on meteorology and evaporation which were the effect of sun's heat. Leonardo da Vinci in late 1400s wrote *Where there is life, there is light and where vital heat is, there is movement of vapour*. Pan evaporation developed in late seventy's has remained to be same till today. In 1795, Dalton constructed lysimeter for runoff and drainage. Fitzgerald[2] (1886) identified many quantities and variables related to pan and lake evaporation. Rohwer[3] (1931) worked on evaporation from free water body. Thornthwaite and Holzman[4] (1942) measured evaporation from land and water surface. Robinson and Johnson[5] (1961) had published a bibliography on evaporation and evapotranspiration of the United States since 1960.

---

[1] Biswas, A.K., *History of Hydrology*, North Holland Publishing, Amsterdam, 1972.
[2] Fitzgerald, D., *Evaporation*, Trans, ASCE, 15, pp. 581–646, 1886.
[3] Rohwer, C., Evaporation from Free Water Surface, USDA, Tech. Bulletin, No. 271, 1931.
[4] Thornthwaite, C.W. and Holzman, B., Measurement of Evaporation from Land and Water Surface, USDA, Tech, Bulletin, No. 817, 1942.
[5] Robinson, T.W. and Johnson, A.I., Selected Bibliography on Evaporation and Evaporation US Geo., Sur. Water Supply Paper 1539R, 1961.

Theoretical work by Penman[6] (1948) in which he combined vertical energy budget with horizontal wind surface and lysimeter studies by Harrold and Dreilbelbis[7] (1958, 1967) are significant contributions. Tanner and Fucks[8] (1968), Van Bavel[9] (1966), Monteith[10] (1965), Rijtema[11] (1965) modified Penman Model by direct net radiation estimate and wind profile theory. Hillel and Talpez[12] (1976) commented that plant roots and soil moisture have received considerable attention in recent years. Wartena[13] (1974) has provided a useful summary of the past century research on evapotranspiration (ET). According to him basic problems of effect of turbulent air and influence of soil water content are yet to be explored. ET from vegetal cover is the result of several processes like radiation exchanges, vapour transport and biological growth operating within a system that involves temperature of atmosphere, plants and soil. Examples of such integrated system approach have been reported by Ritchie[14] (1972), Van Keulen[15] (1975), Van Bavel and Ahmed[16] (1976), Brutsaert[17] (1982) and others. Researchers like Tanner[18] (1957), Goodell[19] (1966), and Penman et al (1967) have provided a good description of ET from vegetal cover, which requires energy input, water availability and transport process from earth's surface to atmosphere.

Thus, lots of investigations have been done by many investigators on evapotranspiration (ET).

5.3 FACTORS AFFECTING EVAPORATION

There are various factors like hydrological, meteorological and physical that affect the rate of evaporation from earth's surface. They are discussed as follows:

---

[6] Penman, H.L., Natural Evaporation from Open Water, Bare Soil and Grass, Proc. Soc. London Ser. A, No. 1032, 193:120–145, 1948.

[7] Harrold, L.L. and Dreilbelbis, F.R., Evaporation of Agricultural Hydrology by Monolith Lysimeters, USDA, Tech, Bulletin, No. 1179, 1958.

[8] Tanner, C.B. and Fucks, M., Evaporation from Unsaturated Surfaces: A Generalized Combination Method, *Jour. Geophysics Res.*, 73 (4), pp. 1299–1304, 1968.

[9] Van Bavel, C.H.M., Potential Evaporation: The Combination Concept and its Experimental Verification, *Water Res. Research*, 12(3), 1966.

[10] Monteith, J.L., Evaporation and Environment in Fogg, G.E. (Edition). The state and movement of water in living organism, Academic Press, New York, 1965.

[11] Rijtema, P.F., An Analysis of Actual Evapotranspiration, *Agr. Res., Rep.*, 689, Pudoc, Wageningen, The Netherlands, 1965.

[12] Hillel, D. and Taplez, H., Simulation of Root Growth and its Effect on the Pattern of Soil Water Uptake by a Non-uniform Root System, *Soil Sci.*, 121 (5), 1976.

[13] Wartena, L., Basic Difficulties in Predicting Evaporation, *Jour. Hydrology,* 23 (Y_2), 1974.

[14] Ritchie, J.T., Model for Predictions of Evaporation from a Row Crops with Incomplete Cover, *Water Res. Research* 8(5), pp. 1204–1213, 1972.

[15] Van Keulen, H., Simulation of water use and herbage growth in region. Simulation Monograph. Pudoc Wageningen, the Netherlands, 1975.

[16] Van Bavel, C.H.M. and Ahmed, J., Dynamic Simulation of Water Depletion in the Root Zone, Ecol. Modeling 2: 109–212, 1976.

[17] Brutsaert, W.H., *Evaporation into the Atmosphere*, D. Reidel Dordrecht, Holland, 1982.

[18] Tanner, C.B., Factors Affecting Evaporation from Plants and Soil, *Jour. Soil and Water Conserv.*, 12(5), pp. 221–227, 1957.

[19] Goodell, B.C., Watershed Treatment Effects on Evapotranspiration, Int. Symp. on Forest Hydrology, Penn. State. Univ. Aug 29 to Sept 10, 1965, Pergamum Press, New York, 1966.

Radiation: Radiation is the most important factor of evaporation. Solar radiation supplies continuous energy, which is essential for evaporation. Evaporation is directly proportional to radiation. Solar energy near the equator is more, therefore, evaporation is much more.

Vapour pressure: Evaporation rate varies directly with difference of vapour pressure between air and water. If E is the rate of evaporation (mm/day) and, e_w and e_a are the vapour pressure in water and in air, then,

$$E = C\,(e_w - e_a) \tag{5.1}$$

Here C is constant. Equation (5.1) is called **Dalton's law of evaporation** in honour of John Dalton[20] (1802), who first proposed it.

Temperature: Increase in air temperature increases evaporation when other factors remaining same yet the high correlation coefficient between the two does not exist. In cold dry season although temperature is less, rate of evaporation is more because some of heat energy absorbed at lower depth in hot weather is released in cold season.

Wind velocity: The increase in wind velocity increases evaporation. Wind removes the evaporated water vapour and thereby creates space for new evaporated water vapour. When there is no wind above the water body where the evaporated water vapour is in still condition, further evaporation ceases to take place. If wind velocity over the water body is high, it does not increase correspondingly the evaporation. There is always a critical velocity of wind beyond which evaporation does not increase. Wind velocity near the ground and water level is effective and influence on evaporation is more if the wind velocity is turbulent.

Atmospheric pressure: If atmospheric pressure is more, according to Dalton's law, e_a is more, hence less evaporation. Thus, decrease in atmospheric pressure can increase evaporation. At higher altitude, atmospheric pressure is low; hence evaporation should have been more. But this is not necessary because temperature at higher altitude is low which reduces evaporation.

Area of water surface: Evaporation is directly proportional to the area exposed. Hence, if area is more, evaporation is more. But there is always a difference of evaporation rate between smaller and larger water surface although total evaporation from larger area will be more. The main cause is that when air moves over a large water body, it holds water vapour, and eventually water vapour holding capacity decreases as it moves further. Of course, it largely depends on the humidity of incoming air.

Quality of water: It also affects the rate of evaporation. If water contains dissolved salts, it reduces the saturated vapour pressure e_s and by Dalton's law, E decreases. Also turbidity of water has some indirect effects.

Nature of evaporating surface: Evaporating surface is classified into three main surfaces: land surface, water surface and snow surface. Temperature remaining the same, evaporation from saturated soil surface is same to that of adjacent water surface. Evaporation decreases when

[20] Dalton, J., Experimental Essay on the Constitution of Mixed Gases; on the Force of Steam or Vapour from Waters and other Liquids, both in Torricellian Vacuum and in Air; on Evaporation and on Expansion of Gases by Heat, *Mem. Proc. Manch. Lit. Phil Soc.,* vol. 5, pp. 535–602, 1802.

the soil surface is dry. Thus, evaporation is dependent on the availability of water on the surface. Further, it is again dependent on pressure or absence of vegetal cover. Evaporation rate decreases in the following order: bare ground, grass and croplands, light forests and dense forest.

Salinity of water: It actually falls under quality of water. Yet if the soil is saline, evaporation decreases.

Depth of water in the water body: If the depth is more, it increases evaporation in winter season. On the other hand, low depth increases evaporation in summer as all water gets warmed up by solar radiation.

Humidity: If humidity is more, water holding capacity of air is less, so less evaporation. If water content is less in air, more evaporation will take place.

5.4 MEASUREMENT OR ESTIMATION OF EVAPORATION

The following methods are employed to measure evaporation:

5.4.1 Empirical Formulae

Based on Dalton's theory various empirical formulae have been developed by different investigators with some modifications to estimate the evaporation from free water surface. Most of them are dependent on wind velocity, temperature and atmospheric pressure. Some of the formulae commonly used are as follows:

Fitzgerald's equation (1886): D. Fitzgerald gave the following equation in 1886:

$$E = (0.4 + 0.124V)\,(e_s - e_a) \tag{5.2}$$

where E is evaporation in mm/day, e_s is saturated vapour at the temperature of water surface in mm of mercury, e_a is actual vapour pressure of air in mm of mercury and V is average wind speed at the surface in km/hr.

Meyer's equation (1915): Another formula given by Meyer in 1915 is:

$$E = C\,(1 + V/16)(e_s - e_a) \tag{5.3}$$

The values of E, V, e_s and e_a are already defined with their magnitude and dimensions. Here e_a and V are measured at a height of 9 m above the surface. C is a coefficient having a valve of 0.36 for large deep water and 0.50 for small and shallow water.

Rohwer's equation (1931): It is C. Rohwer who developed the following equation:

$$E = 0.771(1.465 - 0.000732P_a)(0.44 + 0.0733V)\,(e_s - e_a) \tag{5.4}$$

where E, e_s and e_a have the same meaning like Eq. (5.2), V is mean velocity of wind in km/hr and at 0.6 m above the surface and P_a is mean atmospheric pressure in mm of mercury.

Horton's equation (1939):

$$E = 0.4(2 - e^{-0.124V})(e_s - e_a) \tag{5.5}$$

The values of E, e_s, e_a and V have the same significance as in Eq. (5.2).

Lake Mead's equation:

$$E = 0.0331V\,(e_s - e_a)[1 - 0.03(T_a - T_w)] \tag{5.6}$$

Here T_a and T_w are average temperature in °C of air and water surface respectively.

These empirical formulae or equations can be used for estimation quite quickly. But they must be used with caution. Constants of the equation may vary in different areas or watersheds.

5.4.2 Water Budget Method or Storage Equation

Evaporation E from a reservoir or water body can be determined by the following water budget or storage equation:

$$E = P + 1 - O + O_u + \Delta S \tag{5.7}$$

where P is the total precipitation, I is total inflow, O is total outflow, O_u is total underground inflow or outflow which is positive for inflow and negative for outflow and ΔS is change in storage (+ve for an increase in storage and –ve for a decrease in storage).

All the parameters are converted into same units, preferably in terms of depth of water area for some convenient time interval.

5.4.3 Energy Budget Method

This method is based on the application of the law of the conservation of energy in the form of heat. The energy budget equation from Figure 5.1 may be written as:

or $$Q_i - Q_r - Q_l - Q_c - Q_e = Q_s - Q_a$$

∴ $$Q_e = Q_i - Q_r - Q_1 - Q_c - Q_s - Q_a \tag{5.8}$$

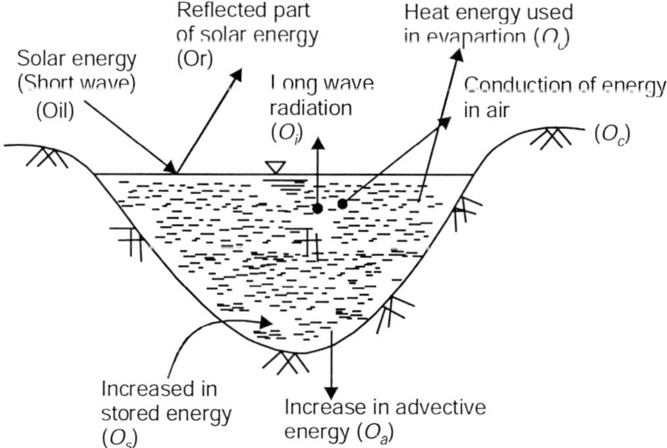

Figure 5.1 Energy balance in a water body to estimate evaporation.

where Q_i is short solar radiation, Q_r is reflected part of solar every, Q_l is long wave atmosphere radiation, Q_c is conduction energy in air, Q_e is heat energy used in evaporation, Q_s is increase is stored energy and Q_a is advective energy.

To calculate Q_c, Bowen ratio[21] (1926) equation is used, which states that

$$r = Q_c/Q_e$$

Since $$Q_e = \rho LE \tag{5.9}$$

where ρ is the density of water, L is the latent heat of evaporation and E is in mm.

Thus, $$r = \frac{Q_c}{Q_e} = \frac{Q_c}{\rho LE}$$

$$= 6.1 \times 10^{-4} P_a \left[\frac{T_w - T_a}{e_w - e_a} \right] \tag{5.10}$$

where P_a is atmospheric pressure in mm of mercury, e_w is saturated water vapour pressure in mm of mercury, e_a is actual vapour pressure in mm of mercury, T_w is temperature of water surface at 0°C and T_a is temperature of air at °C.

Bowen ratio can again be expressed as:

$$r = \frac{Q_c}{Q_e} = \frac{Q_i - Q_r - Q_l - Q_e - Q_s + Q_a}{Q_e}$$

or $$r = \frac{Q_i - Q_r - Q_l - Q_s + Q_a}{Q_e} - 1$$

or $$(1 + r) = \frac{Q_i - Q_r - Q_l - Q_s + Q_a}{\rho LE} \qquad \because Q_e = \rho LE \tag{5.11}$$

∴ $$E = \frac{Q_i - Q_r - Q_l - Q_s + Q_a}{\rho L(1 + r)}$$

which gives the evaporation. When parameters are measured for short period, Q_s and Q_a (negligibly small) may be neglected. Then Eq. (5.11) becomes

$$E = \frac{Q_i - Q_r - Q_l}{\rho L(1 + r)} \tag{5.11a}$$

Equation (5.11) may again be written as:

$$E = \frac{(Q_i - Q_r - Q_1) + (Q_a + Q_s)}{\rho L(1 + r)} \tag{5.11b}$$

or $$E = \frac{Q_{in}}{\rho L(1 + r)}$$

Here Q_{in} is net solar radiation.

---

[21] Bowen, I.S., The Ratio of Heat Losses by Conduction and Evaporation from any Water Surface, *Phys. Rev.* 27: pp. 779–787, 1926.

5.4.4 Mass Transfer Method

When wind flows on the surface, a boundary layer is formed. The method is based on turbulent mass transfer in this boundary layer to calculate the mass of water vapour transferred from the surface to the surrounding atmosphere. Therefore, it is known as **vapour flow approach** or **aerodynamics approach**. Prandtl mixing length theory of boundary layer concept is applied to obtain an expression of evaporation. Assuming logarithmic wind velocity distribution in turbulent flow and an adiabatic atmosphere, E is obtained as:

$$E = \frac{46.08(e_1 - e_2)(v_2 - v_1)}{(T - 273)\log\left(\frac{z_2}{z_1}\right)^2} \tag{5.12}$$

Here z_1 and z_2 are arbitrary lower and upper level (in metre) above the surface, E is in mm/hr, e_1 and e_2 are vapour pressures at z_1 and z_2 respectively, v_1 and v_2 are wind velocity at z_1 and z_2 levels and T is the average temperature of air in °C between z_1 and z_2.

Assume z_1 is in thin layer of saturated vapour and $e_1 = e_S$. e is the vapour pressure where velocity is v, a slight modification of equation (5.12), it may be written in the general form well known Dalton's equation i.e. $E = K\ (e_w - e_e)(a + b)v$: as:

$$E = K'\ (a + bv)\ (e_s - e) \text{ where } K'\ a,\ b \tag{5.13}$$

are some constant, V is average velocity between z, and z_2. e is the air vapour pressure.

Penman[22] (1956) has suggested following equation of evaporation:

$$E_a = 0.35(0.5 + 0.54v)\ (e_s - e_a) \tag{5.14}$$

where E_a is in mm/day, e_s and e_a in mm of mercury and V is in m/sec at height 2 m above the surface.

5.4.5 Combined Energy Budget and Mass Transfer Approach

In energy budget and mass transfer methods, measurement of data makes the use of the methods difficult. To alleviate this difficulty, Penman (1956) gave the following simple equation by both the methods. The data appeared in the following equation is limited number of fairly easily measured meteorological variables:

$$E = \frac{(\Delta Q_{in}/L_e) + (0.00061\ PE_a)}{\Delta + 0.00061P} \tag{5.15}$$

Here E_a is given by Eq. (5.14).

Δ is the slope of the saturated vapour pressure curve at air temperature in mm of mercury/ °C, and is obtained from the following equation:

$$\Delta = \frac{e_s}{T}\left(\frac{679^{0.498}}{T} - 5.02808\right) \tag{5.16}$$

[22] Penman, H.L., "Evaporation: An Introductory Survey", *Jour. of Agri. Science*, 4, pp. 9–29, 1956.

Here T is the air temperature in K, P is the air pressure in mm of mercury, Q_{in} is the net solar radiation in cal/cm^2 and L_e is the latent heat of varporisation of water in cal/gm which varies with temperature and is obtained from:

$$L_e = 597.3 - 0.564t \tag{5.17}$$

where t is air temperature in °C.

5.4.6 PAN MEASUREMENT METHOD

The five methods of estimation (or measurement) of evaporation are not directly applicable in design problem and also in some cases required data are not easily available. Therefore, in most of the design problems, evaporation is measured by evaporation pans which are called **evaporimeters**. Pans are commonly used to measure evaporation, and this method has become increasingly popular.

The different evaporimeters that are in use are discussed as follows:

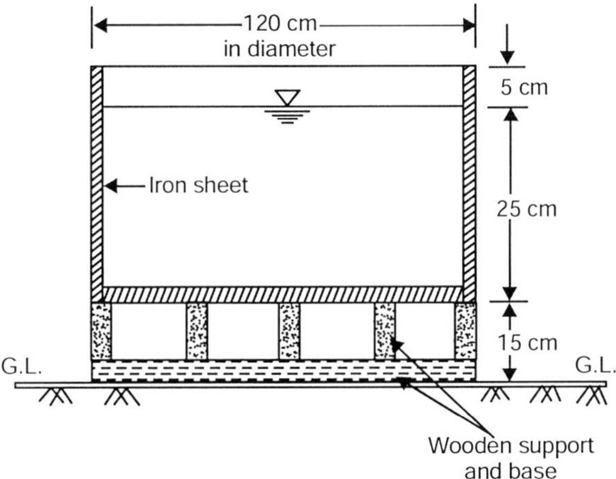

Figure 5.2 US Weather Bureau class A Pan evaporimeter (surface pan).

US weather bureau class A pan (Surface pan): The US weather bureau class A pan or surface pan evaporimeter along with its dimensions is shown in Figure 5.2. It is made of galvanized iron sheet. It is painted white. The pan is kept in wooden platform as shown in the figure so that air can circulate freely. Evaporation is measured by a hook gauge in a stilling well. Water level is measured daily. Water is added every day up to the fixed level. A pan coefficient of 0.7 is taken and thus pan coefficient is:

$$\text{Pan coefficient} = 0.7 = \frac{\text{Lake evaporation}}{\text{Pan evaporation}}$$

Advantages

1. It gives stable pan coefficient (0.6 to 0.8). The average value 0.7 is normally adopted without appreciable error.
2. It is easy for observation, i.e., measurement.
3. It has relative freedom of dirt and trash.
4. Cost of installation is reasonably low.

Disadvantage

1. The pan gives higher rate of evoporation than that of large free water surface.
2. Effects of wind and radiation are more which overestimate the expiration rate.

ISI standard pan (IS5973–1970): In Figure 5.3, ISI standard pan has been shown. It is also known as **modified class A pan.** Dimensions of the pan are shown in the figure. It is placed in the vicinity of the lake or reservoir to determine the evaporation of the lake. It is covered with wire mesh of galvanized iron to protect the water in the pan from birds. Pan is made of copper sheet of 0.9 mm thickness. The pan has a stilling well with a point gauge and thermometer. Amount of water lost can be measured by the point gauge. Water is added to bring it back to the original level. Readings are measured normally twice a day. The annual pan coefficient is 0.7.

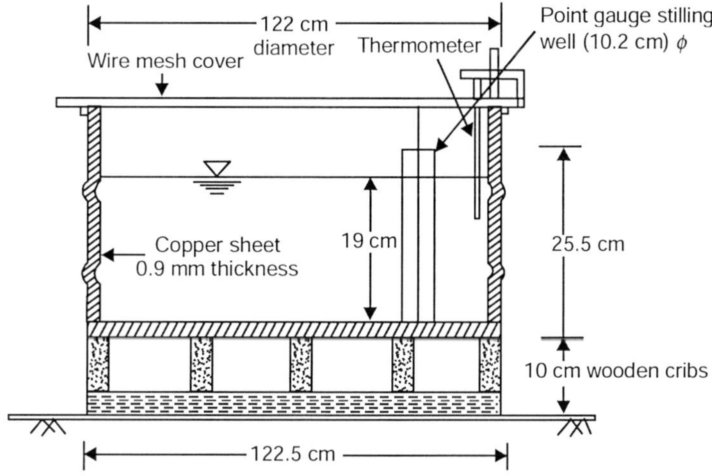

Figure 5.3 ISI standard pan (IMD land pan).

Advantages

1. As the pan is placed over wooden support, air circulation around the pan is alright.
2. Allowance for rainfall can be made if required.
3. Evaporation rates obtained are quite compatible to the rate of large waterbody.

Disadvantages

1. If screened at the top, interference of wind takes place and there by evaporation rate is reduced about 1.144 times as reported from experiment.
2. If left unscreened, birds-bathing in the pan and drinking water from the pan may affect the evoporation.

Colorado sunken pan: The Colorado sunken pan with its usual dimensions is shown in Figure 5.4. It is buried into the ground. Water level in the pan is at ground level.

Advantages

The advantage of this pan is that radiation and aerodynamics effects are similar to lake.

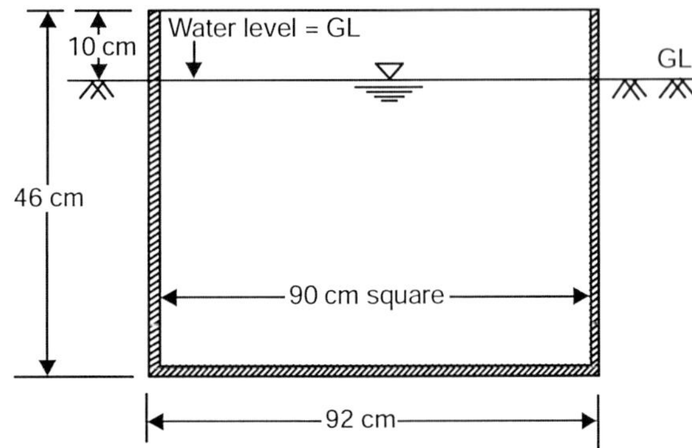

Figure 5.4 Colarodo sunken pan.

Disadvantages

Yet, it has some disadvantage like

1. It is difficult to detect leak, if any.
2. Extra precaution is to be taken for the effect of surrounding tress in the area.
3. Installation is expensive.

US geological survey floating pan: To simulate the characteristics of large body of water, this floating pan is adopted. It is square pan of 90 cm side and 45 cm depth. It is supported by drum floats in the middle of the raft of size 4.25 m × 4.87m. The water level in the pan is kept at the same level of the lake with rim of 7.5 cm (see Figure 5.5). To prevent wave action of the lake, diagonal baffles are provided in the pan so that surging in the pan is reduced.

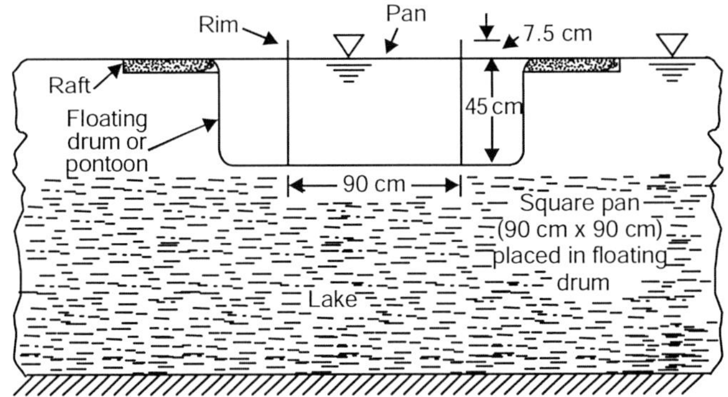

Figure 5.5 US geological survey floating pan.

Advantages

1. The pan floats in take and environment for evaporation is same as that of the lake.
2. Evaration rates obtained is very close to lake evaporation.

Disadvantages

1. High cost of installation is there.
2. Maintenance cost is also more.
3. Measurement of evaporation is the main disadvantage.

Pitche evaporimeter or atmometer: Atmometer is another device to measure evaporation. These atmometers are provided with some special surfaces which are kept wet and from which water loss by evaporation is recorded. Pitche atmometer or evaporimeter is a graduated glass tube of 1.5 cm in diameter and 30-cm long and one open end. It is filled with water and its open end is covered with a dry paper, held in position by metal clip. The tube is held in inverted position so that water wets the paper from which evaporation takes places. Water loss in the tube is recorded as a measure of evaporation. Atmometer is shown in Figure 5.6.

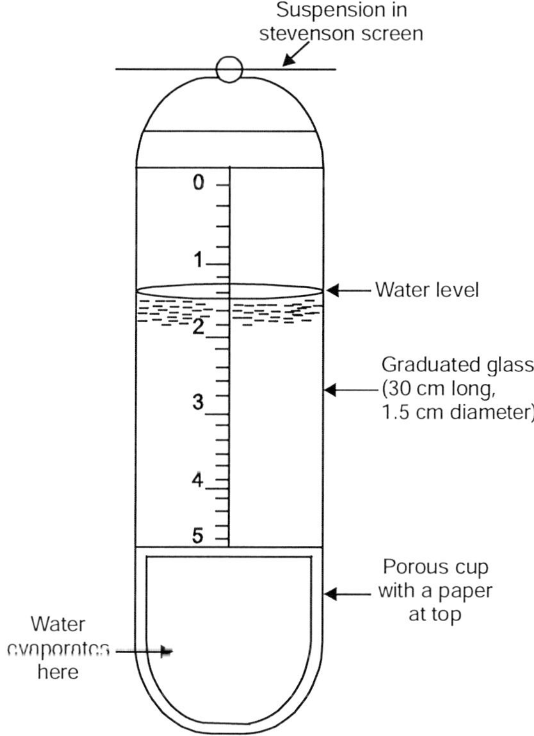

Figure 5.6 Pitche evaporimeter or atmometer.

Advantages

1. It is easy to handle as the size is small and reading are obtained directly from graduated scale.
2. For quick and rough estimate, it is better.

Disadvantages

1. This atmometer does not provide good measurement.
2. The readings obtained are more erratie them other standard pan.
3. Because of its small size, rates of evaporation are in excess of the rates of other evaporimeters.
4. Values are usually higher than those obtained by U.S. Weather Bureau class A pan.

5. Readings tend to overestimate due to wind effects and underestimate due to radiation effects.
6. Evaporating surface is often subject to contamination by dust and other foreign materials.

5.5 METHODS TO REDUCE RESERVOIR EVAPORATION

Evaporation from water body is a continuous process. In Indian conditions, annual evaporation varies from 1.5 m to 2.0 m. This value may even increase in arid region. Multipurpose projects are always with a big reservoir with bigger surface area, so loss due to evaporation is more. Various methods to reduce reservoir evaporation are discussed as follows:

Reduction of surface area of reservoir: Evaporation is directly proportional to water surface area of the reservoir. In selecting the site for reservoir, due consideration should be given for deep reservoir with less surface area. Reservoirs in deep gorges are preferable.

Wind breakers: Some has advocated for wind breakers as wind velocity over the surface increases the evaporation. If tall trees like causerina are allowed to grow on windward side of the reservoir, it may act as natural wind breakers. Besides it may help in cooling the water surface obstructing solar radiation at least for a part of the day. Wind breakers are effective only in small reservoir.

Mechanical covers: Small reservoirs are sometimes entirely covered to reduce evaporation. Cooley[23] (1970) suggested floating cover and Meyer and Frazier[24] (1970) advocated floating granular materials. Such methods are effective but expensive to apply.

Monomolecular films: Extensive research has been undertaken by Mansfield[25] (1955), Gunaji[26] (1968), Pushkarev and Levchenko[27] (1967) for use of monomolecular films by cetylalcohol to reduce evaporation from surface of bodies of water. Linsley[28] et al (1982) said that this chemical is very effective in evaporation, but for big reservoir this is not the case. Therefore, it appears that any hope for achieving a reliable, practical reduction in evaporation from large reservoir lies in finding a material which can increase the reflectivity of water surface without causing undesirable side effects.

---

[23] Cooley, K.P., Energy Relationships in the Design of Floating Covers for Evaporation Reduction, *Water Resource. Res.*, vol. 6, pp. 717–727, 1970.

[24] Meyer L.E. and Frazier, G.W., Evaporation Reduction with Floating Granular Materials, *J. Irrg. Drain. Div*, ASCE, vol. 96, pp. 425–436, December 1970.

[25] Mansfield, W.W., Influence of Monolayer on the Natural Rate of Evaporation of Water, *Nature*, vol. 175, 1955.

[26] Gunaji, N.N., Evaporation Investigation at Elephant Butte Reservoir in New Mexico, Int. Assoc. Sci. Hydrology Pub, 78, pp. 308–325, 1968.

[27] Pushkarev, V.F. and Levchenko, G.P., Use of Molecular Films to Reduce Evaporation from the Surface of Bodies of Water, Tr. GGI 142, pp. 84–107 (Sov. Hydrology. Pap. No. pp. 33, 253–272) 1967.

[28] Linsley, R.L., Kohler, M.A. and Paulhus, J.L.H., *Hydrology for Engineers,* McGraw-Hill, Auckland, 1982.

5.6 ESTIMATION OF EVAPOTRANSPIRATION (OR CONSUMPTIVE USE)

Consumptive use *Cu* may be defined as the amount of water used in evapotranspiration from an area under vegetation plus the water used by the plants in their process for building up the plant tissues. The amount of water required in building up the metabolic process is quite insignificant compared to evapotranspiration and hence, evapotranspiration *ET* is practically equal to consumptive use *Cu*.

There are several methods of estimating this *ET* or *Cu*. The following are some of the methods:

1. **Empirical Equations**
   - Blaney-criddle method (1966) or equation
   - Thornthwaite equation (1964)
   - Penman equation (1948)
   - Christiansen equation (1968)
2. **Field measurement methods**
   - Lysimeters
   - Field plots
   - Soil-moisture depletion studies
   - Water balance method
   - Evaporation index method

Blaney-Criddle method (1966): H.F. Blanay and W.D. Criddle[29] (1966) while studying the consumptive use *Cu* (or *ET*) of water derived an equation of *Cu* (or *ET*). They initially derived the equation in FPS unit. When the equation is converted to MKS units, it is written as:

$$Cu = ET = \sum \frac{kp(4.61t + 81.3)}{100} \tag{5.18}$$

Here *Cu* or *ET* is in cm, *t* is mean monthly temperature in °C, *k* is monthly consumptive use coefficient determined from experimental data, *p* is monthly percentage of hours of bright sunshine and Σ refers the summation of all months of the season.

Hence if *t*, *p*, *k* are known, evapotranspiration *ET* or Consumptive use *Cu* can be estimated.

Thornthwaite equation (1964): Thornthwaite[30] first defined the monthly heat index *i* as:

$$i = \left(\frac{t}{5}\right)^{1.514} \tag{5.19}$$

Here *t* is the mean monthly temperature in °C. Monthly heat indices are then obtained taking the summation of *i* for 12 month, i.e.,

[29] Balney, H.F. and Criddle, W.D., Determining Consumptive Use for Water Developments, ASCE, *Irrig. and Drain. Spec. Conf. Proc.*, pp. 1–34, November 2–4, 1966.

[30] Thornthwaite Assoc., Average Climate Water Balance Data of the Continents, United States, Publication in Climatology, 17(3), 1964.

$$I = \sum i = \sum_{m=1}^{12} \left(\frac{t_m}{5}\right)^{1.514} \tag{5.20}$$

Monthly potential evapotranspiration is calculated from:

$$PET = 1.6b\left(\frac{10t}{I}\right)^{a} \tag{5.21}$$

where $$a = 67.5 \times 10^{-8} I^3 - 77.1 \times 10^{-6} I^2 + 0.01791 I + 0.492 \tag{5.22}$$

and $$b = \frac{12 \text{ months'hour}}{12 \times 30}$$

PET is in cm/month.

Penman equation (1948): Penman equation of evaporation is given by Eq. (5.15), i.e.,

$$E = \frac{(\Delta.Q_{in}/Le) + (0.00061 PE_a)}{\Delta + 0.00061P}$$

Equation (5.15) is based on theoretical approach. When Eq. (5.15) is multiplied by a coefficient *K*, it will yield potential evaporation *PET*.

$$\therefore \qquad PET = KE \tag{5.23}$$

where *E* is given by Eq. (5.15), and the value of *K* depends on types of crop.

Christiansen equation (1968): The Christiansen[31] equation for estimation of potential evaporation is:

$$PET = 0.473\, Q_0 C \tag{5.24}$$

Here Q_0 is the solar radiation at the top of the atmosphere converted to mm of equivalent evaporation and C is a coefficient derived from series of climatic measurements like temperature, humidity, wind, sunshine, elevation, etc.

Lysimeters: Lysimeter is an evapotransporimeter. Many observations of evapotranspiration are made by this lysimeter. Mather[32] (1954), in the John Hopkins University laboratory has nicely described its operation and comparative usefulness. Lysimeter is a tank circular in cross-section with pervious bottom whose diameter may be extended to even 5 m (Pruit and Laurence[33], 1966). Mather (1954) states:

[31] Christiansen, J.E., Pan Evaporation and Evapotranspiration from Climatic Data, ASCE, *J. Irrig and Drain. Div.*, 94, pp. 243–265, 1968.

[32] Mather, J.R. (Edition), The Measurement of Potential Evapotranspiration, John Hopkins Univ. Lab Climatology, Seabrook, N.J. Publication, Climatology, vol. 7, No. 1, 1954.

[33] Pruit, W.O. and Laurence, F.J., Test of Aerodynamics: Energy Budget and other Evaporation Equations over Grass Surface in Investigation of Energy Momentum and Mass Transfer near the ground, U.S. Army Electronics Command Atmospheric Lab., Ft. Huachuca, Ariz, pp. 37–63, 1966.

"The properly operated evapotransporimeter, i.e., when watered sufficiently so that there is no moisture deficiency and no appreciable moisture surplus in the soil of the lysimeter and when exposed homogenously within the protected buffer area of the proper size to eliminate the effect of moisture advection, is an instrument which should give reasonably reliable values of potential evapotranspiration".

Great care must be taken in the operation of the instrument. Standardized soil, vegetation, cultivation and watering practices must be maintained on the tanks in order to ensure comparable results from one installation to another. Pruit and Laurence (1966)'s recommendation of larger size allows the growth of root zone of the crop within the lysimeter. Unless a great care is taken in operation, the difference may exist between lysimeter and natural conditions of soil profile surrounding it.

Field plots: A field plot is selected. The elements or all parameters of water budget in a known interval of time are measured. The parameters that are required to measure in the plot of given period are precipitation, supply of irrigation, runoff, increase in soil storage and ground water loss. The ground water loss due to deep percolation is very difficult to measure. It can however, be minimized maintaining the moisture condition in the plot at the field capacity. Thus, ground water loss is assumed to be negligible,

$$ET = \text{Precipitation} + \text{Irrigation water} - \text{Runoff-increase in soil moisture} \tag{5.25}$$

The method produces quite good results; only limitation of the method is the measurement of ground water flow.

Soil moisture depletion studies: Throughout the growth period of a crop, measurement of soil moisture at frequent intervals from various depths are to be made. If there is again a ground water flow that takes place at deep root zones, problem may arise in correct assessment of *ET*. If root does not penetrate to ground water, the method provides a good assessment. The evapotranspiration *ET* for any time period between two successive samplings may be obtained as:

$$ET = \sum_{i=1}^{n} \left[\frac{M_{1i} - M_{2i}}{100} \right] G_i D_i \tag{5.26}$$

where M_{1i} is soil moisture percentage at the time of first sampling in the *i*th layer, M_{2i} is soil moisture percentage at the time of second sampling in the *i*th layer, G_i is apparent specific gravity of the *i*th layer of the soil, D_i is depth of *i*th layer, in mm within the root zone and *n* is number of soil layer considered in the entire root zone.

Water balance method: The reliability of this method largely depends on time increment of the basin. Knox and Nordenson[34] remarked that as a rule normal annual evapotranspiration can reliably computed as the difference between a long time averages of precipitation and

[34] Knox, C.E. and Nordenson, T.J., Average Annual Runoff and Precipitation in the New England-New York Area, U.S. Geolo. Surv. Hydro. Invest. Atlas, HA-7, undated.

stream flow, since the change in storage over a long period of years is irrelevant. Actual evaporation is:

Evapotranspiration (of the basin) = (Surface flow + Sub surface inflow + Imported water) – (Surface outflow + Sub surface outflow + Domestic, municipal and industrial use + Exported water + Increase in surface storage + Increase in ground water storage) (5.27)

The assessment of all the parameters makes the method extremely difficult. But in brief, it can be obtained as the difference of long time average precipitation and stream flow as the change in storage for a long time period is inconsequential according to Knox and Nordenson.

Evaporation index method: It is G.H. Hargreaves[35] who in 1966 presented the study of estimating consumptive use *Cu* (or *ET*) from evaporation. A high degree of correlation between consumptive use *Cu* (or *ET*) and evaporation *E* exists. Therefore, *ET* is expressed in terms of *E* as:

$$ET = kE \tag{5.28}$$

where k is the coefficient giving the ratio of (ET/E) and was found to vary with the stage of growth of the crop. Hargreaves presented a detailed study of coefficient k.

Example 5.1

A class A pan was setup adjacent to a lake. The depth of water in the pan at beginning was 195 mm. In that week, a rainfall of 45 mm came and 15 mm of water was removed from the pan to keep the water level in the specified depth. If the depth of water at end of the week was 190 mm, calculate pan evaporation. Using suitable pan coefficient, estimate lake evaporation in that week.

Solution:

Initial depth of water in the pan = 195 mm

After the rainfall, total water in the pan = (195 + 45 – 15) mm = 225 mm

At the end of the week depth water in the pan is 190 mm.

∴ Evaporation from the pan in the week = (225 – 190) mm = 35 mm

Assume suitable pan coefficient is 0.7.

∴ Evaporation from the lake in the week = (0.7 × 35) mm = 24.5 mm

Example 5.2

A reservoir has an average area of 50 km^2 over a year. Normal annual precipitation is 120 cm and evaporation from class A pan is 240 cm. Assuming the land flooded by the reservoir has

[35] Hargreaves, G.H., Consumptive Use Computation from Evaporation Pan Data, ASCE, *Irrig. and Drain. Spec. Comf. Proc.*, November 2–4, pp. 36–52, 1966.

a runoff coefficient 0.4, estimate the net annual increase or decrease in the stream flow as a result of the reservoir.

Solution:

Total annual rainfall = 120 cm

Runoff coefficient = 0.4

Effective rainfall = (120 – 0.4 × 120) = 72 cm

∴ Evaporation (actual) = 240 – 72 = 168 cm = 1.68 m

and reservoir level falls = (168 cm – 72 cm) = 96 cm = 0.96 m

Now, decrease in volume of stream flow = Area of reservoir × 0.96 m

$$= 50 \times 10^6 \times 0.96$$

$$= 48 \times 10^6 = 48 \text{ mm}^3$$

EXAMPLE 5.3

A reservoir has average surface area of 20 km^2. In the month of June, mean rate of inflow is 10 m^3/sec, mean outflow is 15 m^3/sec, rainfall is 10 cm and change of storage is 16 million m^3. Assuming surface losses to be 1.8 m, estimate the evaporation.

Solution:

Writing water budget equation as per the problem:

Inflow + Precipitation = Outflow + Surface outflow + Change in storage + Evaporation (i)

Now,

$$\text{Inflow} = 10 \text{ m}^3/\text{sec} = \frac{10 \times 24 \times 30 \times 60 \times 60}{20 \times 10^6} = 1.296 \text{ m}$$

$$\text{Precipitation} - 10 \text{ cm} - 0.1 \text{ m}$$

$$\text{Outflow} = 15 m^3/\text{sec} = \frac{10 \times 24 \times 30 \times 60 \times 60}{20 \times 10^6} = 1.944 \text{ m}$$

$$\text{Surface outflow} = 1.8 \text{ cm} = 0.018 \text{ m}$$

$$\text{Storage} = \frac{16 \times 10^6}{20 \times 10^6} = 0.8 = -0.8 \text{ m}$$

Storage – ve, since Outflow > Inflow

Substituting the values in Eq. (i),

$$1.296 + 0.1 = 1.944 + 0.018 - 0.8 + E$$

$$\therefore \quad E = 1.296 + 0.1 + 0.8 - 1.944 - 0.018 = 0.234 \text{ m} = 23.4 \text{ cm}$$

The evaporation in June is 23. 4 cm

EXAMPLE 5.4

For a particular place in November, the p.c. of sunshine hours is 7.2 and mean temperature is 18°C. If the consumptive use coefficient of the crop is 0.7 for that month, find the consumptive use or evapotranspiration of the crop in mm/day by Balney-Criddle method.

Solution:

Blaney-Criddle equation is expressed as:

$$Cu = ET = \sum \frac{kp(4.61t + 81)}{100}$$

where Cu in cm.

Here $\quad k = 0.7,\ p = 7.2$ p.c., $t = 18°C$

Then
$$Cu = ET = \frac{0.7 \times 7.2(4.61 \times 18 + 81)}{100} \text{ cm/month}$$
$$= 8.264592 \text{ cm/month}$$
$$= 2.7548 \text{ mm/day}$$

EXAMPLE 5.5

Determine evapotranspiration in December if the pan evaporation of the month is 8.50 cm. Assume the growing season of the crop from November to February. The consumptive use coefficient at 40% stage is 0.52.

Solution:

By evaporation index method,

$$ET = RE$$

Crop season is November to February = 30 + 31 + 31 + 28 = 120 days

From November to middle December (30 + 15.5) = 45.5 days

$$= \frac{45.5}{120} = 40\%$$

$$ET = 0.52 \times 8.5 = 4.42 \text{ cm/month}$$
$$= \frac{4.42}{30} = 1.4733 \text{ mm/day}$$

EXAMPLE 5.6

Estimate evaporation by (i) Meyer's (ii) Rohwar's equations and (iii) Fitgeralds's equation with the help of the given data. Comment on the answers also.

Reservoir area = 3 km^2

Water temperature = 25°C and e_s at this temperature = 23.75 mm of mercury

Wind velocity at surface V = 10 km/hr

Barometric reading = 750 mm of mercury

Relative humidity = 45%

Find daily evaporation and volume of water evaporated per week.

Solution:

(i) Meyer's Equation is:

$$E = C\left(1+\frac{V}{16}\right)(e_s - e_a)$$

Here C = 0.36 for large reservoir, e_s = 23.75 mm of mercury,
V = 10 km/hr, e_a/e_s = Relative humidity = 0.45 and E is in mm/day.

$\therefore$ $$e_a = e_s \times 0.45 = 23.75 \times 0.45 = 10.6875 \text{ mm of } Hg$$

and $$E = 0.36\left(1+\frac{10}{16}\right)(23.75 - 10.6875)$$

$$E = 7.64156 \text{ mm/day}$$

$\therefore$ Volume of water evaporated per week $= \left(\dfrac{7.64156}{1000}\times 7\times 3\times 10^6\right)\text{m}^3$

$= 160472.76 \text{ m}^3$

(ii) Rohwar's equation is:

$$E = 0.771(1.465 - 0.000732\ P_a)(0.44 + 0.0733V)(e_s - e_a)$$

Here P_a = 750 mm of mercury

And $V_{0.6}$ (i.e., at height 0.6 m above the surface is assured) = 10 km/hr.

Then $E = 0.771(1.465 - 0.000732)(0.44t\ 0.0733 \times 10)(23.75 - 10.6875)$

$$E = 10.8211 \text{ mm/day}$$

$\therefore$ Volume of water evaporated per week $= \left(\dfrac{10.8211}{1000}\times 7\times 3\times 10^6\right)\text{m}^3$

$= 227873.1 \text{ m}^3$

(iii) Fitzgerald's equation is:

$$E = (0.4 + 0.124V)\ (e_s - e_a)$$

or

$$E = (0.4 + 0.124 \times 10)(23.75 - 10.68875)$$

$$E = 21.4225 \text{ mm/day}$$

$\therefore$ Volume of water evaporated per week $= \left(\dfrac{21.4225}{1000}\times 7\times 3\times 10^6\right)\text{m}^3$

$= 449872.5 \text{ m}^3$

Comment on evaporation by these three equations:

E in Meyer's equation = 7.64156 mm/day

E in Rohwar's equation = 10.8211 mm/day which is 1.4116 times more than Meyer's equation

E in Fitzgerald's equation = 21.4225 mm/day which is 2.8 times more than Meyer's equation.

Thus, estimations vary quite widely by these empirical equations. This is the main limitations of empirical equations.

We may also check by Horton's equation, i.e.,

$$E = [0.4\ (2 - e^{-124V})\ (e_s - e_a)]$$
$$= 0.4\ (2 - e^{-124\times10})\ (23.75 - 10.6875)$$
$$= 8.965 \text{ mm/day}$$

which is 1.173 times more of Meyer's equation.

It appears Meyer's and Horton's equations give almost the same results. Rohwar's and Fitzgerald's equations overestimate evaporation.

5.7 CONCLUSION

Evaporation and evapotranspiration starting from historical review by different investigators and estimation of evaporation by six different methods, and estimation of evapotranspiration by nine different methods are presented. Estimation of evaporation has been checked by an example with different empirical formulae which have shown the limitation of use of empirical formulae in the estimation of such meteorological parameters.

EXERCISES

5.1 The following are the monthly evaporation data (in cm) in a certain year (January to December) in the vicinity of lake.

Month	J	F	M	A	M	J	J	A	S	O	N	D
Evaporation	15.7	14.1	16.9	24.0	27.5	21.4	15.7	16.2	16.2	20.5	19.7	15.4

The water spread area in the lake in the January was 3.2 km^2 and in December 2.6 km^2. Calculate the loss of water due to evaporation in that year. Assume pan coefficient of 0.71

(***Ans:*** 4.51345 mm^3)

5.2 For a particular place in December, the percentage of sunshine is 7 and mean temperature is 16°C. If the consumptive use coefficient of the crops is 0.65 for that month, find consumptive use or evapotranspiration of the crop in mm/day by Balney-Criddle method.

(***Ans:*** 2.346 mm/day)

5.3 Determine the consumptive use in June if the pan evaporation of the month is 20 cm. Assume the growing season of the crop from April to September. The consumptive use coefficient at 33% growth stage is 0.5.

(***Ans:*** 10 cm/month or 3.333 mm/day)

5.4 The evaporation in a place is to be estimated by four different empirical formulae: (i) Horton (ii) Meyer (iii) Rohwar (iv) Fitzgerald.

Data of the place are:

Water temperature = 20°C

e_s at 20°C = 17.54 mm mercury

Wind velocity = 12 km/hr

Atmosphere pressure = 70 mm of mercury

Relative humidity = 50%

Find daily evaporation in mm/day by these formulae. Compare the results. Comment where empirical formulae have limitations.

SUGGESTED FURTHER READINGS

Chow, V.T (Ed.), *Handbook of Applied Hydrology*, McGraw-Hill, New York, 1964.

Harrold, L.L. and Dreldelbis, Evaporation of Agricultural Hydrology by Monolith Lysimeters, USDA, *Tech. Bulletin*, No. 1367, 1967.

Horton, R.E., Analysis of Runoff Plots Experiment with Varying Infiltration Capacity, *Trans. Am. Geophysics* – Union Part, IV, 1939.

Lingere, E.T., Climate and Evaporation from Crops, *Jour. Irrig. Drain. Div. Proc.*, ASCE, 93 (IRA), pp. 61–79, December 1967.

Penman, H.L., Estimating Evaporation, *Am. Geophysics Union*, 4(i): 9–29, 1956.

Prescott, J.A., Evaporation from a Water Surface in Relation to Solar Radiation, *Trans. Royal Society of Australia*, 64, pp. 114–116, 1940.

Chapter 6
Runoff

6.1 INTRODUCTION

Runoff is one of the important components of hydrologic cycle. The runoff is defined as that part of precipitation which is not evapotranspirated.

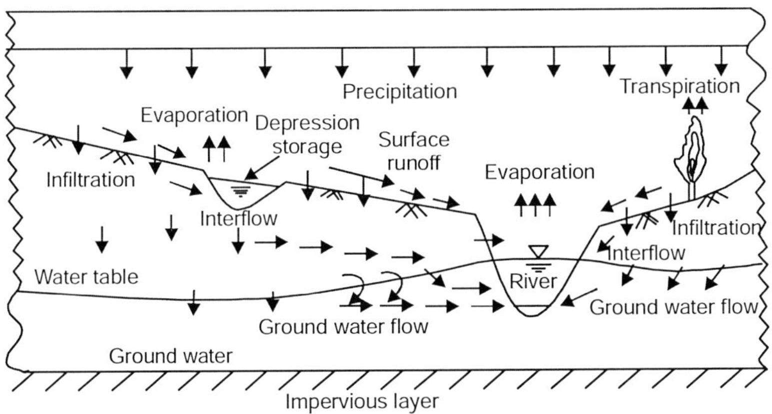

Figure 6.1 Runoff after precipitation.

Figure 6.1 shows how runoff in different forms occurs after precipitation. A part of precipitation goes to atmosphere by way of evaporation and transpiration. The remaining part goes to the stream or river of the catchments as:

1. Surface water flow or overland flow
2. Interflow or sub surface flow
3. Underground flow

The major part of rainfall flows as a thin sheet of water over the surface, which is called **surface flow** or **overland flow** or **direct surface runoff** (DSR). A portion is intercepted by small depression of ground, which is known as **depression storage**. If impermeable strata exists in the soil, a part of the infiltrating water moves laterally below the surface towards the stream. This part is called **interflow** or **sub surface runoff**. If the permeable soil exists just below, major portion of infiltrating water percolates into the ground as deep seepage and builds the ground water table (WT) joining ground water storage. This part of ground water contributes to the stream if the level of stream is lower than the water table.

Overland flow and interflow or sub surface flow when combined together, it may be called **direct runoff (DR)**. Thus, total runoff is the combination of direct surface runoff, interflow and ground water flow. Various investigators have presented different runoff modelling. Contributions of Linsley and Ackermam[1], Cook[2], Kohler and Linsley[3], Milton and Paulhus[4], Chow[5], Betson[6] and Bevin[7] are worth to be referred.

6.2 FACTORS AFFECTING RUNOFF

The runoff from a watershed or catchment or drainage basin depends on the following climatic and physiographic factors.

Type of precipitation: The runoff is quick and immediate if the precipitation falls as rainfall. If it falls as snow, it cannot produce any runoff initially. When melting of snow takes place in warm period, it can contribute to runoff sometimes heavily if the snow precipitation is large enough.

Intensity of precipitation: Runoff is directly proportional to the intensity of precipitation. If the rainfall intensity exceeds the infiltration capacity, runoff increases. Runoff increase rate is although directly proportional to intensity of rainfall; storage of the basin has some influence on runoff and intensity of rainfall.

Duration of rainfall: Duration of rainfall affects the runoff to a great extent. If this duration is greater than the critical concentration time t_c (i.e., the time taken by the raindrop falling at the farthest point to reach the outlet), whole of the rainfall in the basin contribute to runoff as infiltration ceases to take place, and the soil surface is completely saturated due to high duration of the rainfall. If the duration of rainfall is quite small, runoff may not occur due to infiltration and interception.

Arial distribution of rainfall: Uniform rainfall over the whole catchment is rarely observed. If rainfall occurs at the upstream of the outlet, runoff takes time to reach the outlet and peak runoff at outlet occurs after a considerable time. On the other hand, if rainfall occurs close to the outlet, runoff is quick and higher.

---

[1] Linsley, R.K. and Ackermann, W.C., Method of Predicting the Runoff from Rainfall, *Trans. ASCE*, vol. 107, pp. 825–846, 1942.

[2] Cook, H.L., The Infiltration Approach to the Calculation of Surface Runoff, *Trans. Am. Geophy. Union*, vol. 27, No. 5, pp. 726–747, October, 1946.

[3] Kohlcr, M.A. and Linsley, R.K., Predicting the Runoff from Storm Rainfall, U.S. Weather Bur. Res. Paper, 34, 1951.

[4] Milton, J.F. and Paulhus, J.L.H., Rainfall-Runoff Relation for small Basins, *Trans. Am. Geophy. Union*, vol. 38, No. 2, pp. 216–218, April 1957.

[5] Chow, V.T., Maidment, D.R. and Mays, L.W., *Applied Hydrology*, McGraw-Hill, 1988.

[6] Betson, R.P., What is Watershed Runoff?, *Jour. Geophysics. Res.*, vol. 69, No. 8, pp. 1552, 1964.

[7] Bevin, K.J., *Rainfall-Runoff Modelling*, John Wiley & Sons, West Sussex, England, 2000.

Direction of storm movement: If the storm moves upstream of the outlet, runoff at outlet occurs slowly at much later time; but if the storm occurs near to the outlet, runoff is quick and higher in magnitude. Again the storm crossing the basin in transverse direction, runoff produced will be in between the above two cases.

Antecedent rainfall: Runoff is greatly affected by the soil moisture condition. If the soil moisture condition or relation is quite low before the rain comes, most of the rainfall is lost as infiltration and hence runoff is low. If the soil moisture condition is quite high, rainfall produces high runoff. Even ground water level at this condition contributes to higher runoff.

Other climatic conditions: Other climatic or hydro meteorological conditions like wind velocity, temperature, atmospheric pressure, humidity, radiation and vapour pressure affect the runoff. Their influences are already discussed in Chapter 2 (hydrometeorology).

Land use: In urban areas, scope of infiltration is very small, so runoff is more. In deep forest, it absorbs a lot of rainfall; transpiration takes place and the land with thick grass and plant offers greater resistance to flow producing little runoff. In non-forest areas, interception, infiltration, evaporation is less so runoff is more.

Type of soil: Sandy soil has high infiltration producing less runoff. Clay soil tends to produce high runoff as infiltration is less.

Area of catchment: Every catchment is surrounded by a topographic divide (ridge line), which fixes the area of catchment from which the surface runoff takes place. Similarly, there is an underground or phreatic divide from which groundwater contributes to runoff. When these two divide lines do not coincide as shown in Figure 6.2, watershed leakage area occurs in one catchment which contributes to the total runoff although in many cases, this part of runoff due to watershed leakage is taken to be unimportant.

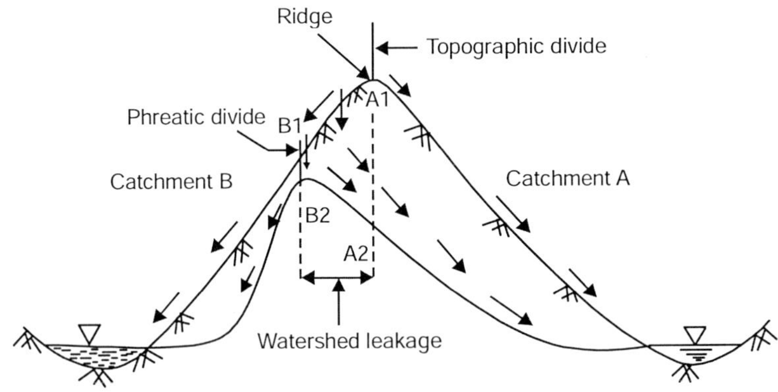

Figure 6.2 Watershed leakage between catchments A and B.

Figure 6.2 shows the watershed leakage between two catchments. In the figure, surface runoff in catchment B takes place from A_1 to B_1. But the interflow or subsurface water after infiltration in catchment B flows to catchment A due to non-coincidence of topographic and phreatic divide lines. Naturally runoff in catchment A will be more. It shows in such situation, only topographic area of catchment is not only the criterion for runoff at the outlet.

Shape of catchment: Figure 6.3 shows the different shapes of catchment which affect the runoff volume at the outlet. Intensity of runoff at outlet is different for different catchment shape. For fern or elongated shape [Figure 6.3(b)] runoff at outlet is of lesser intensity. Runoff from nearest tributaries is already drained out before the floods of farthest tributaries arrive at the outlet. Thereby, time of draining out the runoff is more and hence peak of runoff less. In broad shape catchment [Figure 6.3(c)], runoff from different tributaries meets the main stream at the middle almost at the same time resulting high peak runoff at outlet. For fan shaped [Figure 6.3(a)], runoff peak is in between the elongated and broad shaped.

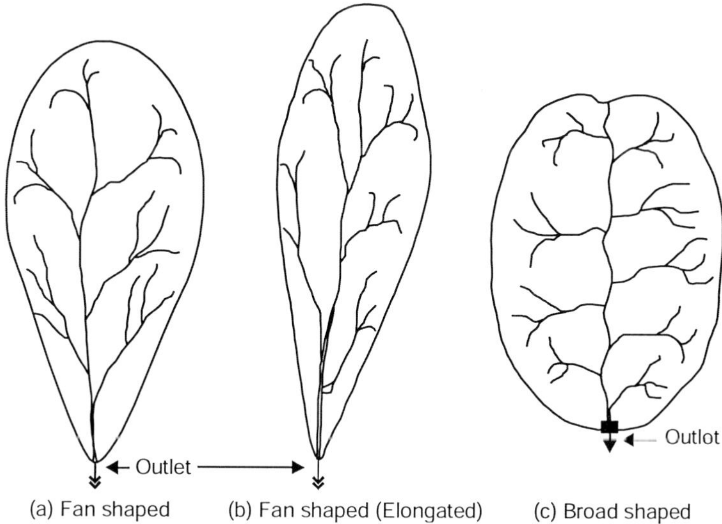

Figure 6.3 Different shapes of catchment.

Elevation of catchment: Variation of elevation of the catchment causes the variation of temperature, humidity, vapour present, wind surface velocity, which in turn may increase or decrease (depending on elevation) evaporation, transpiration and even rainfall also. Thus, change of runoff takes place due to rise and fall of surface level, i.e., elevation.

Slope of the catchment: The slope of the watershed is one of the important factors that controls the time of overland flow and concentration of runoff in the main stream. If slope is more, velocity of surface runoff or overland flow is more and infiltration is less. Runoff generated by rainfall quickly arrives at outlet producing more runoff peak.

Orientation of catchment: If the basin is oriented most of the time towards the sunrays, temperature increases due to heat received from the sun. Increased temperature accelerates the rate of evaporation, which affects the runoff due to rain. Also if the precipitation is snow and basin gets the heat by sun rays, runoff increases due to melting of snow.

Drainage network: Natural drainage channels are formed in the catchment area. Their formation, direction, capacity, etc. mainly depend on topography. If the drainage systems are well developed flowing towards the main stream at an inclination, runoff takes less time to arrive at outlet of the main stream to produce more runoff. The characteristics of drainage network may be judged from (i) order of stream (smaller to bigger sizes), (ii) stream density (number of perennial stream/unit area), etc.

Indirect drainage: A part of rainwater flows through indirect drainage and thereby affects the surface runoff in the main streams or drains. These indirect drains are formed in very pervious surface with flat topography and in the regions formed by soluble rocks like limestone. It has undulating surfaces with conical hillocks and almost circular sinks. Runoff from such areas enters the ground through the sinkholes and rock fissures and its way through underground drainage to adjacent lakes and streams. Thus, this drainage affects the normal runoff in the rivers.

Artificial drainage: These are mainly man made drain like open ditches, tile drains, storm sewer drains in urban areas and portion of areas covered by pavement, building etc. with little or no infiltration. Runoff through such artificial drainages is accelerated, but this local effect has no significant increase in peak runoff in large stream.

Depressions: Depressions on ground can hold quite a lot of rainwater. This stored water is eventually lost by evaporation and infiltration affecting direct surface runoff (DSR).

Pool, ponds and lakes: It also similarly affects runoff by evaporation and infiltration.

Check dams: Check dams are constructed much upstream of the stream to arrest sediments and silts and also in hilly area for soil conservation. These dams store a part of the surface water and thus they affect the normal runoff.

Upstream reservoirs and lakes: Upstream reservoirs and lakes arrest a considerable portion of stream flow and it mitigates the magnitude of runoff at the outlet.

Ground water storage: Major portion of infiltration seeps to ground water. Only a small portion flows to stream as inert flow so runoff is likely to be affected.

6.3 ESTIMATION OF RUNOFF

Following are some of the available methods of estimating runoff or measurement of runoff:

Rainfall-runoff relation: If R is rainfall, P is the precipitation, then:

$$R = kP \tag{6.1}$$

where k is the runoff coefficient. Its values range from 0.5 to 0.95.

Again annual runoff R and precipitation P may be related by the regression analysis as:

$$R = aP - b \tag{6.2}$$

In Eq. (6.2), a and b are constants estimated by regression analysis of R and P of a catchment data for few years.

Empirical formulae: The types of empirical formulae are:

$$R = C_1P + C_2 \tag{6.3}$$

$$R = C_3P^n \tag{6.4}$$

$$R = \frac{C_4 P^{n_1}}{T^{n_2} A^{n_3}} \tag{6.5}$$

where C_1, C_2, C_3 and C_4 are constants, n, n_1, n_2 and n_3 are exponents of the catchment and T is average temperature in °C. Inglis formulae for Bombay-Decan catchment are:

(For Ghat areas) $$R = 0.85P - 30.5 \tag{6.6}$$

(For Plains) $$R = \frac{(P - 17.8)P}{254} \tag{6.7}$$

A.N. Khosla's formula is of the following type:

$$R = PC^T, \quad T > 4.5°\text{C} \tag{6.8}$$

where C is regression constant ranging from 0.952 to 0.971.

Another formula of Khosla is:

$$R = Pe^{T/a} \tag{6.9}$$

Again a is a regression constant

Khosla's another formula for Northern India is:

$$R = P - \frac{T}{3.74} \tag{6.10}$$

Some of the Empirical Formulae of India for different basins are:

Ganga basin $$R = 2.14P^{0.64} \tag{6.11}$$

Yamuna basin in Delhi $$R = 0.14P^{1.1} \tag{6.12}$$

Rihand basin in UP $$R = P - 1.17P^{0.86} \tag{6.13}$$

Chambal basin in Rajasthan $$R = 120P - 4945 \tag{6.14}$$

Towa basin in MP $$R = 90.5P - 4800 \tag{6.15}$$

Tapti basin in Gujarat $$R - 135P - 17200 \tag{6.16}$$

In these formulae, R is annual runoff in cm, P is annual average rainfall in cm and T is the temperature in C°.

Thus, different empirical formulae are developed. But all the equations have their limitations of constants and exponents to be used in different catchments.

Infiltration method: From the infiltration capacity curve [Figure 6.4 (a)] and ϕ-index shown in Figure 6.4(b) by dotted areas, total infiltration can be obtained. Then net runoff is obtained by deducting areas of infiltration from hyetograph. Runoff volume is obtained shown by shaded area. The volume divided by area of catchment gives the runoff in depth.

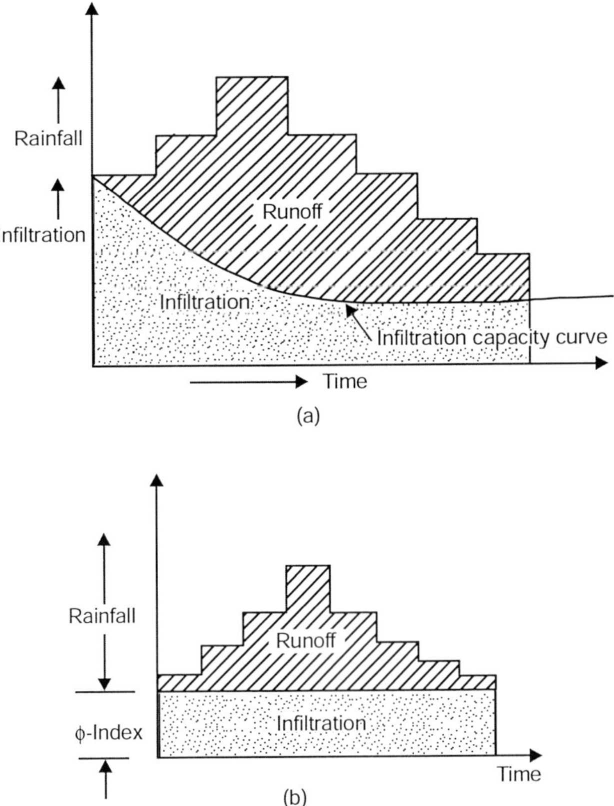

Figure 6.4 Runoff from Infiltration.

Runoff curves: Some curves of rainfall vs runoff area available. When the measured rainfall is known, corresponding runoff is obtained from the curves once the constants a and b of the catchment are known.

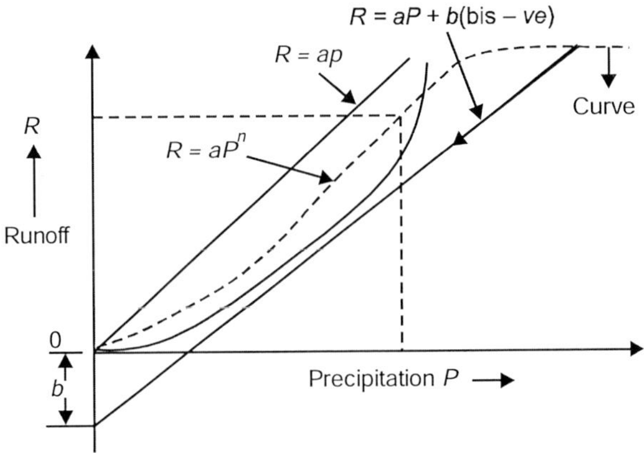

Figure 6.5 Runoff precipitation curves.

Runoff tables: Some tables are of the following types presented separately by Binnie, Strange and Barlow:

Rainfall in increasing order	% Runoff
x_1	60%
x_2	62%
x_3	64%
x_4	70%
x_5	80%
....	....
....	....

From those tables runoff can be computed from the measured rainfall. These tables are also not much reliable like empirical formulae.

Hydrograph analysis: After a storm, discharges are measured at outlet at different time interval (1 hr or 2 hr or 4 hr). The measured discharges Q when plotted with corresponding time, graph obtained is called **hydrograph** (Details will be presented in Chapter 7). The river or stream has some initial flow before the storm, which is called **base flow**. This base flow is separated as shown by dotted area. The area shown by shaded lines is the total volume of runoff due to that storm. This volume divided by area of the catchment gives the runoff in depth.

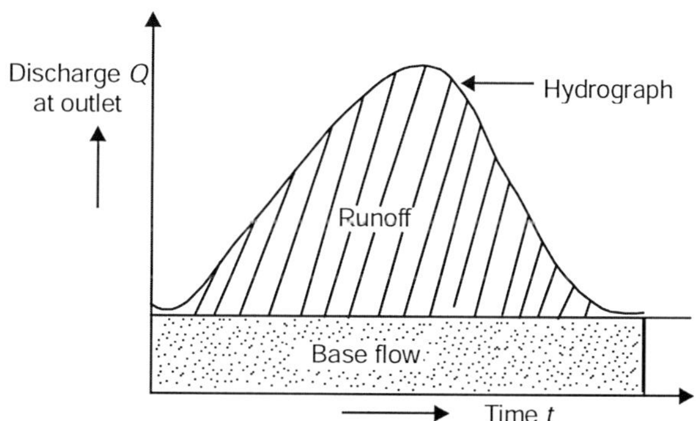

Figure 6.6 Hydrograph analysis to compute runoff.

Unit hydrograph (UH) method: Unit hydrograph will be discussed in details in Chapter 7. (Hydrographs).

It is a hydrograph due to unit depth (1 cm) of runoff (or rainfall excess or net rainfall after deducting losses). Area under the unit hydrograph shown in Figure 6.7 represents the total volume of rainfall due to 1 cm of net rainfall. If a storm produces x cm net rainfall, then volume of runoff = (area of UH $\times$ x).

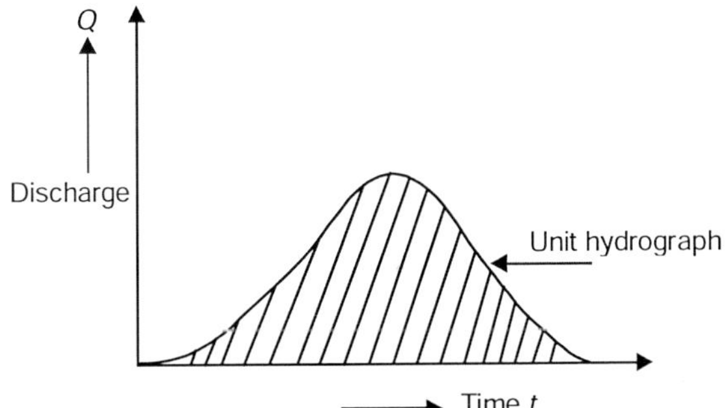

Figure 6.7 Unit hydrograph method.

Curve number method: This method was developed in 1972 by the Soil Conservation Service[8] of the USA to determine the runoff (rainfall excess). The method is based on the concepts of a reservoir operation. This method has been presented by Das, Ghansyam[9] (2000).

Rational method: This method is developed by Mul Vaney in 1851 for determination of peak flow rate. Although this method was developed in 19th century, Pilgrim (1986), Linsley (1986) claimed that it is still widely used method.

In metric unit, it is expressed as:

$$Q_p = \frac{1}{3.6} CIA \tag{6.17}$$

where Q_p is peak runoff rate in (m^3/sec), I is intensity of precipitation (mm/hr) for duration equal to time of concentration t_c, A is the area of drainage basin in km^2 and C is coefficient (varies from 0.3 to 0.82 based on soil cover and soil type).

6.4 RUNOFF CYCLE

Hydrologic cycle is the term applied to the general circulation of water in its various states from sea and earth's surface to the atmosphere and back to sea and earth's surface.

Runoff cycle is the descriptive term applied to that portion of hydrologic cycle between incident precipitation over land areas and subsequent discharge through streams and direct return to the atmosphere through evapotranspiration.

Figures 6.8(a), (b), (c) and (d) are the conditions of runoff cycle at four different stages. Figure 6.8 (a) is after an extended dry period and just prior to the beginning of rainfall. Figure 6.8(b) is shortly after beginning of rainfall, but before interception and depression storages have been satisfied. Figure 6.8(c) is nearly at the end of rainfall but, rainfall still continuing. Figure 6.8(d) is after the end of a protected heavy storm, but before channel and retention storages have been depleted.

---

[8] U.S. Soil Conversation Service, Hydrology, Nat'l Engg. Handbook, Section 4, Washington D.C., 1972.

[9] Das, Ghanasyam, *Hydrology and Soil Conservation Engineering,* Prentice-Hall of India, New Delhi, 2000.

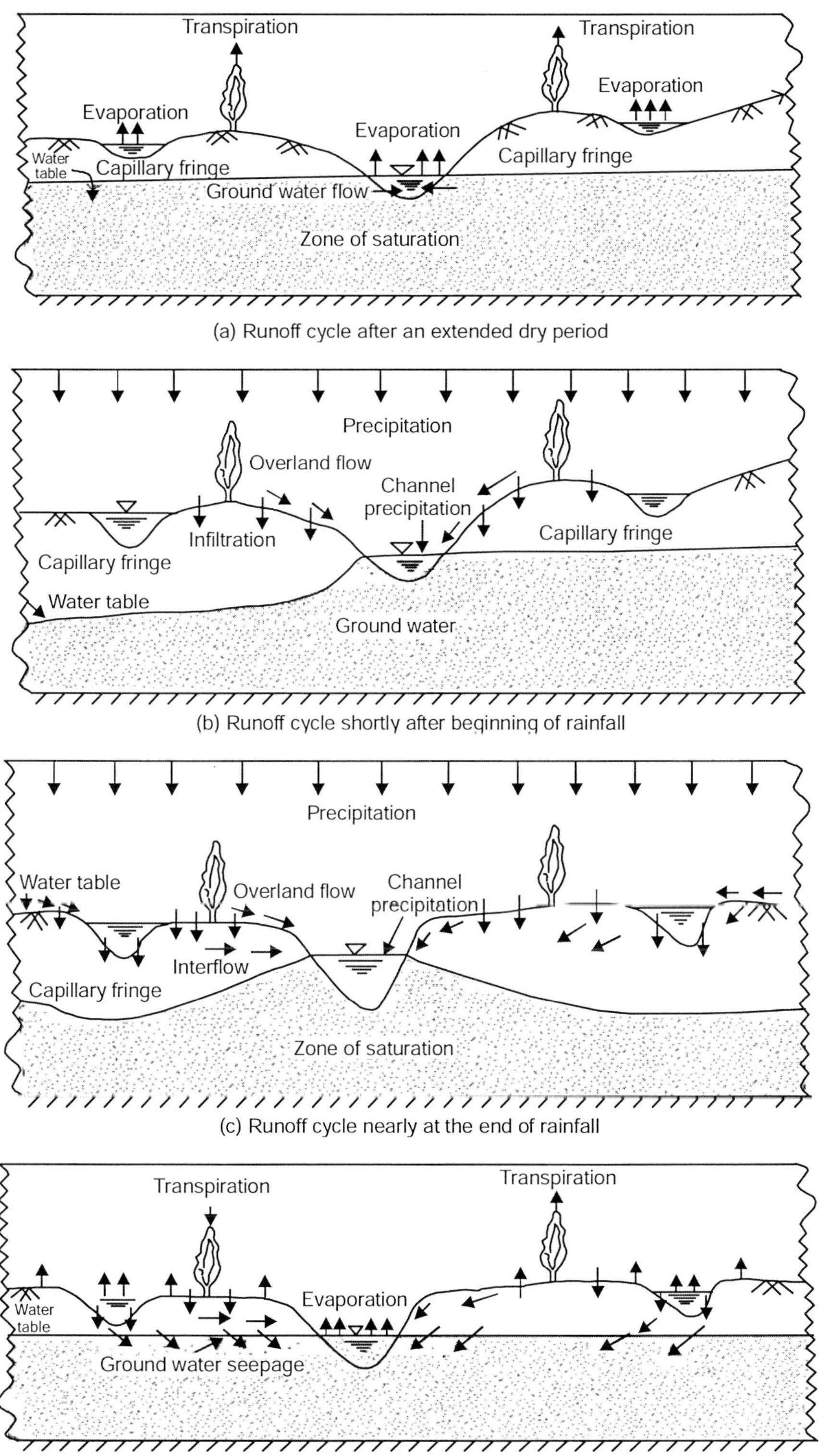

Figure 6.8 Four states of runoff cycle.

6.5 CONCLUSION

Definition of runoff, factors affecting runoff and different methods of runoff estimation are discussed systematically along with runoff cycle. The concept of runoff will help in the analysis of the hydrographs and floods.

EXERCISES

6.1 What is run off? Discuss the factors in details that affect the runoff.

6.2 Explain the different methods of estimating runoff.

6.3 What is runoff cycle? Discuss the four stages of runoff cycle with figures.

SUGGESTED FURTHER READINGS

American Society of Civil Engineers (ASCE) and Water Environment Federation, Urban Runoff Quality Management, WEF Manual and Report on Engineering Practice, No. 87, New York, 1998.

Betson, R.P., Tucker, R.L. and Haller, F.M. Using Analytical Methods to Develop a Surface-Runoff Model, *Water Resource. Res.*, vol. 5, No. 1, pp. 103–111, February, 1969.

Chow, V.T., Maidment, D.R. and Mays, L.W., *Applied Hydrology*, McGraw-Hill, New York, 1964.

Dugal, K.N. and Soni, J.P., *Elements of Water Resources Engineering,* New Age International, New Delhi, 1996.

Freeze, R.A., Role of Subsurface Flow in Generating Surface Runoff, *Water Resource*, vol. 8, No. 5, pp. 1272–1283, 1972.

Kuichling, E., Relationship between the Rainfall and Discharge of Sewers in Populous Districts, *Trans.*, ASCE, vol. 20, 1889.

Linsley, R.L., Kohler, M.A. and Paulhus, J.L.H., *Hydrology for Engineers*, 3rd ed., McGraw-Hill International, Auckland, 1982.

Murteja, K.N., *Applied Hydrology*, Tata McGraw-Hill, New Delhi, 1990.

Nemec, J., *Engineering Hydrology*, McGraw-Hill, New York, 1972.

Petersen, M.S., *River Engineering,* Prentice Hall, Upper Saddle River, N.J., 1986.

Raghunath, H.N., *Hydrology: Principle, Analysis, Design,* New Age International, New Delhi, 1996.

Reddy, P.J., *A Textbook of Hydrology,* Lakshmi Publication, Delhi, 1986.

Sing, V.P., *Elementary Hydrology*, Prentice Hall Inc., Upper Saddle River, N.J., 1992.

Subramanya, K., *Engineering Hydrology,* Tata McGraw-Hill, New Delhi, 2000.

Varshney, R.S., *Engineering Hydrology,* Nem Channel & Bros., Roorkee, 1979.

Viessman, W.G. and Lewis, G.L., *Introduction to Hydrology*, Harper Collins, New York, 1996.

Walesh, S.G., *Urban Surface Water Management*, John Wiley & Sons, New York, 1989.

Ward, A.D. and Elliot, W.J., *Environmental Hydrology,* Lewis Publishers, New York, 1995.

Chapter 7

Hydrographs

7.1 INTRODUCTION

Hydrographs are graphical representations of flow parameters against time. When a storm falls in a catchment, after initial losses and loss due infiltration, rainfall excess eventually flows to the main stream. If the discharge Q, average velocity V and depth or stage Y at outlet point of the stream is measured from beginning of the storm to the effect of storm in the catchment and if they are plotted against time, they may be called **hydrograph**. For example,

1. If Q is plotted against time t, it is called flow or **flood hydrograph**.
2. If V is plotted against t, it is known as **velocity hydrograph**.
3. If depth Y or stage H is plotted against time t, it is called **depth** or **stage hydrograph**.

But most commonly in hydrology, hydrograph means the graph of flow rate Q against time t.

7.2 HYDROGRAPHS OF SINGLE AND COMPLEX STORMS

When a single storm occurs [Figure 7.1(a)], the hydrograph has a single peak as shown in Figure 7.1(b).

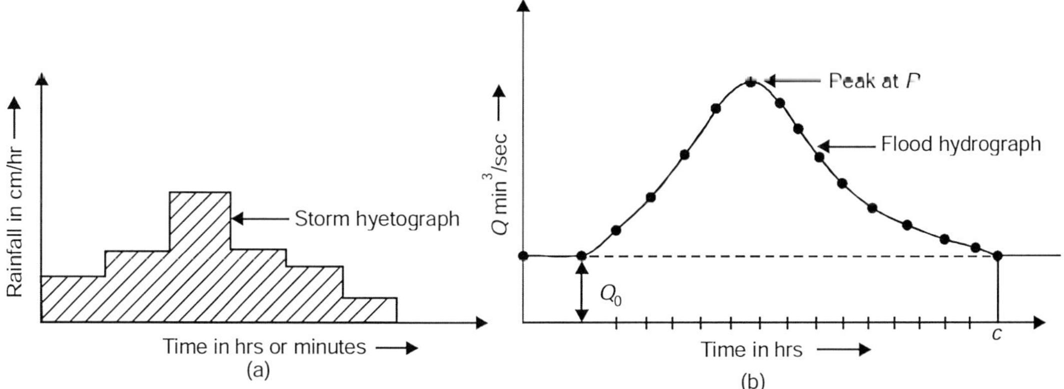

Figure 7.1 Flood hydrograph (b) due to single storm in (a).

Figures 7.2(a) and (b) shows the hyetograph and corresponding hydrograph of three different storms with different peak flows.

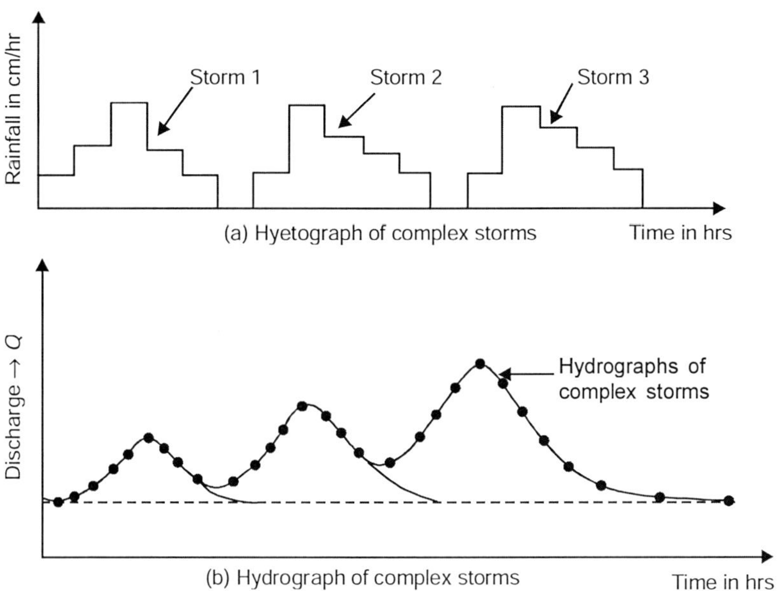

Figure 7.2 Hyetograph and hydrograph of complex storms.

7.3 COMPONENTS OF HYDROGRAPH OF ISOLATED STORM

From Figure 7.3, the components of hydrograph are:

1. *AB* is approach segment.
2. *BC* is rising limp or concentration curve.
3. *CPD* is crest segment or peak limb.
4. *DE* is falling limb.
5. Lower portion *DE* is called ground water depletion curve.
6. *P* is the point of crest or peak.
7. t_B is time of rise.
8. t_P is the time of peak.
9. Area under *ABE* is the volume of base flow (shown by shaded line). This volume is assumed to have no contribution of storm.
10. Area between *BPE* and *BE* is the volume of direct surface runoff (DSR) shown by dots.
11. *OA* is the value of discharge at initial time $t = 0$, and this *OA* may be assumed to be the base flow.
12. *C* is the first point of inflexion and *D* is the second point of inflexion.
13. OE_1 is the base period *T* of the hydrograph.

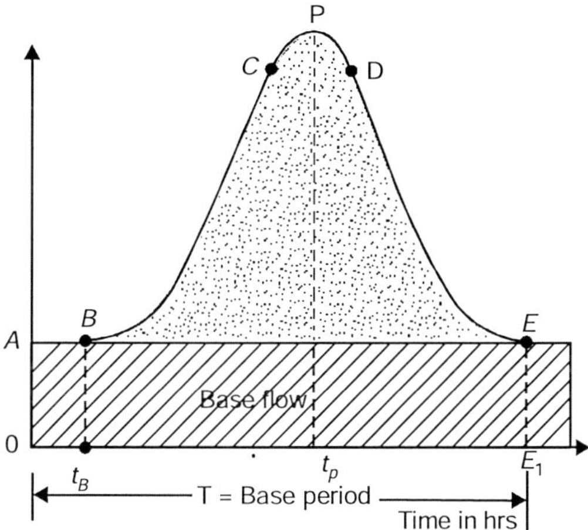

Figure 7.3 Hydrograph of isolated storm.

Out of these 13 components, the main components are:

Rising limb or segment or concentration curve: In the initial time of rainfall, losses cause the discharge to rise slowly. Building up of storage is gradual. If the storm is continuous, discharge from upstream reaches the outlet as infiltration. Absorption also decreases with time. Of course, shape of basin and storm characteristics affect the rising limp.

Peak limb or crest segment: It is the most important part of hydrograph. Peak outflow is essential for design of hydraulic structures in the catchment. When runoffs from all sources reach the outlet, peak flow occurs. In large catchment, peak flow may even occur after the storm. Estimation of peak flow and its time of occurrence are very important in hydrology.

Recession limb or depletion curve: The recession limb represents the withdrawal of water from storage after all inflows to the channel have ceased. Therefore, it is time independent of rainfall or infiltration and is essentially dependent on channel features alone.

F.R. Hall[1] (1968) advocated that number of functions have used to describe recession curve or depletion curve and the one in general use being:

$$Q_t = Qk_r^{-t} \tag{7.1}$$

where Q is the flow at the start of period, Q_t is the flow at the end time t, and k_r is recession content less than unity.

Equation (7.1) can be written in the more general form as:

$$Q_t = Qe^{-\alpha t} \tag{7.2}$$

where $\alpha = -\log_e k_r$

As Q_t is the derivation of storage S_t with respect to time t,

$$-dS_t = Q_t\, dt$$

Integrating

$$S_t = -\int Q_t\, dt$$

[1] Hall, F.R., Base Flow Recessions, A Review, Water Resource Research, vol. 4, No. 5, pp. 973–983, 1968.

or $$S_t = -\int (Qe^{-\alpha t})\, dt$$

or $$S_t = (Q_t/\log_e k_r) \tag{7.3}$$

Hence discharge at any time is proportional to water remaining in storage.

or $$\frac{Q}{Q_t} = \frac{S}{S_t} \tag{7.4}$$

Taking $\log_e$ on both sides of Eq. (7.1)

$$\log_e Q_t = \log_e Q \quad t \log_e k_r \tag{7.5}$$

which is straight line in the form $y = mx + c$ where $y = \log_e Q_t$, $m = -\log_e k_r$, $x = t$, $C = \log_e Q$ on semi-log paper. If continuous stream flow data record is available for number of years, hydrograph of recession part can be plotted in a semi-log paper.

7.4 FACTORS AFFECTING FLOOD HYDROGRAPH

Factors affecting hydrograph are:

1. Climatic factors are movement, infiltration, intensity and duration of rainfall.
2. Physiographic factors depend on:
   (a) Shape of basin
   (b) Size of basin
   (c) Slope of basin
   (d) Nature of basin
   (e) Drainage density
   (f) Elevation of catchment
   (g) Land use and cover
   (h) Soil type and geological condition
   (i) Lakes, storages, swamps
   (j) Channel characteristics

Climatic factors: It includes movements of storm direction, its intensity and duration. It also depends on initial loss and evapotranspiration. If the storm moves from upstream to downstream or from downstream to upstream, effects on hydrographs are shown in Figures 7.4(a) and 7.4(b).

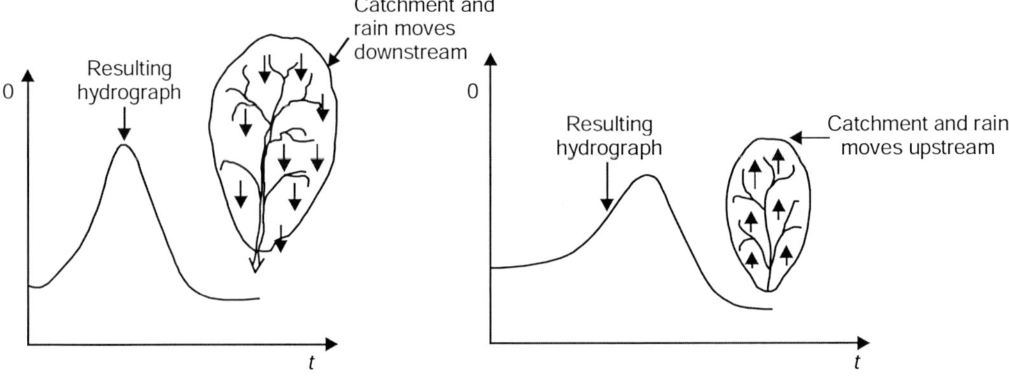

Figure 7.4 Hydrographs affected by movement of rainfall.

Physiographic factors: The affects on resulting hydrographs on the different shapes of catchment are shown in Figures 7.5(a), (b) and (c). Size of basin has predominant effect on peak discharge, overland flow, base of the hydrograph, etc. Usually $Q \propto A^n$ where n is less than unity and time of occurrence is also proportional to A^m where m is even less than 0.5. Overland flow is predominant in small size catchment. In large catchment channel, phase is important.

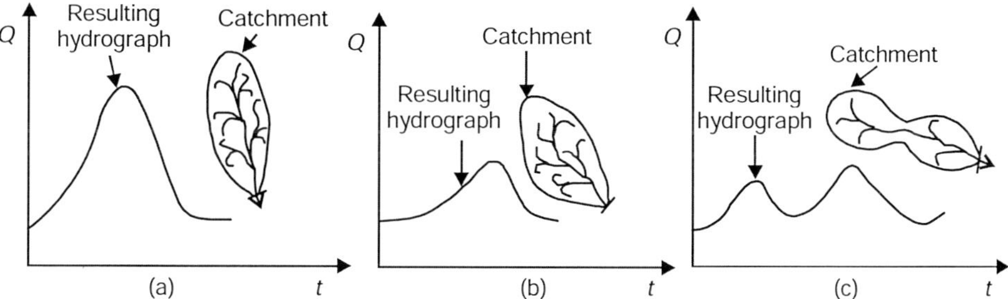

Figure 7.5 Effect on hydrograph by shape of catchment.

If the slope is more, floodwater drains out quickly, hence hydrograph has a small time base. The falling limb of the hydrograph becomes steep for bigger slope of the catchment. In small basin, overland flow is important and in such basin higher slope results in higher peak discharge.

Density of drainage has pronounced effect on peak of the hydrograph. If drainage density is higher, i.e., catchment has quite a number of drainages or channels to drain out the excess rainfall, peak is more and if drainage density is low, peak is lower. Other factors like land use and cover, elevation, soil type, stages in the catchment, channel characteristics have affected the hydrograph.

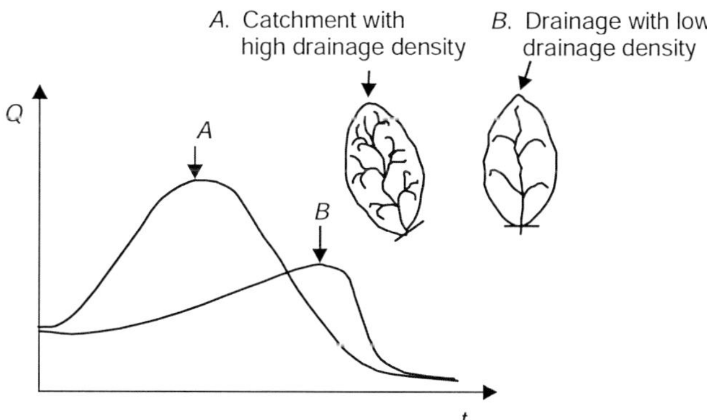

Figure 7.6 Drainage density affects the hydrographs of catchments *A* and *B*.

Vegetative covers retard the flow and increase infiltration and interception, thereby reduce the peak flow. High elevation upstream causes quick flow downstream. Type of soil like sand absorbs quite a larger part, so flood is less, whereas clay soil produces more flow. Geologic conditions also affect the runoff. If there are some cracks on rocks joining surface water with ground water, a part of water flows quickly to underground. Lakes, storages, swamp hold a part of excess rainfall so peak is less. If the channel has lot of meandres, travel time of flow is long, as it has to move a longer distance, so peak may be low.

7.5 BASE FLOW SEPARATION

Base flow is the initial flow of the stream before the rain comes. It is the sustained or dry weather flow of the stream resulting from the outflow of perennial or almost permanent ground water flow that reaches the channel. This base flow of such channel is more or less assumed to be constant.

There are some methods of separating this base flow from the total recorded flows of the hydrograph.

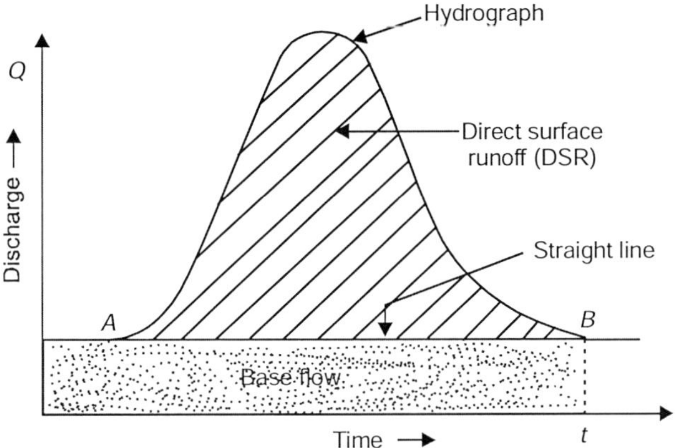

Figure 7.7 Separation of base flow by straight line.

Straight line method: In this method, separation is achieved by a straight line joining from a point A (Figure 7.7) to a point B on the recession limb representing the end of surface runoff. This method is approximate, but very easy to separate. Error is not much.

Fixed base method: Linsley et al[2] (1982) are of the opinion that for application to the unit hydrograph theory, time base of direct surface hydrograph (DSR) should remain relatively constant for storm to storm. In using method of separation, this point should be kept in mind. This is usually provided by terminating the direct runoff at a fixed time after the peak of the hydrograph. As a rule of thumb, time in days N may be approximated by

$$N = 0.8A^{0.2} \tag{7.6}$$

where A is in km^2. Figure 7.8 shows the reasonable and unreasonable values for N.

Variable slope methods: Here base flow curve existing prior to the commencement of surface runoff is extended till it intersects the ordinate drawn at point D. D is joined with a straight line DB. The volume below ADB is volume of base flow [Figure 7.9(a)]. For Figure 7.9(b), base flow recession curve after depletion of the floodwater at B is extended back-

[2] Linsley, R.L., Kohler, M.A. and Paulhus, J.L.H., *Hydrology for Engineers* 3rd ed., McGraw-Hill International, 1982.

wards till it intersects the ordinate *DB* at point of inflection. The points *A* and *D* are joined by straight line. The volume under *ADB* represents the volume of base flow. Chow[3] (1964) and Wilson[4] (1974) gave description of base flow separation. Tallaksen[5] (1955) gave a nice review of base flow recession analysis.

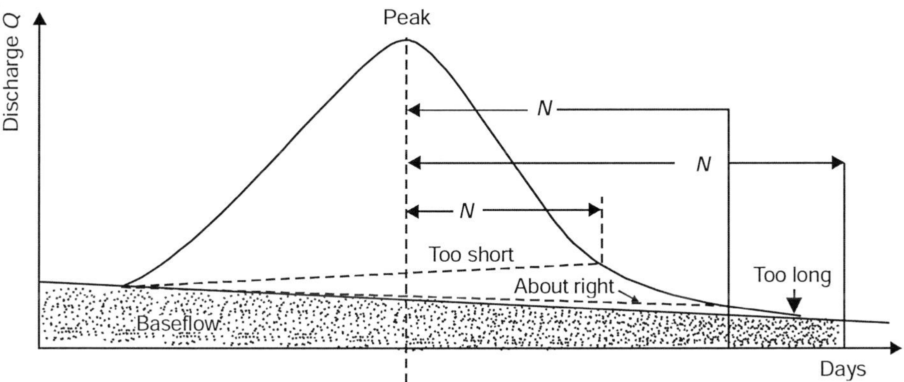

Figure 7.8 Selection of time base for the direct surface runoff hydrograph.

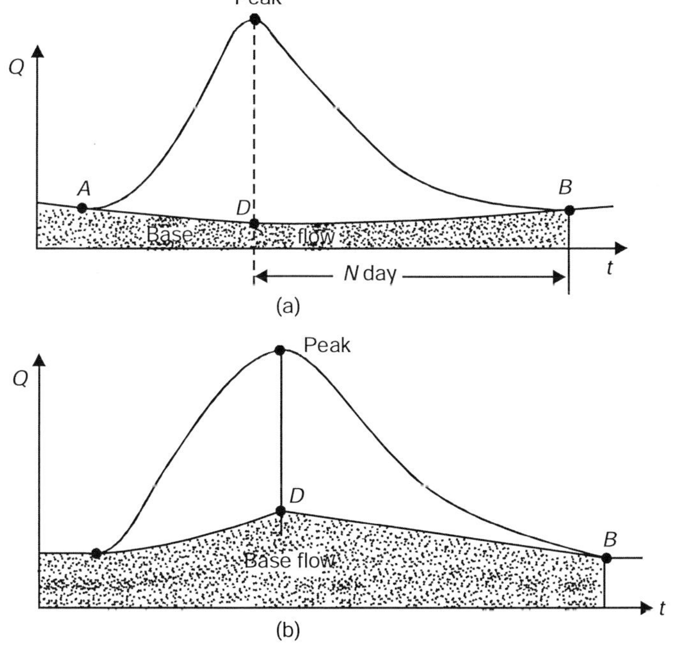

Figure 7.9 Base flow assessment by variable slope method (a) & (b).

[3] Chow, V.T., *Handbook of Applied Hydrology*, McGraw-Hill, New York, 1964.
[4] Wilson, E.M., *Engineering Hydrology,* 2nd ed., Macmillan, London, 1974.
[5] Tallaksen, L.M., A Review of Base Flow Recession Analysis, *Jour. of Hydro.*, vol. 165, 1955.

7.6 COMPUTATION OF DIRECT SURFACE RUNOFF (DSR) FROM FLOOD HYDROGRAPH

Figure 7.10 shows the flood hydrograph of a storm with flood ordinates at different time interval Δt. From the flood hydrograph, separate the base flow by any one of the methods described in Section 7.5. Let the base flow is separated by simple straight line method and direct surface runoff (DSR) hydrograph is drawn in Figure 7.11.

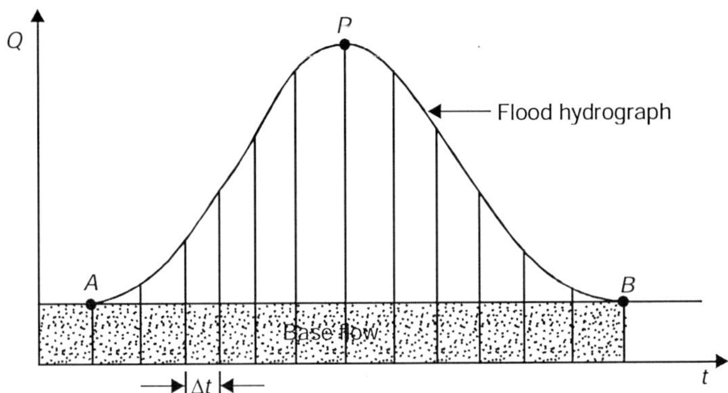

Figure 7.10 Food hydrograph of a storm.

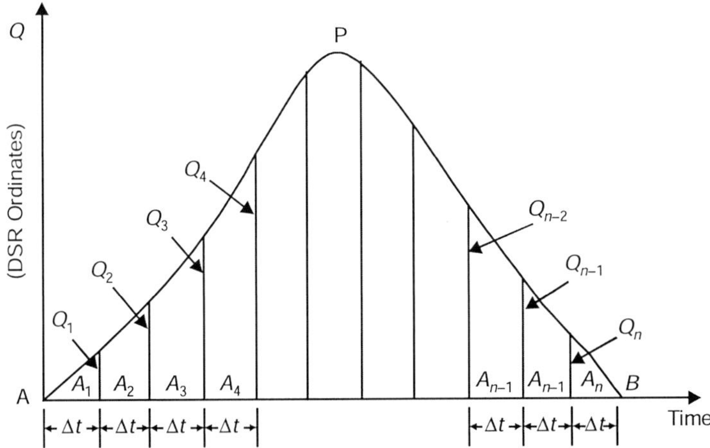

Figure 7.11 Direct surface runoff (DSR) hydrograph.

Let Q_1, Q_2, Q_3, Q_4, $\cdots$, Q_{n-2}, Q_{n-1}, Q_n are the ordinates of direct surface runoff hydrograph at different time interval Δt.

To find the area under the DSR hydrograph, which represent the volume of DSR:

$$\text{Area under the curve } APB = \text{Volume of } DSR$$

$$= A_1 + A_2 + A_3 + A_4 + \cdots + A_{n-2} + A_{n-1} + A_n$$

$$= \frac{1}{2}Q_1.\Delta t + \left(\frac{Q_1 + Q_2}{2}\right)\Delta t + \frac{Q_2 + Q_3}{2}\Delta t + \frac{(Q_3 + Q_4)}{2}\Delta t + \cdots + \frac{(Q_{n+2} + Q_{n+1})}{2}\Delta t$$

$$+ \frac{(Q_{n-1} + Q_n)}{2}\Delta t + \frac{1}{2}\Delta t$$

$$= \left(\frac{1}{2}Q\Delta t + \frac{Q_1}{2}\Delta t\right) + \left(\frac{Q_2}{2}\Delta t + \frac{Q_2}{2}\Delta t\right) + \left(\frac{Q_3}{2}\Delta t\right) + \left(\frac{Q_4}{2}\Delta t + \frac{Q_4}{2}\Delta t\right) \cdots + \left(\frac{Q_{n-2}}{2}\Delta + \frac{Q_{n-2}}{2}\Delta t\right)$$

$$+ \left(\frac{Q_{n-1}}{2}\Delta t + \frac{Q_1}{2}\Delta t\right) + \left(\frac{Q_n}{2}\Delta t + \frac{Q_n}{2}\Delta t\right)$$

$$= (Q_1 + Q_2 + Q_3 + Q_4 + \cdots Q_{n-2} + Q_{n-1} + Q_n)\Delta t$$

$$= \sum Q\Delta t$$

= Total volume of direct surface runoff

If Q in m^3/sec, and Δt in hr, then

$$\text{Volume of DSR} = \sum Q\Delta t \frac{\text{m}^3}{\text{sec}} \times (60 \times 60)\,\text{sec}$$
$$= \sum Q\Delta t \times (60 \times 60)\,\text{m}^3 \qquad (7.7)$$

If A is the area of the catchment in km^2,

$$A = A \times 10^6 \text{ m}^3$$

$$\therefore \text{DSR in cm} = \left\{\left[\frac{(60 \times 60)\sum Q\Delta t}{A \times 10^6}\right] \times 100\right\}\text{cm}$$

Thus, DSR in depth = Rainfall excess or net rainfall

Example 7.1

The flood data and base flow in a storm are estimated for a storm in a catchment area of 600 km^2. Estimate the rainfall excess.

Time in days	0	1	2	3	4	5	6	7	8	9
Discharge (m^3/sec)	20	63	151	133	90	63	44	29	20	20
Base flow (m^3/sec)	20	22	25	28	28	26	23	21	20	20

Solution:

Ordinates of DSR hydrograph after the separation of the base flow are: 0, 41, 126, 105, 62, 37, 21, 8, 0, 0. Then

$$\Sigma Q = (41 + 126 + 105 + 62 + 37 + 21 + 8) = 390 \text{ m}^3 \text{ day}$$

$$\Sigma\Delta Qt = (400 \times 1 \times 24 \times 60 \times 60)\ \text{m}^3$$

$$\therefore \quad \text{Runoff depth} = \text{Rainfall excess} = \left(\frac{400 \times 24 \times 60 \times 60}{600 \times 10^6}\right)$$

$$= 0.0576 \text{ m} = 5.76 \text{ cm}$$

7.7 UNIT HYDROGRAPH

The theory of unit hydrograph was proposed first by L.K. Sherman[6] (1932). Since then it has gone many refinement. Sherman developed the theory based on the idea of base flow separation by Folse (1929) and on the report of Boston Society of Civil Engineers (1931) that the base of flood hydrograph of a catchment appears to be same.

Unit hydrograph (UH) is defined as the hydrograph of surface runoff of a catchment area resulting from unit depth (usually 1 cm) of rainfall excess or net rainfall occurring uniformly over the basin and at uniform rate for a specified duration. Unit hydrograph is a linear model of the catchment which is used to find out the volume of DSR due to 1 cm of direct surface runoff or 1 cm of rainfall excess. This is always constant for the catchment since area of the catchment is constant. If rainfall comes to the catchment producing 2 cm of rainfall excess, the ordinates of the DSR will be twice as great of the unit hydrograph (UH) ordinates and volume of DSR will be two times the volume of DSR of unit hydrograph.

Therefore, UH is said to be linear model of the catchment usually employed to determine the volume of runoff or flood of any storm occurring in the catchment.

7.8 ASSUMPTIONS IN DERIVATION OF UH THEORY

1. The rainfall is of spatially uniform intensity within its specified duration.
2. The effective rainfall is uniformly distributed throughout the whole area of drainage basin.
3. The base of time duration of hydrograph of direct runoff due to effective rainfalls of unit duration is constant. Base period of hydrograph of different rainfall intensities remain approximately same. This is represented in Figure 7.12.

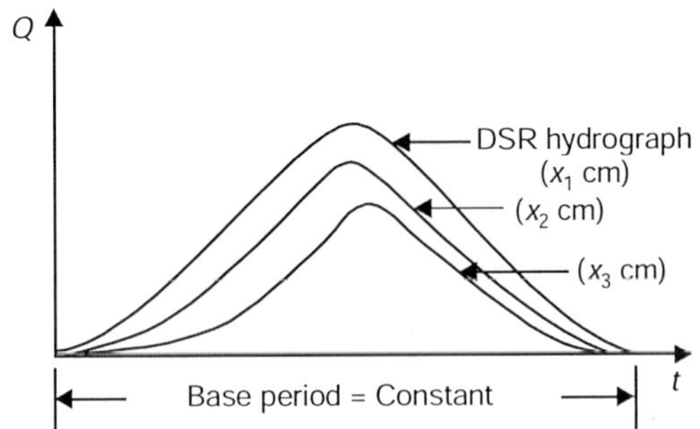

Figure 7.12 Base period is assumed to be constant.

[6] Sherman, L.K., Stream Flow from Rainfall by Unit Hydrograph Method, *Eng. News-Rec.*, vol. 108, pp. 501–505, 1932.

4. The ordinates DSR hydrographs due to net rains of different intensities but same duration are proportional.
5. A unit hydrograph reflects all the combined physical characteristics of the basin.

7.9 DERIVATION OF UH FROM A SIMPLE FLOOD HYDROGRAPH OF ISOLATED STORM

Different steps required to derive UH are:

Step 1 From the given flood hydrograph, separate the base flow by any one of the methods. Most commonly used method to draw a straight line without much error for simplicity (Figure 7.13).

Step 2 Determine the volume of DSR hydrograph by the formula:

$$\text{Volume of DSR} = \Sigma Q\Delta t$$

Step 3 Divide this volume by known area of catchment to get DSR in depth in cm, i.e., net rainfall or rainfall excess.

Step 4 Divide the ordinates of DSR by the depth of DSR hydrograph to obtain ordinates of UH.

Step 5 Plot the ordinates of UH against time to get the UH of the catchment as shown in Figure 7.13.

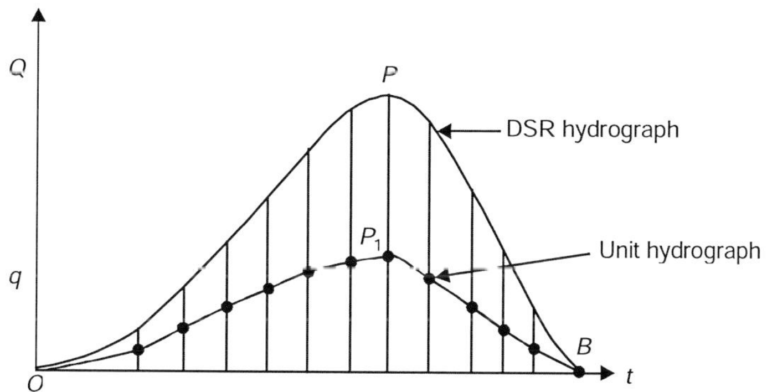

Figure 7.13 DSR hydrograph and unit hydrograph.

EXAMPLE 7.2

The following are the ordinates of the flood hydrograph from a catchment area of 780 km^2 due to 6 hr storm. Derive the 6 hr unit hydrograph of the basin.

Time (hrs)	6	12	18	24	6	12	18	24	6	12	18	24	6
Discharge (m^3/sec)	40	64	215	360	405	350	270	205	145	100	70	50	40

Solution:

Assume a base flow of 40 m^3/sec. Then

ΣQ for direct surface runoff $= (64 - 40) + (215 - 40) + (360 - 40) + (405 - 40) + (350 - 40) + (270 - 40) + (205 - 40)(145 - 40) + (100 - 40) + (70 - 40) + (50 - 40)$

$= (24 + 175 + 320 + 365 + 310 + 230 + 165 + 105 + 60 + 30 + 10)$

$= 1794 \text{ m}^3/\text{sec}$

$$\therefore \qquad \text{DSR in depth} = \left(\frac{1794 \times 6 \times 60 \times 60}{780 \times 10^6} \times 100 \right) \text{cm} = 4.968 \text{ cm}$$

Therefore, the ordinates of UH are obtained by dividing the ordinates of DSR hydrograph by rain excess 4.968 cm to get ordinates of UH.

Time (hrs)	6	12	18	24	6	12	18	24	6	12	18	24	6
Ordinates of UH	0	4.83	35.22	64.42	74.47	62.4	46.29	33.21	21.13	12.077	6.04	2.01	0

Example 7.3

The following data represent the ordinates of hourly interval of one unit hydrograph:

Time (hrs)	0	1	2	3	4	5	6	7	8	9	10	11	12	13
UH ordinates m^3/sec.	0	58	110	96	53	26	14	8	5	4	3	1.5	1	0

Compute storm hydrograph resulting from three-hour storm rainfall as:

Time	1st hr	2nd hr	3rd hr
Rainfall depth (cm)	4	3	2.5

Take ϕ-index as 2 cm/hr and assume a base flow of 2 m^3/sec.

Solution:

Net rainfall or rainfall excess or DSR in depth:

1st hr $= 4 - \phi$ index $= 4 - 2 \times 1 = 2$ cm

2nd hr $= 3 - 2 \times 1 = 1$ cm

3rd hr $= 2.5 - 2 \times 1 = 0.5$

Time (hrs)	0	1	2	3	4	5	6	7	8	9	10	11	12	13	14	15
Ordinates of UH (m^3/sec)	0	58	110	96	53	26	14	8	5	4	3	1.5	1	0		
Ordinates of DSR hydrograph due to 1st storm, i.e., 2 cm	0	58×2 =116	110×2 =220	96×2 =192	53×2 =106	26×2 =52	14×2 =28	8×2 =16	10	8	6	3	1.5	0		

Ordinates of DSR hydrograph due to 2nd storm, i.e., 1 cm	0	0 × 1 = 0	58 × 1 = 58	110 × 1 = 110	96 × 1 = 96	53 × 1 = 53	26 × 1 = 26	14 × 1 = 14	8	5	4	3	1.5	0.75	0.5	0
Ordinates of DSR hydrograph due to 3rd storm, i.e., 0.5 cm	0	0	0	58 × 0.5 = 29	110 × 0.5 = 55	48	26.5	13	7	4	2.5	2	1.5	0.75	0.5	0
Add ordinates of 3 rainfalls to obtain ordinates of DSR (1) + (2) + (3)	0	116	278	331	257	153	80.5	43	25	17	12.5	8	5	1.75	0.5	0
Add base flow 2 m³/sec to get the ordinates of UH (4 + 2 m³/sec)	0	118	280	333	259	155	82.5	45	27	19	14.5	10	7	3.50	3	0

Storm hydrograph	Time (Hrs)	0	1	2	3	4	5	6	7	8	9	10	11	12	13	14	15
	Q	2	118	280	333	259	155	82.5	45	27	19	14.5	10	7	3.50	3	0

7.10 LIMITATIONS AND USES OF UH THEORY

Limitations:

1. Similar rainfall distribution from storm to storm over a large area is rare. Hence UH theory is suited to catchment area under about 500 km³ (≅ 2000 sq. miles).
2. Odd shaped basins particularly those which are long and narrow, commonly have very uneven rainfall distribution and hence UH theory for such basins is not much suitable.
3. In mountainous regions, subject to orographic rainfall, aerial distribution is very uneven, but the pattern tends to remain the same from storm to storm, and UH theory may not be successfully applied.
4. The unit hydrograph method cannot be applied when an appreciable portion of the storm precipitation falls as snow.
5. The catchment having large storage like reservoir, lake, low areas, etc. affect the linear relationship and hence theory cannot be applied.
6. If the variation of base period and peak flow vary more than ± 20% and + 10%, theory applied is not generally accepted.

Uses:

1. As the UH is a linear model of the catchment, it is used to determine runoff hydrograph of the catchment even for extreme magnitude to determine peak flow for design of hydraulic structures.
2. It can be used for flood forecasting and warning.
3. Based on rainfall records, it is used for extension of flood flow records.

7.11 DERIVATION OF UH FROM COMPLEX STORM

Discussion and derivation of UH from an isolated single storm has been shown in previous sections. In practice, it is not always possible to have such isolated storms. Storms having different rainfall excess like R_1, R_2, and R_3 may occur. Such storms of different rainfall excess are called **complex storm**.

Assume 3 storms which can produce rainfall excess R_1, R_2 and R_3. The hydrograph of these 3 storms (after deducting base flow) is shown in Figure 7.14.

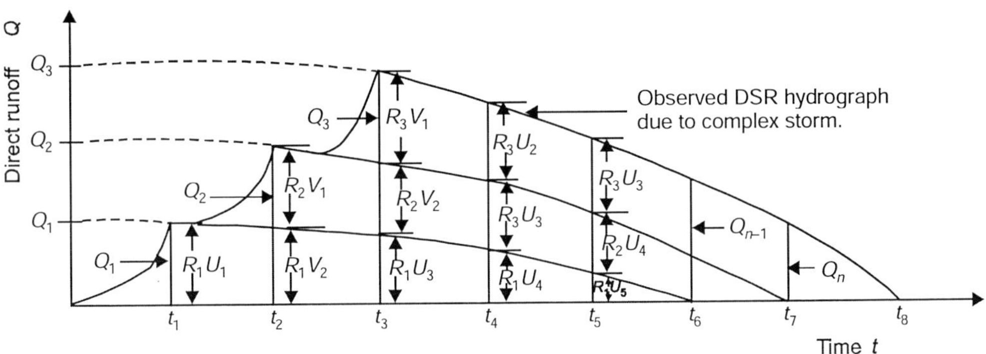

Figure 7.14 Derivation of UH from complex storm.

Let Q_1, Q_2, Q_3, ..., Q_n are the known ordinates of complex storm in Figure 7.14, and U_1, U_2, U_3, ..., U_n are ordinates of unit hydrograph which are to be determined. R_1, R_2 and R_3 are known rainfall excesses due to complex storm, which are also known. Then

$$Q_n = R_nU_1 + R_{n-1}U_2 + R_{n-1}U_3 + \cdots, \tag{7.8}$$

When

$n = 1$, $Q_1 = R_1U_1$, ($\because$ Q_1 and R_1 are known, U_1 is determined.)

$n = 2$, $Q_2 = R_2U_1 + R_1U_2$ ($\because$ Q_2, R_2, U_1, R_1 are known, U_2 is determined.)

$n = 3$, $Q_3 = R_3U_1 + R_2U_2 + R_1U_3$ ($\because$ Q_3, R_3, U_1, U_2, R_1, R_2 are known, U_3 is determined.)

$n = 4$, $Q_4 = R_4U_1 + R_3U_2 + R_2U_3 + R_1U_4$

R_4U_1 is zero, since no 4th rainfall is considered.

Since Q_4 R_3 R_2 R_1, U_2, U_3 are known, U_4 can be determined.

Thus, all the ordinates of unit hydrographs U_1, U_2, U_3, U_4, ..., U_n can be determined and resulting unit hydrograph (UH) can be obtained.

EXAMPLE 7.4

Example 7.3 may be solved by the explained in Section 7.11. The unit hydrograph given in Example 7.3 is:

Time (hrs)	0	1	2	3	4	5	6	7	8	9	10	11	12	13
Ordinates of UH (m^3/sec)	0	58	110	96	53	26	14	8	5	4	3	1.5	1	0

Solution: Here

$U_1 = 58$, $U_2 = 110$, $U_3 = 96$, $U_4 = 53$, $U_5 = 26$,
$U_6 = 14$, $U_7 = 8$, $U_8 = 5$, $U_9 = 4$, $U_{10} = 3$,
$U_{11} = 1.5$ and $U_{12} = 1$

Let $Q_1, Q_2, Q_3, Q_4, \ldots, Q_{12}$ are ordinates of DSR hydrograph without base flow.

Then
$$Q_n = R_n U_1 + R_{n-i} U_2 + R_{n-2}\, U_3 + R_{n-3} U_4 + \cdots$$

When $n = 1$,
$$Q_1 = R_1 U_1 = 2 \times 58 = 116 \text{ m}^3/\text{sec}$$

$n = 2$,
$$Q_2 = R_2 U_1 + R_1 U_2 = 1 \times 58 + 2 \times 110 = 278 \text{ m}^3/\text{sec}$$

$n = 3$,
$$Q_3 = R_3 U_1 + R_2 U_2 + R_1 U_3 = 0.5 \times 58 + 1 \times 110 + 2 \times 96 = 331 \text{ m}^3/\text{sec}$$

$n = 4$,
$$Q_4 = \overset{0}{\cancel{R_4}} U_1 + R_3 U_2 + R_2 U_3 + R_1 U_4 = 0.5 \times 110 + 1 \times 96 + 2 \times 53 = 257 \text{ m}^3/\text{sec}$$

$n = 5$,
$$Q_5 = \overset{0}{\cancel{R_5}} U_1 + \overset{0}{\cancel{R_4}} U_2 + R_3 U_3 + R_2 U_4 + R_1 U_5 = 0.5 \times 96 + 1 \times 53 + 2 \times 26 = 153 \text{ m}^3/\text{sec}$$

Similarly, $Q_6, Q_7, \ldots$, and all other ordinates of DSR hydrograph are obtained. Adding the base flow 2 m^3 sec, exactly same answer is obtained just like in Example 7.3.

7.12 UNIT HYDROGRAPH OF DIFFERENT DURATIONS

Unit hydrograph may be of different duration D. If a unit hydrograph of 2 hrs duration available, unit hydrograph of 4 hrs duration, 6 hrs duration, 8 hrs duration, etc. may be obtained, i.e., the unit hydrograph to be derived is an integral multiple. In such cases, durations from 2 hrs to 4 hrs, 6 hrs etc. can be done by method of superposition. But if the duration of unit hydrograph to be derived is less than duration of known unit hydrograph, this method of superposition will not hold good. Thus, unit hydrograph of different duration may be derived by the following two methods:

1. Method of superposition when duration of UH to be derived is 2 times, 3 times or 4 times of the duration of known UH.
2. S-curve method (Summation hydrograph method) when duration of UH to be derived is less than the duration of the known UH.

Method of superposition:

1. As shown in Figure 7.15, D-hr UH is plotted (Curve 1).
2. Also plot the UH lagged D hrs (Curve 2).
3. Add the two unit hydrographs and plot it (Curve 3).
4. Divide Curve 3, i.e., DSR hydrograph by $2D$ duration to get the curve shown by dotted line, i.e., Curve 4.

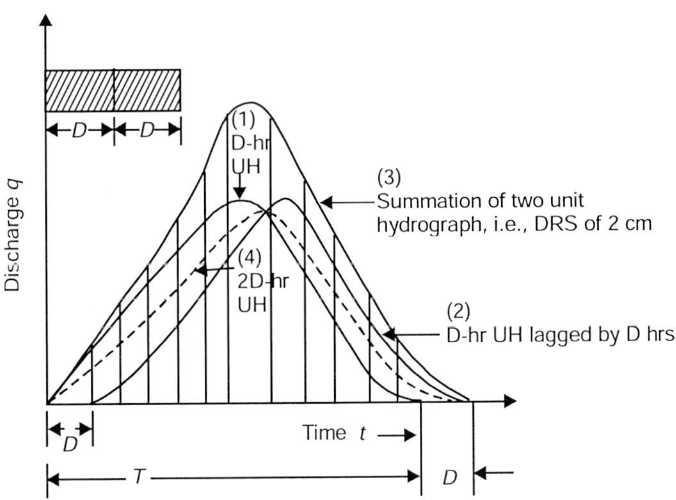

Figure 7.15 Method of superposition.

It is noted that base period of 2-*D*-hr UH shown by dotted line increases by *D*-hrs from the given UH of *T* hrs base period. Therefore, the peak of 2-*D*-hrs UH will decrease as the base period is increased from *T* to $(T + D)$ hrs.

EXAMPLE 7.5

A 4-hr UH is given as:

Time (hr)	0	2	4	6	8	10	12	14	16	18	20
4-hr UH ordinates	0	9	19	20	14	12	8	5	3	1	0

Derive 8-hr unit hydrograph.

Solution:

Time (hrs) (1)	4 *hrs* UH (2)	*Lagged* 4 *hrs* (3) UH	*Combined DSR hydrograph* (4)	*Ordinates of* 8 *hr* UH (5) = (4)/(2)
0	0	–	0	0
2	9	–	9	4.5
6	16	0	16	8
8	14	16	30	15
10	12	(20) peak	32	(16) peak
12	8	14	22	11
14	5	12	17	8.5
18	1	5	6	3
20	0	3	3	1.5
22	–	1	1	0.5
24	–	0	0	0

Column (5) gives the ordinates of 8-hr unit hydrograph derived from column (4) and unit UH in column (2). The peak of the 8-hr UH is reduced to 16 m^3/sec from 20 m^3/sec as the base period is increased from 20 to 24 hrs.

S-curve method: S-curve (summation curve), also known as **S-hydrograph** is a hydrograph produced by a continuous effective rainfall of unit depth at a constant rate for a unit period. It is a curve obtained by summation of an infinite series of unit hydrographs each lagged by *t*-hrs or *D*-hrs with respect to preceding one. It is shown that when a UH of n D duration is required where n is whole number like 2, 3, 4, etc., method of superposition serves the purpose. If n is fraction, i.e., to derive UH of smaller duration from the available duration, then S-curve method is required. As an example, from UH of 4 hrs duration to develop UH of 2 hrs duration cannot be developed by the previous method of superposition. In such situation, S-hydrograph based on principle of superimposition is constructed. Then it is used to find the UH of desired duration. As shown in Figure 7.16, S-hydrograph is continuously a rising curve which ultimately attains a constant value when equilibrium discharge is reached.

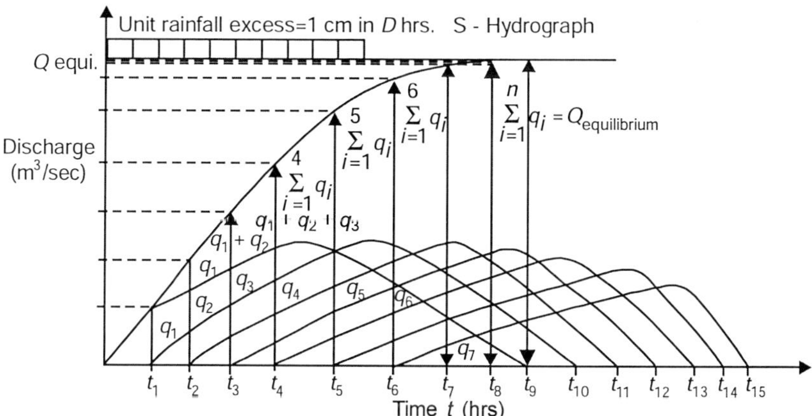

Figure 7.16 S-hydrograph or S-curve.

From the known UH ordinates of q_1, q_2, q_3, ..., q_n of *D*-hrs duration S-curve is plotted (Figure 7.16).

First ordinate of S-curve = q_1

Second ordinate of S-curve = $q + q_2$

Third ordinate of S-curve = $q_1 + q_2 + q_3$

....

....

*n*th ordinate of S-curve = $(q_1 + q_2 + q_3 + \ldots, q_n)$

$$= \sum_{i=1}^{n} q_i$$

= Equilibrium discharge Q equi.

Since S-curve is a summation resulting from continuous summation of unit storms producing 1 cm in D-hrs, so equilibrium discharge is:

$$Q_e = \frac{\text{Volume due to 1 cm rainfall}}{\text{Duration in } D \text{ hrs}}$$

$$= \frac{A \times 1}{D}$$

$$= \frac{\left(A \times 10^6 \times \frac{1}{10}\right) \text{m}^3}{(D \times 60 \times 60) \text{ sec}} \quad \text{where } A \text{ is in km}^2 \text{ and } D \text{ is in hrs}$$

$$= \left(2.7778 \frac{A}{D}\right) \text{m}^3\text{/sec}$$

$$\therefore \quad Q_e = \left(2.7778 \frac{A}{D}\right) \text{m}^3\text{/sec} \tag{7.9}$$

S-curve can be used

1. As a check for uniform distribution of effective rainfall.
2. In deriving UH of shorter duration from UH of larger duration.

UH can be derived as follows:

Step 1 Construct the S-curve from given UH of known time duration D-hrs.

Step 2 Then advance or offset the position of S-hydrograph for a period equal to the desired duration of D_0 hrs of unknown UH. Name this S-hydrograph as offset hydrograph.

Step 3 Find the difference of the ordinates of original S-curve and offset hydrograph.

Step 4 Divide this each difference by (D_0/D) to get the ordinates of new UH of D_0 hrs duration.

The following numerical example has been solved to demonstrate the method:

Example 7.6

Given below is ordinate of 4 hrs UH for a basin. Derive 2 hrs UH from it using S-curve method.

Time (hrs)	0	1	2	3	4	5	6	7	8	9	10	11	12	13	14	15	16	17
UH ordinates (m^3/sec)	0	20	36	60	80	112	120	105	73	40	24	14	9	7	5	3	2	0

Solution:

Here $D = 4$ hr, $D_0 = 2$ hr.

$$\therefore \quad \left(\frac{D_0}{D}\right) = \left(\frac{2}{4}\right) = \frac{1}{2} = 0.5$$

Time (hrs) (1)	*UH ordinates* (2)	*S-curve additions unit storm after every 4 hrs* (m³/sec) (3)				*S-curve ordinates* 2+(i)+(ii) +(iii)+(iv) = 4	*S-curve lagged by 2 hrs* = (5)	(4) – (5) = (6)	2 *hrs UH* (6)/0.5 (7)
		(i)	(ii)	(iii)	(iv)				
0	0	After				0	–	0	0
1	20	4 hrs	After			20	–	20	40
2	36		2nd 4 hrs	After 3rd		36	0	36	72
3	60		i.e.	4 hrs		60	20	40	80
4	86	0	After	i.e.	After	86	36	50	100
5	112	20	8 hrs	After12hrs	4th	132	60	72	144
6	120	36			4 hrs	156	86	70	140
7	105	60			i.e.	165	132	33	66
8	73	86	0		After	159	156	3	6
9	40	112	20		16 hrs	172	165	13	26
10	24	120	36			180	159	21	42
11	14	105	60			179	172	7	147
12	9	73	86	0		168	180	–12	–24
13	7	40	112	20		179	179	0	0
14	5	23	120	36		185	168	17	34
15	3	14	105	60		182	179	3	6
16	2	9	73	86	0	168	185	–17	–34
17	0	7	40	112	20	179	182	–3	–6

Column (7) gives the ordinates of 2-hrs unit hydrograph. A slight adjust is required at tail end of the unit of hydrograph. This adjustment usually required at the tail end by this method.

In Figure 7.17, this adjustment at tail end are entered and shown in the following table.

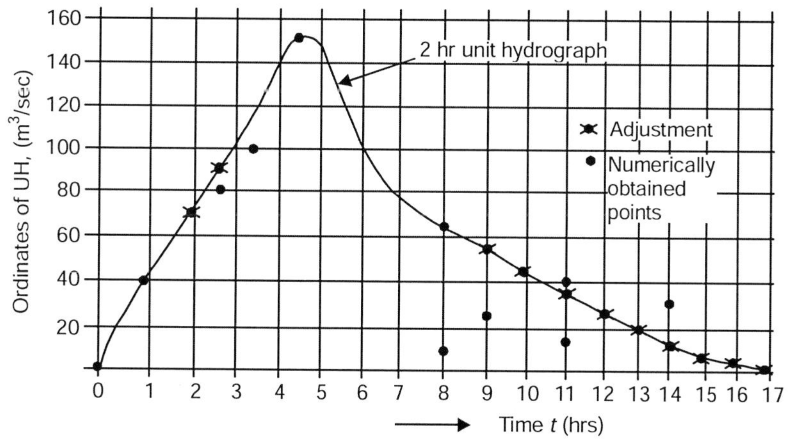

Figure 7.17 Adjusted 2-hr unit hydrograph.

Time (hrs)	0	1	2	3	4	5	6	7	8	9	10	11	12	13	14	15	16	17
Ordinates of UH (m^3/sec)	0	40	72	88	105	144	140	66	57	43	38	27	21	17	11	7	3	0

7.13 SYNTHETIC UNIT HYDROGRAPH (SUH)

Synder[7] (1938) pioneered the concept of synthetic unit hydrograph followed by McCarthy[8] (1938), and Clark[9] (1945). Synder's concept is based on the analysis of gauged watersheds, which can be applied to ungauged basins using only the measurement of basin characteristics. He studied 20 watersheds located in the Appalachian Highlands of eastern United States with areas ranging from 25.9 to 25, 900 km^2. His method is widely used in the USA and other countries. He developed some empirical formulae which can be used to develop unit hydrograph of the catchment in which required parameters to develop UH are not available. Unit hydrograph developed for such ungauged catchment with the help of Synder's formula is called **synthetic unit hydrograph** (SUH).

Synder selected three parameters for the development of SUH. They are:

(i) Base time width T.

(ii) Peak discharge Q_p.

(iii) Lag time, i.e., Basin lag time t_p.

These three parameters used by Synder have been shown in Figure 7.18(b). He proposed the following three equations for these three parameters from his study of 20 watersheds:

$$\text{Lag time } t_p = C_t(LL_{ca})^{0.3} \tag{7.10}$$

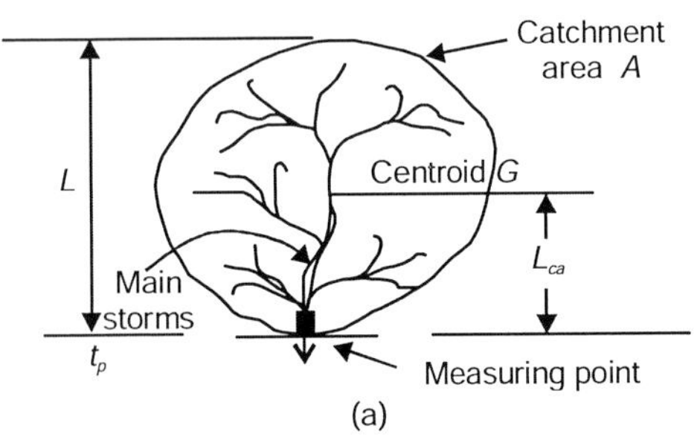

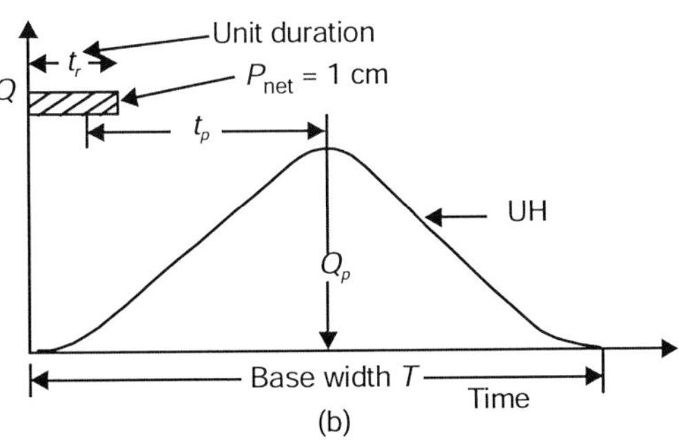

Figure 7.18 Unit hydrograph to define T_i Q_p and t_p of Synder unit hydrograph.

where t_p in hours, C_t is coefficient reflecting slope, land use, associated storage characteristics of the basin, L is the length of longest water course [shown in Figure 7.18(a)] up to measuring

[7] Synder F.F., *Synthetic Unit Hydrography*, Trans. Am. Geophys. Union, vol. 19, Pt. 1, pp. 447–454, 1938.

[8] McCarthy, G.T., The Unit Hydrograph and Flood Routing, Unpublished manuscript presented at a conference of North Atlantic Div. U.S Army Corps of Engineering, June 24, 1938.

[9] Clark, C.O., Storage and the Unit Hydrograph, Trans. ASCE, 110, pp. 1419–1446, 1945.

point and L_{ca} is length from centroid G of catchment to measuring point. L and L_{ca} shown in Figure 7.18(a) are in miles. C_t varies from 1.8 to 2.2, average being 2. When L and L_{ca} are in km, C_t varies from 1.35 to 1.65, average being 1.5 and t_p in hours.

Now,

$$T = 3 + \frac{t_p}{8}\,(\text{days})$$
$$T = (72 + 3t_p)(\text{hrs}) \tag{7.11}$$

$$Q_p = 2.78\left(\frac{c_p A}{t_p}\right) \tag{7.12}$$

where Q_p is in m^3/sec, A is in km^3, C_p in Synder's study varies from 0.56 to 0.69. Others got this value from 0.4 to 0.8. Synder used the equation of standard duration t_r (or D hr) in hour for the unit hydrograph as:

$$t_r = D\text{ hr} = \frac{t_p}{5.5} \tag{7.13}$$

If a synthetic unit hydrograph of any other duration is required, then lag time t_{pr} is given by

$$t_{pr} = t_p + \frac{D' - t_r}{4} = t_p + 0.25(D' - t_r) \tag{7 14}$$

Example 7.7

Derive 3-hr synthetic unit hydrograph of basin with following data with a catchment area of 2500 km^3:

Length of main stream = 120 km

Distance from central outlet = 80 km

C_t and C_p of the catchment are assumed to be 1.5 and 0.6 respectively. Use Synder's method.

Solution:

$$t_p = C_t\,(LL_{ca})^{0.3}$$

Here C_t = 1.5, L = 120 km, L_{ca} = 80 km

$\therefore$ t_p = 23.484 hrs.

$$\text{Standard duration } t_r = D = \left(\frac{t_p}{5.5}\right) = \left(\frac{23.484}{5.5}\right) = 4.27\text{ hrs}$$

Required duration D' = 3 hrs

$$\text{Lag time } t_{pr} = t_p + \frac{D' - t_r}{4} = t_p + 0.25(D' - t_r)$$

or $$t_{pr} = 23.484 + 0.25(3 - 4.27)$$
$$= 23.16 \text{ hrs}$$

Now, $$Q_p = 2.78\left(\frac{0.6 \times 2500}{23.16}\right) = 180 \text{ m}^3/\text{sec}$$

and $$T = 72 + 3 \times 23.16 = 141.48 \text{ hours}$$

With these values Q_p, T and t_{pr}, a smooth curve may be drawn.

To plot the smooth synthetic unit hydrograph (SUH), US Army Corps Engineers[10] (1959) gave the width of SUH as:

$$W_{50} = \frac{5.87}{\left(\dfrac{Q_P}{A}\right)^{1.08}}$$

$$W_{75} = \frac{3.35}{\left(\dfrac{Q_P}{A}\right)^{1.08}}$$

where W_{50} and W_{75} are widths in hrs of SUH at 50% and 75% of Q_p.

∴ $$W_{50} = \frac{5.87}{\left(\dfrac{180}{2500}\right)^{1.08}} = 100.62 \text{ hrs}$$

and $$W_{75} = \frac{3.35}{\left(\dfrac{180}{2500}\right)^{1.08}} = 57.42 \text{ hrs}$$

Hence the synthetic unit hydrograph (SUH) may be drawn as shown in Figure 7.19.

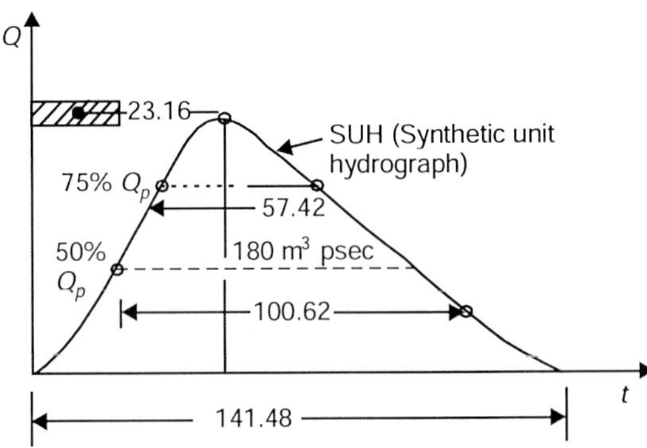

Figure 7.19 Answer of Example 7.7.

[10] US Army Corps of Engineers, Flood Hydrograph Analysis and Computations, *Engg. and Design Manual*, EM 1110-2-1405, US Govt. Printing Office, Washington, D.C., August 31, 1959.

EXAMPLE 7.8

Two catchments x and y are considered meteorologically homogenous. The characteristics of the two are:

x	y
$L_x = 50$ km	$L_y = 80$ km
$L_{ca\,x} = 30$ km	$L_{ca\,y} = 60$ km
$A_x = 5000$ km^2	$A_y = 1000$ km^2

For catchment x, UH developed is of 2 hr duration and peak discharge $Q_{px} = 80$ m^3 sec. The time of peak from beginning of rainfall in x is 13 hrs. Using synder's method, develop SUH for catchment y.

Solution: For catchment x, t_{px} is obtained as:

$$13 = t_{px} + \frac{t_r}{2} = t_{px} + \frac{2}{2}$$

∴ $$t_{px} = 12 \text{ hrs}$$

Now, $$t_{px} = C_t(L_x L_{ca})^{0.3}$$

$$12 = C_t(50 \times 30)^{0.3}$$

∴ $$C_t = 1.3376$$

Again $$Q_{px} = 2.78\left(\frac{C_p A_x}{t_{px}}\right) = 80 = 2.78\left(\frac{500}{12}\right)^{cp}$$

∴ $$C_p = 0.69$$

Since x and y are homogenous, the values of C_t and C_p for x may be used to develop SUH for catchment y.

∴ $$t_{py} Ct(L_y . L_{cay})^{0.3} = 1.3376(80 \times 60)^{0.3}$$

or $$t_{py} = 17.00 \text{ hrs}$$

∴ $$\text{Standard duration } t_{ry} = D_y = \frac{t_{py}}{5.5} = \frac{17.00}{5.5} = 3.09 \text{ hrs}$$

$$\cong 3 \text{ hrs}$$

Now, $$\text{Lag time } t_{pry} = t_{tpy} + \frac{D_y - t_{rx}}{4}$$

or $$t_{pry} = 17.00 + \frac{3-2}{4} = 17.25 \text{ hrs}$$

∴ $$Q_{py} = 2.78\left(\frac{C_p + A_y}{t_{pry}}\right)^{2.78} = \left(\frac{0.69 \times 1000}{17.25}\right) = 111.2 \text{ m}^3\text{/sec}$$

$$T_y = 72 + 3\left(\frac{t_{pry}}{8}\right) = 72 + 3\left(\frac{17.25}{8}\right) = 78.84 \text{ hrs}$$

$\therefore$ $$Q_p = 40 \text{ m}^3 \text{ sec},\ T = 78.84 \text{ hrs},\ t_p = 17.25 \text{ hrs}$$

Now, $$W_{50} = \frac{5.87}{\left(\dfrac{Q_P}{A}\right)^{1.08}}$$

$$= \frac{5.87}{\left(\dfrac{115.2}{1000}\right)^{1.08}} = 62.93 \text{ hrs}$$

and $$W_{75} = \frac{5.87}{\left(\dfrac{111.2}{1000}\right)^{1.08}} = 35.91 \text{ hrs}$$

Answer of Example 7.8 is given in Figure 7.20.

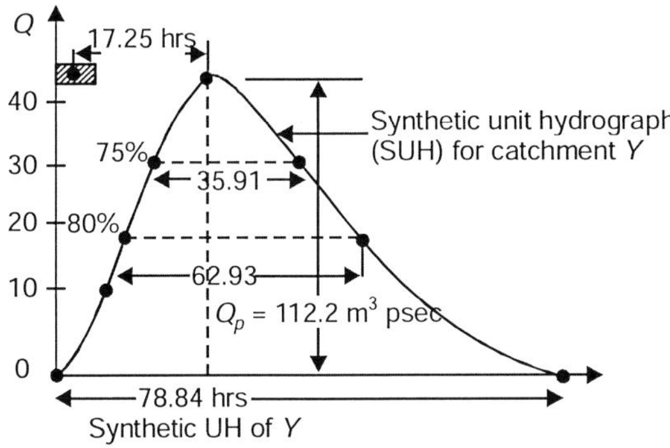

Figure 7.20 Answer of Example 7.8.

7.14 MUTREJA'S EMPIRICAL EQUATION IN INDIAN CONDITIONS

Mutreja[11] (1986) has suggested the following equations for Indian conditions:

$$t_p = 1.13\frac{(LL_{ca})^{0.2769}}{\sqrt{s}} \tag{7.15}$$

$$Q_{ps} = 0.315A^{0.93}s^{0.53} \tag{7.16}$$

[11] Murtreja, K.N., *Applied Hydrology*, Tata McGraw-Hill, New Delhi, 1984.

$$W_{50} = 2.18\left(\frac{Q_{\text{peak}}}{A}\right)^{-1.12} \tag{7.17}$$

$$W_{75} = 0.81\left(W_{50}^{0.72}\right) \tag{7.18}$$

$$\text{Width of rising side of UH } W_{R\text{-}50} = 0.69(W_{50})^{0.69} \tag{7.19}$$

$$\text{Width of rising side of UH } W_{R\text{-}75} = 0.605(W_{R\text{-}50})^{0.95} \tag{7.20}$$

$$T = t_b = 4.3\frac{(LL_{ca})^{0.28}}{s^{1/2}} \tag{7.21}$$

Here s is the slope of the drainage channel.

7.15 TRIANGULAR UNIT HYDROGRAPH

This method was developed by Mockus[12] (1957). Consider Figure 7.21. The area triangular unit hydrograph gives volume Q_v, i.e.,

$$Q_v = \frac{1}{2}(t_{\text{peak}} + t_b)Q_{ps} \tag{7.22}$$

or

$$Q_{ps} = \frac{2Q_v}{t_{\text{peak}} + t_b} \tag{7.22a}$$

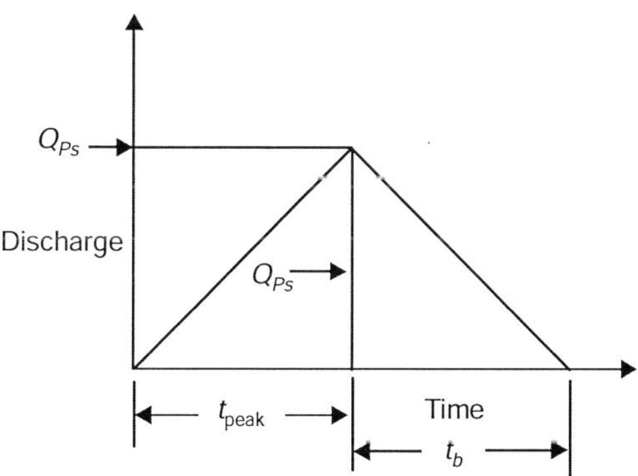

Figure 7.21 A triangular UH.

According to US soil conservation service,

$$t_{\text{peak}} = 0.7t_c$$

or

$$t_{\text{peak}} = \frac{\text{Rainfall effective time}}{2} + 0.6t_c \tag{7.23}$$

[12] Mockus, V., Use of Storm and Watershed Characteristics in Synthetic Unit Hydrograph Analysis and Applications, United States Soil Conversation Service (USSCS), 1957.

Here t_c is the time of concentration.

Then $$\text{Time lag} = 0.6t_c \tag{7.24}$$

$$T = 1.67t_{\text{peak}} \tag{7.25}$$

Therefore, $$Q_{ps} = \frac{2Q_v}{2.67t_{\text{peak}}} = \frac{0.75Q_v}{t_{\text{peak}}} \tag{7.26}$$

When area A is in km^2,

$$Q_{ps} = \frac{0.208AQ_v}{t_{\text{peak}}} \text{ m}^3\text{/sec} \tag{7.27}$$

7.16 NON-DIMENSIONAL UNIT HYDROGRAPH

This non-dimensional unit hydrograph has been developed by US Soil Conservation Service (USSCS). Non-dimensional discharge

$$Q_n = (Q_t/Q_p)$$

where Q_t is the discharge at any time, Q_p is the peak discharge.

Non-dimensional time $T_n = (t/t_{\text{peak}})$, where t is the time at any instant and t_{peak} is the time of Q_p. Mockus (1957) in USSCS gave the non-dimensional UH as tabulated in Table 7.1.

Table 7.1 Non-dimensional Discharge vs Non-dimensional time

T_n	0	0.25	0.50	0.75	1.00	1.25	1.50	1.75	2.00	2.25	2.50	
		2.75	3.00	3.25	3.50	3.75	4.00	4.25	4.50	4.75	5.0	
Q_n	0	0.12	0.43	0.83	1.00	0.88	0.66	0.45	0.32	0.22	0.12	0.105
		0.075	0.053	0.036	0.026	0.018	0.012	0.009	0.006	0.004		

Plotting Q_n against T_n, non-dimensional unit hydrograph is obtained. Q_p is given by USSCS as:

$$Q_p = \frac{5.36A}{t_{\text{peak}}} \tag{7.28}$$

where A is in km^2, Q_p is in m^3/sec and t_{peak} is in hours.

EXAMPLE 7.9

A unit hydrograph of 2-hr duration is to be constructed by US Soil Conservation Service (USSCS) method and using Table 7.1 by Mockus (1957).

Given data: Catchment area = 350 sq.km, Lag time = 14 hr

Solution:

$$t_{\text{peak}} = \text{Lag time} + \frac{\text{Duration of UH}}{2}$$

$$= 14 + \frac{2}{2} = 15 \text{ hrs}$$

$$Q_{\text{peak}} = \frac{5.36A}{t_{\text{peak}}} = \frac{5.36 \times 350}{15} = 125.06 \text{ m}^3\text{/sec}$$

When $T_n = \dfrac{t}{t_{\text{peak}}} = 1, \dfrac{Q_t}{Q_{\text{peak}}} = 1$, i.e., peak discharge occurs.

At $t = 15$ hrs, $Q_{\text{peak}} = 125.06$ m^3/sec occurs.

When $\left(\dfrac{t}{t_{\text{peak}}}\right) = T_n \cong 5,\ Q_t \cong 0$.

∴ Base period of hydrograph $T = 5 \times t_{\text{peak}}$

$= 5 \times 15 = 75$ hrs

From Table 7.1, we compute Q vs t at different Q_n and T_n and the unit hydrograph (UH) is plotted in Figure 7.22.

(i) When $T_n = 0.50, \dfrac{Q}{Q_{\text{peak}}} = 0.43$.

∴ $t = 0.5 \times 15 = 7.5$ hrs, $Q = 0.43 \times 125.06 = 53.77$ m^3/sec

(ii) $T_n = 1$, $Q = 125.06$ m^3/sec (already determined)

(iii) $T_n = 1.5$, $t = 1.5 \times 15 = 22.5$ sec, $Q = 0.66 \times 125.06 = 82.54$ m^3/sec

(iv) $T_n = 2.0$, $t = 2 \times 15 = 30$ sec, $Q = 0.32 \times 125.06 = 40.02$ m^3/sec

(v) $T_n = 2.5$, $t = 2.5 \times 15 = 37.5$ sec, $Q = 0.22 \times 125.06 = 23.11$ m^3/sec

(vi) $T_n = 3.0$, $t = 3 \times 15 = 45$ sec, $Q = 0.75 \times 125.06 = 9.38$ m^3/sec

(vii) $T_n = 4.0$, $t = 4 \times 15 = 60$ sec, $Q = 0.018 \times 125.06 = 2.25$ m^3/sec

(viii) $T_n = 5.0$, $t = 5 \times 15 = 75$ sec, $Q = 0.004 \times 125.06 = 0.50$ m^3/sec

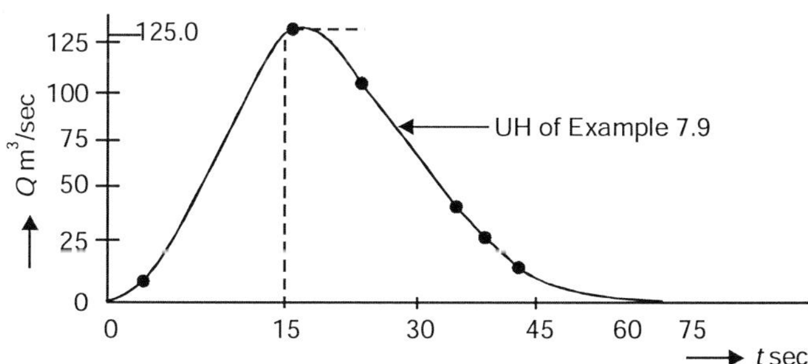

Figure 7.22 UH of Example 7.9.

7.17 CLARK SYNTHETIC UNIT HYDROGRAPH

On the concept of routing a time-area relationship through a linear reservoir, Clark[13] (1943)

[13] Clark, C.O., Storage and the Unit Hydrograph, Proc. ASCE, vol. 9, pp. 1333–1360, 1943.

developed another method to develop synthetic unit method, which is called **Clark method.** The basic concept used to develop the Clark Method may be described as follows:

Water covering the catchment to a depth of 1 cm is released instantaneously and allowed to run of the basin. A time-area relationship represents the translation hydrograph of this runoff as influenced by watershed characteristics, such as size, shape and surface roughness. The translation hydrograph is routed through a linear reservoir to capture additional storage effects of the watershed. The resulting unit hydrograph represents the runoff from a rainfall of zero duration. To obtain a unit hydrograph for a rainfall of duration *D*-hr, two instantaneous unit hydrographs lagged by *D*-hr are combined, and then divided by 2 to reduce the resulting two points, *D*-hr duration hydrograph to a unit hydrograph.

Figure 7.23 shows the isochrone lines drawn on the watershed map to show the loci of points of travel time to the outlet. The travel time for any point in the watershed refers to the time required for a parcel of water to travel from that point to the outlet. It is the time of concentration t_c which is the travel time from the most remote point to the outlet. A travel-time against watershed sub-area relationship indicates the contributing area as a function of time; with entire watershed contributing flow at the outlet at t_c. A relationship between travel time T and contributing watershed area may by developed based on travel time T. Travel time is a fraction of time of concentration t_c, i.e., $t_c = \Sigma\ T_i$ and T is function of flow length and velocity V, i.e.,

$$T = \frac{L}{D}$$

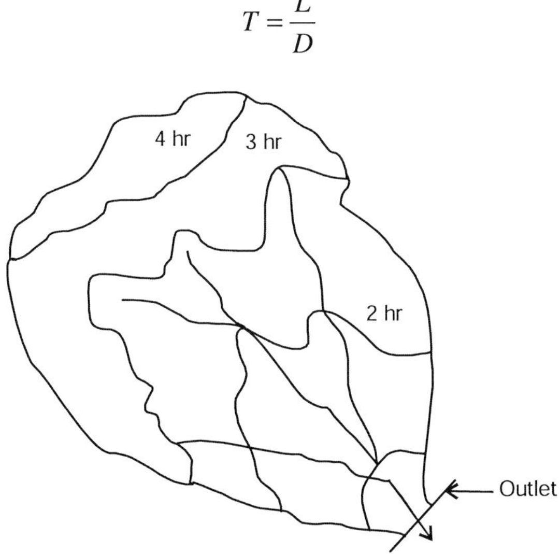

Figure 7.23 The isochrone map shows travel times from watershed outlet.

The determination of V in this case is not very simple like the velocity of an open channel. Investigator like McCuen[14] (1998) used the modified Manning's equation of type $V = Ks^{1/2}$, where value K is different for different surfaces of watershed. V may be used to determine this overland flow based on Chow's table[15] (1972). However, estimation of t_c, and development of

[14] McCuen, R.H., *Hydrologic Analysis and Design,* 2nd ed., Prentice Hall, Upper Saddle River, N.J., 1998.
[15] Chow, V.T., *Open Channel Hydraulics*, McGraw-Hill International, Auckland, 1972.

time-area relationship is difficult. To simplify the Clark method, Hydrologic Engineering Centre[16] (HEC, 1998) developed the following empirical relationship based on analysis of many watersheds:

$$\left(\frac{A_c}{A}\right) = 1.414T^{1.5} \qquad \text{(For } 0 \le T \le 0.5\text{)} \tag{7.29}$$

$$\left(\frac{A_c}{A}\right) = 1 - 1.414(1-T)^{1.5} \qquad \text{(For } 0.5 \le T \le 1.0\text{)} \tag{7.30}$$

where A_c is the contributing area of total area A, T is the fraction of concentration time t_c.

A translation hydrograph based on the time area relationship is routed through a linear reservoir. Hydrologic routing continuity equation may be written as:

$$\left(\frac{I_1 + I_2}{2}\right) - \left(\frac{Q_1 + Q_2}{2}\right) = \left(\frac{S_2 - S_1}{\Delta t}\right) = \frac{\Delta S}{\Delta t} \tag{7.31}$$

or

$$\bar{I} - \left(\frac{Q_1 + Q_2}{2}\right) = \frac{\Delta S}{\Delta t} \tag{7.31a}$$

where I is inflow, Q is outflow and S is storage.

Equation of linear reservoir is:

$$S = KQ \tag{7.32}$$

Here K is a storage coefficient.

Substituting Eq. (7.32) in Eq. (7.31a), Routing equation may be obtained as:

$$Q_2 = \left(\frac{2\Delta t}{2K + \Delta t}\right)\bar{I} + \left(1 - \frac{2\Delta t}{2K + \Delta t}\right)Q_1 \tag{7.33}$$

Initially Q_1 is known, hence Q_2 is computed. Then Q_2 is known, Q_3 is computed. Thus, all outflows are routed and hydrograph obtained by this routing represents the runoff for an instantaneous (Zero duration) rainfall event. This instantaneous hydrograph is converted to 1-hr rainfall duration unit hydrograph lagged by 1 hr, combining and dividing by 2 to represent average flow at beginning and end of each Δt to get the final 1-hr unit hydrograph. Example 7.10 is given to understand the whole procedure:

EXAMPLE 7.10

Using Clark method, develop a UH for rainfall duration of 1 hr for a catchment of 75 km^2. Take t_c = 8 hrs. The storage coefficient of linear reservoir is 3.2 hours. Use Hydrologic Engineering Centre (HEC) synthetic time-area relationship.

Solution: Here t_c = 8 hrs, A = 75 km^2, K = 3.2 hrs and Δt = 1 hr.

[16] Hydrologic Engineering Centre (HEC-1), Flood Hydrograph and Package Users, Manual, US Army Corps of Engrs., Davis, CA, June 1998.

Time t (hrs) (1)	$T = t/t_c$ (2)	A_c/A (3)	$A_c = (3) \times A$ (4) km^2	ΔA (5)	*Translated* (6) $2.778 \times (5)$	0-*hr UH* (m^3/sec) (7)	1-*hr UH* (m^3/sec) (8)
0	0	0	0	0	0	0	0
1	0.125	0.062	4.65	4.65	12.9	3.5	1.75
2	0.250	0.176	13.2	8.55	23.7	8.9	6.2
3	0.375	0.324	24.3	11.1	30.8	17.9	13.2
4	0.500	0.500	37.5	13.2	30.6	21.3	19.6
5	0.625	0.675	50.625	13.125	36.4	25.4	23.4
6	0.75	0.823	61.725	11.100	30.8	26.8	26.1
7	0.875	0.937	70.275	8.550	23.7	25.9	26.3
8	1.0	1.0	75	4.725	13.1	22.4	24.1
9	1.125				0	16.4	19.4
10	1.250					11.9	14.2
11	1.375					8.7	10.3
12	1.500					6.3	7.5
13						4.6	5.4
14						3.4	4.0
15						2.4	2.9
16							2.1
17							1.5
18							|
19							|
20							|
21							|
22							|
23							|
24							|
25							|
26							|
27							|
28							|
29							|
30							|

Using $A_c/A = 1.414\ T^{1.5}$ when $0 \le T \le 0.5$
and $A_c/A = 1 - 1.414(1 - T)^{1.5}$ when $0.5 \le T \le 1$
Then Column (4) = (3) × A = [(3) × 75] km^2

} in Column (3), i.e., (7.29) and (7.30).

An example at t = 5 sec, considered,

$$\text{Column (2)}\ \frac{t}{t_c}=\frac{5}{8}=0.625\quad \frac{A_c}{A}=0.675 \qquad \text{[By Eq. (7.30) in Column (3)]}$$

and

$$A_c = 0.675 \times A$$

$$= 0.675 \times 75 = 50.625\ \text{km}^2 \text{ in Column (4)}$$

Now, Column (5), ΔA = 50.625 – 37.5 =13.125 km^2

Column (6), Translated = 2.778 × (5) = 36.4

Then

$$Q=\frac{1\ \text{cm}\times 13.125\ \text{km}^2}{1\ \text{hr}}$$

$$=\frac{0.01\ \text{m}\times 13.125\times 10^6\ \text{m}^2}{(1\times 60\times 60)\text{sec}}=(2.778\times 13.125)\ \text{m}^3/\text{sec}$$

$$=36.4\ \text{m}^3/\text{sec}$$

In Column (7), use Eq. (7.33).

$$Q_5 = 0.27(36.4) + 0.73(21.3) = 25.4\ \text{m}^3/\text{sec}$$

$$\text{Column (8)},\ \left(\frac{21.3+25.4}{2}\right)=23.4\ \text{m}^3/\text{sec}$$

Column (8) in m^3/sec vs Column (1) gives the 1-hr UH.

7.18 INSTANTANEOUS UNIT HYDROGRAPH

In 1945, Clark[17] for the first time introduced the concept of instantaneous unit hydrograph (IUH). When the duration of effective rainfall becomes infinitesimally small, the resulting hydrograph is called **instantaneous unit hydrograph (IUH).**

The major advantage of IUH in comparison with UH is that the IUH independent of duration of effective rainfall thereby eliminating one variable in hydrograph analysis, i.e., when duration of UH tends to zero, the resulting hydrograph is IUH.

The use of IUH is better suited for the needs of theoretical investigations on the rainfall and runoff relationship in drainage basins.

The determination of IUH is numerically tedious than that of unit hydrograph, but it can be simplified through the use of high speed computer. (Figure 7.24).

7.19 CONVOLUTION PROCESS

Here $Q(t)$ is the ordinate of DSR hydrograph, $U(t)$ is the ordinate of IUH, $I(\lambda)$ is rate of rainfall at time λ and D_0 is the total effective duration of rainfall. Rainfall input is the summation of successive instantaneous inputs of depth (Figure 7.24).

[17] Clark, C.O., Storage and Limit Hydrograph, *Trans., ASCE*, Vol. 110, pp. 1419–1446, 1945.

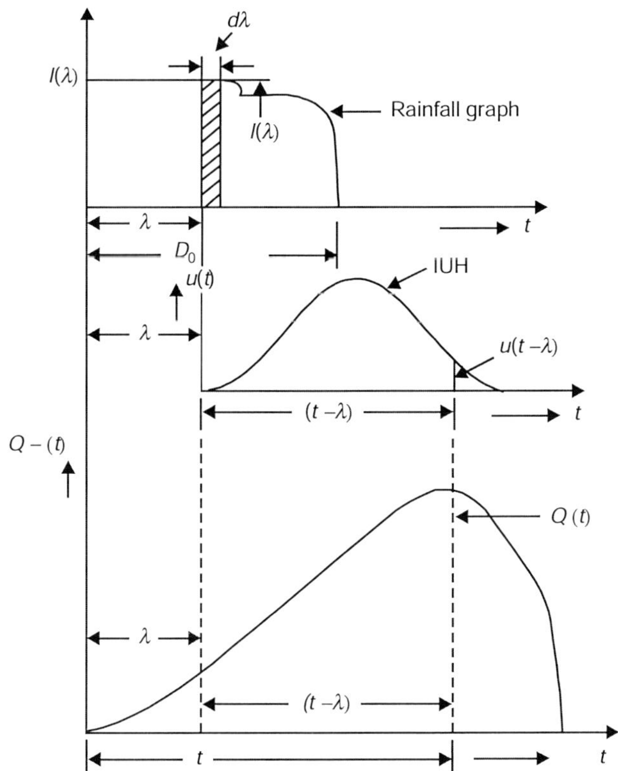

Figure 7.24 Concept of convolution process.

$$I(\lambda) \text{ cm/hr} \times d\lambda(\text{hr}) = I(\lambda)\, d\lambda \text{ cm}.$$

The hydrograph of this rainfall of $I(\tau)\, d\tau$ is the IUH.

The contribution to DSR due to this instantaneous rainfall $I(\tau)d\tau$ is:
Area under IUH = $u\ (t - \lambda)I(\lambda)\ d\lambda$

$$\therefore \qquad Q(t) = \int_{0}^{t \le D_0} u(t - \lambda)\ I(\lambda)d\lambda \tag{7.34}$$

where RHS of Eq. (7.34) is called **convolution integral**, $u(t - \lambda)$ is called **Kernel function** and $I(\lambda)$ is the **input function**.

Equation (7.34) gives the direct surface runoff due to any storm, when IUH of the basin is known.

In Eq. (7.34), $u(t - \lambda)$ is called **Dirac delta function** or **unit impulse function** which is a frequent occurrence in mechanics and in hydrology in IUH based on idea of large parameter acting for a very short time. In mathematics, Dirac delta function or unit impulse function is considered as the limiting form of function (Figure 7.25):

$$\begin{aligned} \delta_e\ (t - a) &= 1/\varepsilon, && a \le t \le a + \varepsilon \\ &= \alpha, && t = a, \text{ i.e., } \varepsilon = 0 \\ &= 0, && t \ne a \end{aligned}$$

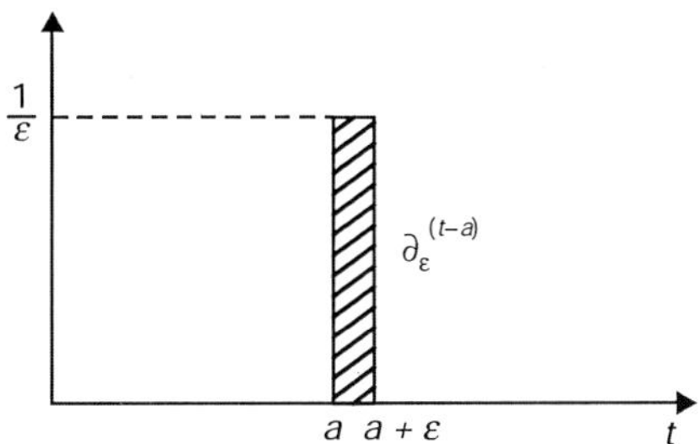

Figure 7.25 Concept of Dirac delta function.

And as $\varepsilon \to 0$, it is clear from Figure 7.25, the height of the strip increases indefinitely and decreases in such a way that its area is always unity.

Thus, the unit Dirac delta functions $\delta(t - a)$ is defined as follows:

$$\delta(t - a) = a, \quad \text{for } t = 0$$
$$= 0, \quad \text{for } t \neq a$$

Such that
$$\int_0^\infty \delta(t - a)\, dt = 1$$

Thus $u(t - 1)$ term in Eq. (7.34) is:

$$u(t - \lambda) = \int_0^\alpha \delta(t - a)\, I(\lambda)\, d\lambda \tag{7.35}$$

7.20 DERIVATION OF IUH FROM S-CURVE

Unit hydrograph of *D*-hr duration is taken to derive a S-curve. Another S-curve lagged by time Δt is also derived from the same *D*-hr duration UH. Then ordinate of the unit hydrograph at time ($\Delta t \to 0$).

$$u(t + \Delta t) = \frac{D}{\Delta t}\left[S(t) - S(t - \Delta t)\right] = D\frac{\Delta S(t)}{\Delta t}$$

where $S(t)$ is S-curve ordinates.

In the limit as Δt tends to zero, one may get the ordinate of IHU as:

$$\text{Lim}_{\Delta t \to 0}\; u(t + \Delta t) = u(t) = D \cdot \frac{ds(t)}{dt}$$

$$\therefore \quad u(t) = D\frac{ds(t)}{dt} \tag{7.36}$$

where ordinate of IUH, $u(t) = D \times$ Slope of S-curve derived from *D*-hr unit hydrograph at *t*. The ordinate $u(t)$ of IUH obtained by this method is approximate since S-curve obtained from a duration unit hydrograph (DUH) cannot be taken to be exact.

7.21 RELATIONSHIP BETWEEN DUH AND IUH

Convolution integral equation already derived is:

$$Q(t) = \int_0^{t \le D_0} u(t - \lambda) I(\lambda)\, d\lambda$$

Let the input of rainfall be $(1/D)$ cm/hr, where D is duration of UH. The corresponding output will be S-hydrograph, i.e., replacing $Q(t)$ by $S(t)$ and $I(\lambda)$ by $\frac{1}{D}$.

$$S(t) = \frac{1}{D} \int_0^t u(t - \lambda)\, d\lambda \tag{7.37}$$

Put $z = t - \lambda$, then $dz = -d\lambda$. Introducing this transformation in Eq. (7.37) with limit of integration as:

$$\lambda = 0, \qquad z = t$$

and

$$\lambda = t, \qquad z = 0$$

$$S(t) = \frac{1}{D} \int_t^0 u(z)(-dz)$$

or

$$S(t) = \frac{1}{D} \int_0^t u(z)\, dz \tag{7.38}$$

Equation (7.38) gives the ordinates of S-hydrograph from an infinite rainfall of constant density $(1/D)$ cm/hr multiplied by a term, which gives the IUH by integrating from zero to t or ordinate of IUH at any time t is obtained by

$$D \times \frac{dS(t)}{dt}$$

This is same as Eq. (7.36).

i.e.,

$$u(t) = D \frac{ds(t)}{dt}$$

7.22 NASH MODEL OF CASCADED LINEAR RESERVOIRS

J.E. Nash[18] (1957) formulated a conceptual model of a catchment having the same hydrologic response as series of n reservoirs, each having the same storage coefficient K for developing IUH. This model is called **Nash model** which is proved to be simple and effective method for deriving catchment IUHs.

[18] Nash, J.E., The Form of the Instantaneous Hydrograph, Intl. Assoc. of Scientific Hydrology Pub., 45(3), pp. 114–21, 1957.

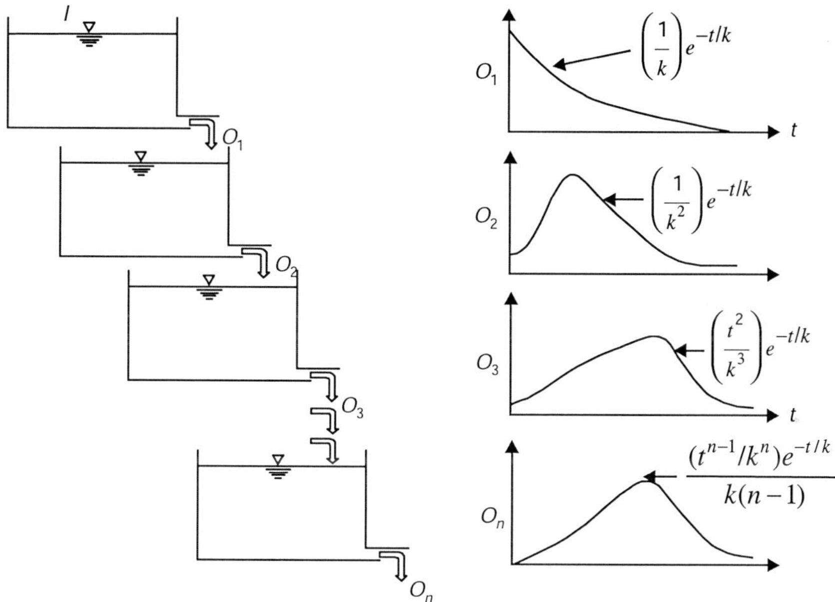

Figure 7.26 Cascaded reservoirs of Nash model.

As shown diagrammatically in Figure 7.26, a hydrograph shape outflow curve is developed and modified by routing the outflow from one reservoir as inflow to the next lower reservoir.

A linear reservoir is a conceptual reservoir in which storage S is directly proportional to outflow O, i.e.,

$$S = KO \tag{7.39}$$

Here K is the proportionality constant, known as **storage coefficient**, having the dimension of time.

Continuity equation gives

$$I - O = \frac{dS}{dt} \tag{7.40}$$

Here I is rate of inflow.

Substituting Eq. (7.39) in Eq. (7.40)

$$I - O = K\frac{dO}{dt} \tag{7.41}$$

or

$$\frac{dO}{dt} + \frac{O}{K} = \frac{I}{K} \tag{7.42}$$

The solution for which is:

$$O = I(1 - e^{-t/K})$$

For equilibrium condition when $t \to \infty$, $O \to I$.

If inflow I stops at time $t = t_0$ at which time outflow $O = O_0$, $I = 0$ and $t - t_0 = \lambda$. Equation (7.41) becomes

$$\frac{dO}{d\lambda}+\frac{O}{k}=0 \tag{7.43}$$

Solution of this ordinary differential Eq. (7.43) is:

$$O=O_0e^{-\lambda/K} \tag{7.44}$$

For an instantaneous inflow which fills a reservoir of storage S_0 in time $t = 0$,

$$O_0=\frac{S_0}{K} \tag{7.45}$$

When Eqs. (7.44) and (7.45) are combined and $t = \tau$,

$$O=\left(\frac{S_0}{K}\right)e^{-t/K} \tag{7.46}$$

In IHU, inflow is instantaneous and storage $S_0 = 1$.

$$\therefore \qquad u(t)=\left(\frac{1}{K}\right)e^{-t/K} \tag{7.47}$$

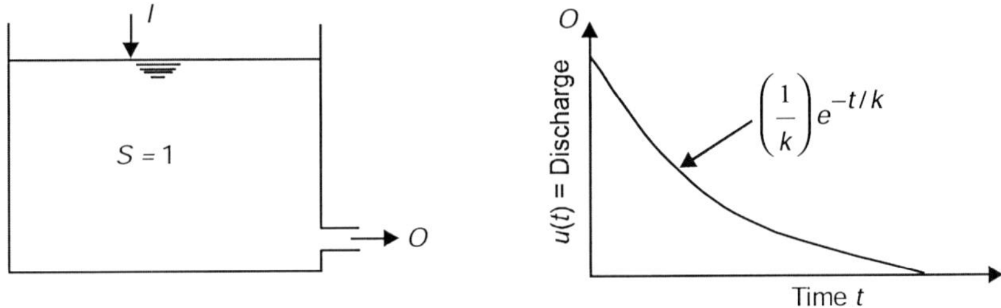

Figure 7.27 IUH for a linear reservoir.

Assuming instantaneous unit output into the initial reservoir, the outflow derived in Eq. (7.47) may be considered as input to the second reservoir of Figure 7.26. Using τ as variable in the convolution integral, outflow from the second reservoir may be obtained as:

$$O_2=\int_0^t I(\lambda)u(t-\lambda)d\lambda$$

$$O_2=\int_0^t \left|\left(\frac{1}{k^2}\right)^{-\lambda/k}\right|\left|\left(\frac{1}{K}\right)e^{-(t-\lambda)/K}\right|d\lambda$$

$$O_2=\left(\frac{t}{K^2}\right)e^{-t/K} \tag{7.48}$$

The repetition of above convolution for a value of n, the generalized equation for IHU of conceptualized drainage basin may be derived as:

$$O_n = u(t) = \frac{1}{K(n-1)!}\left(\frac{t}{K}\right)^{n-1} e^{t/K}$$
$$= \frac{1}{K\Gamma(n)}\left(\frac{t}{K}\right)^{n-1} e^{-t/K} \tag{7.49}$$

where $\Gamma(n)$ is gamma function and K and n are determined by method of moments.

A note on Gamma Function $\Gamma(n)$:

In mathematics $\Gamma(n)$ is defined as

$$\Gamma(n) = \int_0^{\alpha} e^{-x}\, x^{n-1}\, dx \; (n > 0) \tag{7.50}$$

In particular, $\Gamma(1) \int_0^{\alpha} e^{-x}\, x^{1-1} = |e^{-x}|_0^{\alpha} = 1$

Reduction formula of $\Gamma(n)$

Since
$$\Gamma(n + 1) = \int_0^{\alpha} e^{-x} x^n dx$$
$$= |x^n\, e^{-x}|_0^{\infty} + n \int_0^{\alpha} e^{x} x^{n-1} dx$$
$$= n \int_0^{\alpha} e^{x} x^{n-1} dx$$

$\therefore$
$$\Gamma(n + 1) = n\, \Gamma(n)$$

Value of $\Gamma(n)$ *in terms of Factorial*

Using
$$\Gamma(n + 1) = n\, \Gamma(n)$$
$$\Gamma(2) - 1 \times \Gamma(1) = 1!$$
$$\Gamma(3) = 2\, \Gamma(2) = 2 \times 1 = 2$$
$$\Gamma(3) = 2\, \Gamma(2) = 2 \times 1 = 2!$$
$$\Gamma(4) = 3 \times \Gamma(3) = 3 \times 2 \times 1 = 3!$$

– – – – – – – – – – – – – – – – – – –

7.22.1 Determination of *K* and *n*

These two values are determined by the method of moment. The mth movement of any area about the origin is:

$$M_m = \frac{\int_A f(x)dx\, x^m}{\int_A f(x)\, dx}$$

The mth movement of IUH about the origin ($t = 0$) is:

$$M_m = \frac{\int_0^{\alpha} u(t)dt \cdot t^m}{\int_0^{\alpha} u(t)dt} \tag{7.51}$$

Since area of IHU is equal to unity, Eq. (7.51) becomes

$$M_m = \frac{1}{K\Gamma(n)} \int_0^{\alpha} e^{-t/K} \cdot \left(\frac{t}{K}\right)^{n-1} \cdot t^m dt$$

$$= \frac{K^m}{\Gamma(n)} \int_0^{\alpha} e^{-t/K} \cdot \left(\frac{t}{K}\right)^{n-1} \left(\frac{t}{K}\right)^m d\left(\frac{t}{k}\right)$$

and from definition of $\Gamma(n)$,

$$M_m = \frac{K^m \cdot \Gamma(m+n)}{\Gamma(n)} \tag{7.52}$$

Now 1st and 2nd moments are obtained by putting $m = 1$ and 2 respectively, i.e., from Eq. (7.52)

$$M_1 = nK \tag{7.53}$$

$$M_2 = n(n + 1)K \tag{7.54}$$

Let MI_1 and MI_2 be 1st and 2nd moments of effective rainfall and MO_1 and MO_2 are 1st and 2nd moments of direct runoff. It can be shown that the following relationship will hold:

$$MO_1 - MI_1 = nk \tag{7.55}$$

$$MO_2 - MI_2 = n(n + 1)K^2 + 2nk(MI_1) \tag{7.56}$$

Equations (7.55) and (7.56) will yield the values of n and k.

EXAMPLE 7.11

Derive the ordinates of IUH using Nash model and hence ordinates of 2 hr-UH at 2-hr interval for a watershed of area 180 km^2. Take $n = 3$ and $k = 4$ hrs.

Solution:

Given Area $A = 180$ km^2, $n = 3$ and $k = 4$ hrs

$\therefore$ $$\Gamma(n) = \Gamma(3) = 2! = 2 \times 1 = 2$$

$$K\Gamma(n) = 4 \times 2 = 8$$

and $$(n - 1) = (3 - 1) = 2$$

From Eq. (7.49),

$$u(t) = \frac{1}{K\Gamma}\left(\frac{t}{K}\right)^{n-1} e^{-(t/k)}$$

$$= \frac{1}{8}\left(\frac{t}{4}\right)^{2} \overline{e}^{(t/4)} \quad (7.49)$$

The computation are shown in tabular form as follows:

$$K = 4 \text{ hrs},\ u(t) = \frac{1}{8}\left(\frac{t}{4}\right)^{2} e^{-1/4} \text{ cm/sec},\ (6) = (5) \times A = (5) \times 2.778 \times 180 = (5) \times 500$$

Time hr (1)	$^t/_k$ (2)	$e^{-t/k}$ (3)	$(^t/_k)^2$ (4)	$u(t)$ (cm/hr) (5)	$u(t)$ (m^3/sec) (6)	*Ordinates of* 2 hr *UH* (m^3sec) (7)
0	0	1	0	0	0	0
2	0.5	0.6	0.35	0.01875	9.375	4.68
4	1.0	0.37	1.0	0.0426	21. 3	15.34
6	1.5	0.223	2.25	0.0627	31.25	26.27
8	2	0.135	4.0	0.0675	33.75	32.5
10	2.5	0.082	6.25	0.064	32.00	32.875
12	3.0	0.05	9.0	0.05625	28.125	30.065
14	3.5	0.030	12.25	0.046	23.00	25.56
16	4	0.018	16.0	0.036	18.00	20.5
18	4.5	0.011	20.25	0.028	14.00	16.00
20	5.0	0.0067	25.0	0.021	10.5	12.25
22	5.5	0.004	30.25	0.015	7.5	9.00
24	6.0	0.0025	36	0.01125	5.625	6.56
26	6.5	0.0015	42.25	0.008	4.0	4.81
28	7.0	0.00091	49.0	0.0056	2.8	3.4
30	7.5	0.00055	56.25	0.00387	1.935	2.37
32	8	0.000335	64	0.00268	1.34	1.64
34	8.5	0.0002	72.25	0.00018	0.09	0.715
36	9	0.000123	81.00	0.0001245	0.06225	0.076

Sample calculation: Take time 2 hrs in Column 1, then

$\frac{t}{K} = \frac{t}{4}$ (Column 2)

$e^{t/4}$ (Column 3)

$\left(\frac{t}{K}\right)^2 = \left(\frac{t}{4}\right)^2$ (Column 4)

$u(t) = \frac{1}{8} \times (4) \times (3)$ (Column 5)

(5) × 2.7778 × 180 = (5) × 500 (Column 6

$\frac{u(t_1) + u(t_5)}{2}$ = Ordinate of UH at time t_2 (Column 7)

Now, take time 12 hrs in Column (1), then

$\frac{t}{K} = \frac{12}{4} = 3.0$ (Column 2)

$e^{-12/4} = 0.05$ (Column 3)

$\left(\frac{-12}{4}\right) = 3^2 = 9$ (Column 4)

$\frac{1}{8} \times (4) \times (3) = \frac{1}{8} \times 9 \times 0.05 = 0.0565$ cm/hr (Column 5)

$(5) \times 500 = 0.05625 \times 500 = 28.125$ m³/sec (Column 6)

$\frac{u(10) + u(12)}{2} = u(12) = \frac{32.0 + 28.125}{2} = 30.0625$ m³/sec (Column 7)

Therefore,
Column (5) gives ordinates of IUH in cm/hr
Column (6) gives ordinates of IUH in m³/sec
Column (7) gives ordinates 2 hr duration UH

7.22.2 Empirical Equation of Determination of *K* and *n*

From the study of gauged catchments in United Kingdom, Nash recommended that the catchment parameters like length *L*, area *A* and slope *s* have some relationship with *K* and *n*. He gave the following two equations:

$$\left.\begin{aligned} n &= 2.4L^{0.1} && \text{(when } L \text{ in miles)} \\ &= 2.517L^{0.1} && \text{(when } L \text{ in km)} \end{aligned}\right\} \quad (7.57)$$

and

$$\left.\begin{aligned} K &= \frac{11A^{0.3}}{L^{0.1}s^{0.3}} && \text{(when } A, L \text{ are fps unit)} \\ &= \frac{13.95\,A^{0.3}}{L^{0.1}s^{0.3}} && \text{(when } A = \text{km}^2, L = m) \end{aligned}\right\} \quad (7.58)$$

If the ungauged basin is in the same hydrometeorogically homogenous region, Eqs. (7.57) and (7.58) may be used to determine the values of n and K of the basin.

7.23 CONCLUSION

This chapter deals with the contents in very simple methods to facilitate the students to understand basic knowledge on flood hydrograph, base flow separation, surface runoff hydrograph derivation of unit hydrograph and instantaneous unit hydrograph. Hydrographs in different forms are very essential for design of hydraulic structures. Quite a lot of research works are already done on hydrograph analysis. For further study on hydrograph, readers are referred to Dooge[19] (1977) and Singh[20] (1983b).

EXERCISES

7.1 The ordinates of a flood hydrograph in a catchment area of 770 km^2 due to 6-hr rainfall are given as follows:

Find the net rainfall or rainfall excess.

Time (hrs)	0	6	12	18	24	30	36	42	48	54	60	66	72
Discharge (m^3/sec)	40	65	215	360	400	350	270	205	145	100	70	50	40

(Ans: 5.106 cm)

7.2 The ordinates of 6-hr UH are given as follows:

Time (hours)	0	3	6	9	12	15	18	21	24	27	30	33	36
6-hr UH ordinate	0	15	24	42	58	78	69	58	43	30	17	15	0

A storm has successive 3-hr rainfall of 3, 5 and 4 cm respectively. ϕ-index is 0.2 cm/hr, base flow is 53 m^3/sec. Determine the resulting flow hydrograph.

7.3 A 3-hr UH is given. Using method of superposition, determine 9-hr UH.

Time (hours)	0	3	6	9	12	15	18	21	24	27	30	33	36
6-hr UH ordinate	0	12	75	132	180	210	183	156	135	144	96	87	66

Time (hours)	39	42	45	48	51	54	57	60
6-hr UH ordinate	54	42	33	24	18	12	6	0

[19] Dooge, J.C.I, *Problems Methods of Rainfall-Runoff Modeling*, *In Math Model for Surface Water hydrology*, edited by T.A. Ciriani, U. Mione and J.R. Wallis, pp. 71–108, New York, John Wiley, 1977.

[20] Sing, V.P., A Geomorphic Approach to Hydrograph Synthesis with Potential for Application to Ungauged Watersheds, Tech. Report, Louisiana Water Resource Res. Institute, Louisiana State University, USA, 1983b.

7.4 The ordinates of 2-hr UH are given as:

Time (hours)	0	1	2	3	4	5	6	7	8
UH ordinate (m^3/sec)	0	500	1400	2000	1600	1200	800	400	0

Determine S-curve and obtain 1-hr unit hydrograph.

7.5 If the peak discharge of unit hydrograph is 200 m^3/sec, determine the peak flow rate of surface runoff hydrograph from following data:
Direct surface runoff volume = $(5.0 \times 105)^{m3}$
Catchment area = 20 km^2 (*Ans:* 500 m^3/sec)

7.6 Determine Synder's coefficients C_t and C_p for a drainage basin if the following data are available:
$L = 25$ km, $L_{ca} = 10$ km, drainage area = 300 sq. km, $Q_p = 50$ m^3/sec from the beginning of rainfall excess, $t_p = 9$ hr and $t_r = 2$ hr.
(*Ans:* $C_t = 1.5$, $C_p = 0.48$)

7.7 Characteristics of two catchments A and B are given as:

A	B
$L = 148$ km	$L = 106$ km
$L_{ca} = 76$ km	$L_{ca} = 76$ km
$A = 2718$ km^2	$A = 1400$ km^2

For 6-hr UH of catchment basin of A, the peak discharge is at 37 hr from start rainfall excess and its value is 200 m^3/sec. Assuming catchment A and B are homogenous, determine 6 hr synthetic UH for catchment B by Synder's method.
(*Ans:* $t_r = 6$ hr, $t_p = 27.75$ hr, $Q_p = 126$ m^3/sec, $W_{so} = 79$ hr, $W_{75} = 45$ hr, $t_b = 150$ hr)

7.8 Show the components of hydrograph with a sketch for single storm.

7.9 Discuss the factors with sketches where necessary that affect the flood hydrograph.

7.10 Explain with sketches where necessary the separation of base flow from the flood hydrograph.

7.11 Show how the direct surface runoff (DSR) is computed from flood hydrograph.

7.12 What is unit hydrograph? How it is derived from a complex storm? What do you mean by unit hydrograph of different unit durations?

7.13 What is S-hydrograph? Explain its uses.

7.14 Explain the following:

(i) Synthetic unit hydrograph of Synder
(ii) Triangular unit hydrograph
(iii) Non-dimensional unit hydrograph
(iv) Clark synthetic unit hydrograph

7.15 What is instantaneous hydrograph (IUH)? Explain the convolution process. Derive IUH from S-curve.

7.16 What is Nash model of cascaded linear reservoirs? Explain with sketches.

SUGGESTED FURTHER READINGS

Crawford, N.H. and Linsley, R.K., Digital Simulation in Hydrology: Stanford Watershed Model IV, Tech. Report 39, Dept. of Civil Eng., Stanford University, 1966.

Espey, W.H., Altman Jr., D.G. and Graves, C.B., Nomographs for Ten-minute Unit Hydrograph for Small Urban Watershed, Tech. Memo No. 32, Urban Water Resour. *Proc.*, ASCE, New York, December 1977.

Gray, D.M., Synthetic Unit Hydrograph for Small Watershed, *J.H.Y. Div. Am. Soc. Civ. Eng.,* vol. 87, No. HY. 4, pp. 33–34, 1961.

Kirpich, P.Z., Time of Concentration of Agricultural Watersheds, *Civil Engg.*, vol. 10, No. 6. June 1940.

Linsley, R.K. and Grawford, N.H., Continuous Simulation Models in Urban Hydrology, Geophysics, Res. Letters, pp. 59–62, 1974.

Sing, V.P., *Elementary Hydrology*, Prentice Hall Inc., Upper Saddle River, N.J., 1992.

Chapter 8

Methods of Discharge Measurement

8.1 INTRODUCTION

The science of water measurement in hydrology is called **Hydrometry**. It is actually the measurement of flowing water per second, i.e., flow rate-discharge. The chapter has been started from historical review of flow measurement and all the different methods of stream flow measurement are presented. Theoretical principles of the methods developed by different investigators have been explained with sketches where necessary. The techniques adopted and the instruments used are also discussed.

Measurement of discharges forms the most important data for engineers and hydrologists as the peak discharge in a stream is very important in the design of any water resources project. The measurement is required to develop hydrographs, mass curve, S-curve, flood warnings, equitable distribution of water supply and irrigation among the users, and determination of seasonal and annual variation in the runoff.

8.2 HISTORICAL REVIEW OF DISCHARGE MEASUREMENT

The subject *Hydrometry* is as old as human civilization. "Bibliography of Hydrometry" by Kolupaila[1], is quite famous and comprehensive. The oldest hydrometric evidences are the marking of flood stages of the river Nile cut in steep rock faces. Stage discharge relationship was conceived long back in irrigation system to the Mughals in central Asia. The famous Pitot tube for measuring the velocity in the river Seine was proposed by Henri de Pitot, a French engineer in 1732. The Price current meter which has been extensively used throughout the world to measure the velocity of river was initially devised by T.G. Ellis in 1870 and subsequently redesigned by W.G. Price. Parshall Flume which is extensively used in USA to measure discharge was developed by R.L. Parshall in 1920. Ultrasonic method of velocity measurement was first reported by Swengel in 1955. Based on different hydraulic and electronic principles, hydrometry is developed step by step. Many modern methods and techniques have been developed.

---

[1] Kolupaila, S., Early History of Hydrometry in United States, ASCE, Civ. Engrs, *Jour. Hy. Div.*, Vol. 86, No. Hy1, January, 1960.

8.3 DIFFERENT METHODS OF STREAM FLOW MEASUREMENT

Discharge is obtained when area of flow is multiplied by the average velocity. Therefore, in the different methods described below either discharge Q is obtained directly or average velocity or stage of the flow is measured. Following are the methods of measurement available in literature.

8.3.1 Current Meter Method

It is the most popular and widely used method to measure the average velocity of river. The most common current meter is the Price meter shown in Figure 8.1(a), which consists of six conical cups rotating about a vertical axis. Electric contacts drives the cups close to a circuit through a battery, and wires of the supporting cable cause a click for each revolution in headphones worn by the operator. Electrical counting devices are also used. For measurement in deep water, the meter is suspended from a cable as shown in Figure 8.1(a). The tail vanes keep the meter facing into the current and heavy weight keeps the meter vertical. Special boat and crane arrangement is available to support the meter over the surface of water. A calibration curve for the meter is available in the form of $V = a + bN$ where N is number of revolutions per second of the cup meter and a, b are constants of the meter available from the manufacturer. This current meter is also called **vertical axis current meter**. Propeller type [Figure 8.1(b)] current meter (also called **horizontal axis meter**) employs a propeller turning about a horizontal axis. The

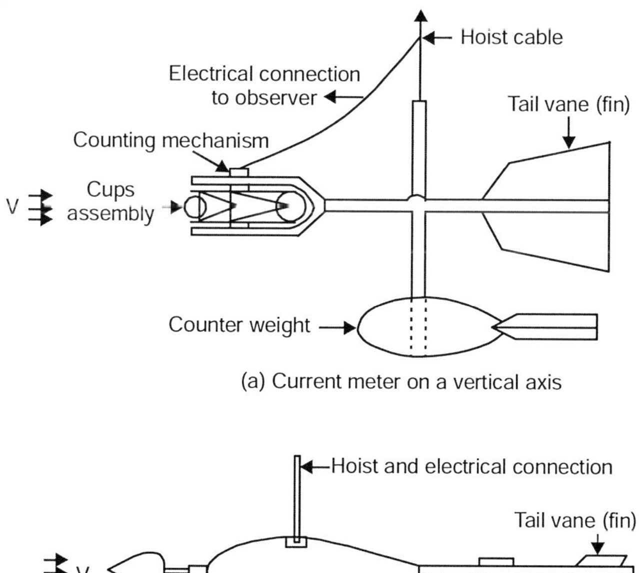

Figure 8.1 Types of current meter to measure average velocity in stream.

contacting mechanism of this type is similar to that of the Price meter, and similar suspensions are used. Velocity is measured normally in two depths, i.e., 0.2*D* and 0.8*D*. The average velocity is then obtained as:

$$V = \frac{V_{0.2D} + V_{0.8D}}{2} \tag{8.1}$$

Here velocity at two depths is obtained from

$$V = a + bN \tag{8.2}$$

where a, b and N are already defined, and D is the total depth of the river. This is in brief the procedure to follow to measure the velocity. But there lies a hydraulic principle why depths are taken at 0.2*D* and 0.8*D*. This principle can be obtained by making an analytical solution starting from Reynold's stress equation and Prandtl mixing length theory and applying velocity defect law. Analytical solution proves that average velocity is equal to the half of the velocities measured at 0.16*D* and at 0.84*D*. If velocities are measured at 0.3*D* and 0.7*D*, the hydraulic theory fails. Therefore, for practical purpose, depths are taken at 0.2*D* and 0.8*D* in place 0.16*D* and 0.84*D*. Velocity thus obtained is used to find the discharge Q as:

$$Q = AV \tag{8.3}$$

where A is the area of the river which is obtained by dividing whole river section into number subsections and velocity is measured as shown in Figure 8.2.

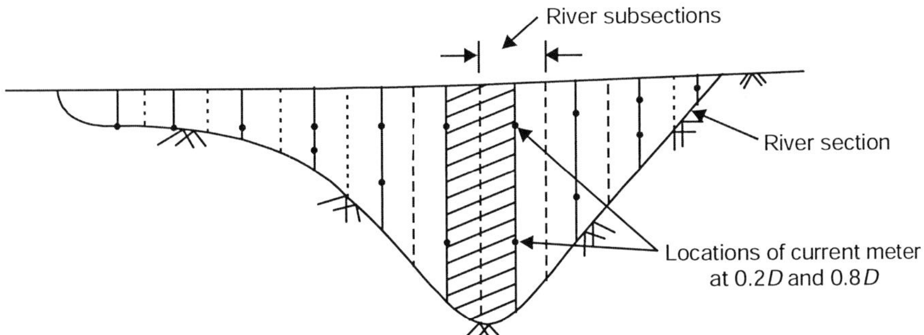

Figure 8.2 Stream division in subsections showing location of current meter.

At the middle of every subsection, velocity is measured at 0.2*D* and 0.8*D*. Area of the subsection is obtained by multiplying the width of the subsection and the average depth. This area of the subsection multiplied by the average velocity at 0.2*D* and 0.0*D* gives the discharge through this subsection. Thus, velocity and area of each subsection is determined. Final discharge is the summation of discharges through all the subsections.

8.3.2 Area Velocity Method

It is based on continuity equation (8.3). The whole river section is divided into number subsections as shown in Figure 8.2. The methods of measurement of area are same. Only velocity is measured by different methods like float at surface, Pitot tube, current meter at 0.2*D*, 0.4*D*, 0.6*D*, 0.8*D* and on surface at different subsections. Velocity profiles along cross-section and vertical are plotted and shown in Figure 8.3. From the velocity profile along vertical, average velocity *V* is determined. Thus, *Q* is obtained by

$$Q = \sum_{i=1}^{n-1} \Delta Q_i \tag{8.4}$$

where ΔQ_i is the discharge in the *i*th sub-section, and is given as:

ΔQ_i = Depth at *i*th subsection × (1/2 Width to the left + 1/2 Width at right) × Average velocity at the *i*th vertical (Verticals drawn in Figure 8.2).

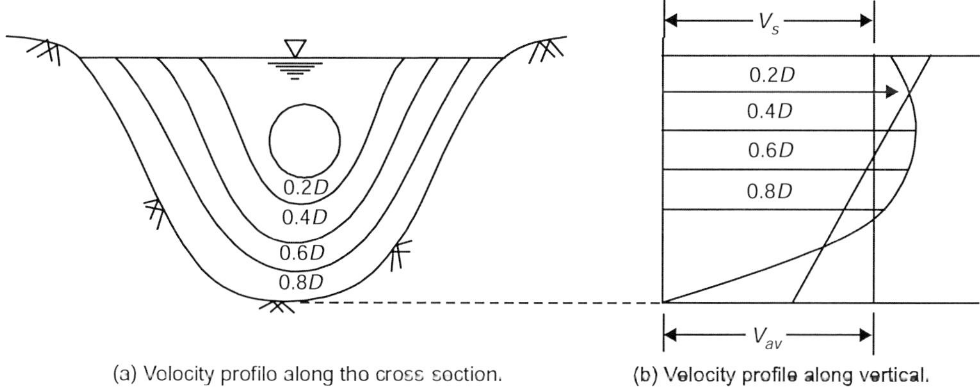

(a) Velocity profile along the cross section. (b) Velocity profile along vertical.

Figure 8.3 Velocity profiles.

8.3.3 Pitot Tube Method

It is a very simple device. The method is named in honour of its originator Henri de Pitot, a French engineer, who developed the method for measuring velocity in the river Seine. The basic principle used in this method is that the velocity at the stagnation point *S* (Figure 8.4) is reduced to zero. Kinetic energy at that point is converted to pressure energy and liquid, therefore, water

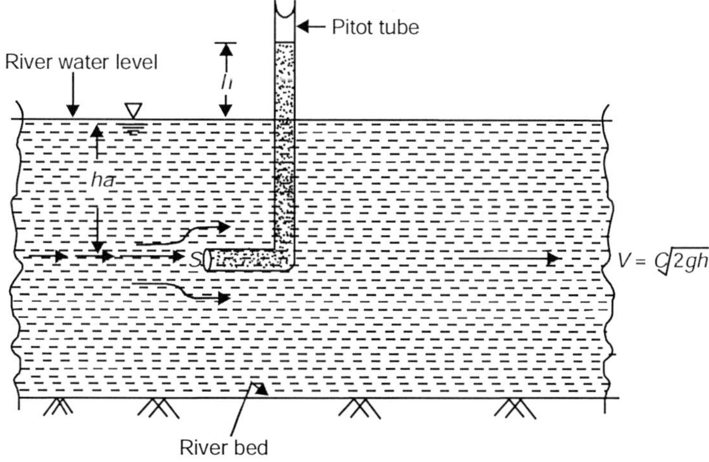

Figure 8.4 Pitot tube method.

in the tube rises to a height h above the general level of water of river. Applying Bernoulli's equation at upstream point (where velocity is V) and at stagnation point S, the average velocity V is obtained as:

$$V = C\sqrt{2gh} \tag{8.5}$$

Here C is the tube coefficient having the effect of loss of kinetic energy due to turbulence at entry. If the tube diameter is smaller, the surface tension may also affect the rise of h in the tube. Most acceptable value of C is 0.8, but actual value is to be calibrated. The open tube is bent as shown in the figure.

8.3.4 Float Method

It is the most simple and quick method of measuring surface of velocity V_S. Different types of float used have been shown in Figure 8.5. If L is the distance in metre moved by the float in T seconds, then

$$V_s = \frac{L}{T} \text{ m/sec} \tag{8.6}$$

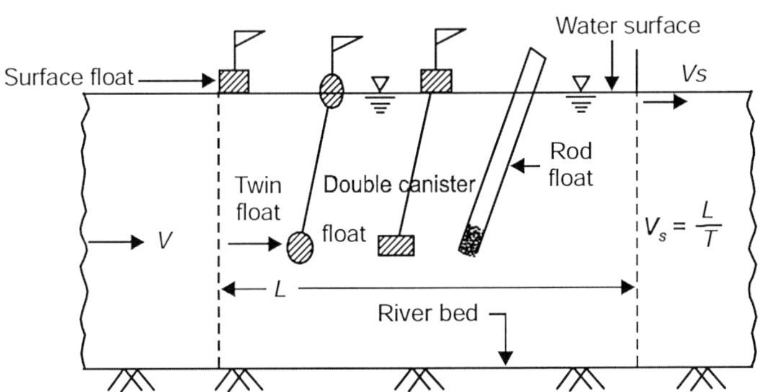

Figure 8.5 Float method.

When this surface velocity is multiplied, a reduction factor (varies from 0.79 to 0.95, acceptable one = 0.85), average velocity of river may be obtained. Mysore Engineering Research station has given the following relation between average V and V_S:

$$V = 0.8529V_s + 0.0085 \tag{8.7}$$

8.3.5 Slope Area Method

This method is due to Benson[2] (1968) and Darlymple and Benson[3] (1967). It is used in the reach

[2] Benson, M.A., Measurement of Peak Discharge by Indirect Method, World Meteoro. Organi. Tech., Note 90, Geneva, 1968.

[3] Dalrymple, T. and Benson, M.A., Measurement of Peak Discharge by Slope Area Method, U.S. Geolo. Surv. Tech. Water. Res. Inv. bk3, Chap. A2, 1967.

of the river in a length L when the reach is approximately straight and bed slope is almost constant. If h is the difference water level between upstream and downstream of the river of reach length L as shown in Figure 8.6.

$$\text{Slope } s = \frac{h}{L} \tag{8.8}$$

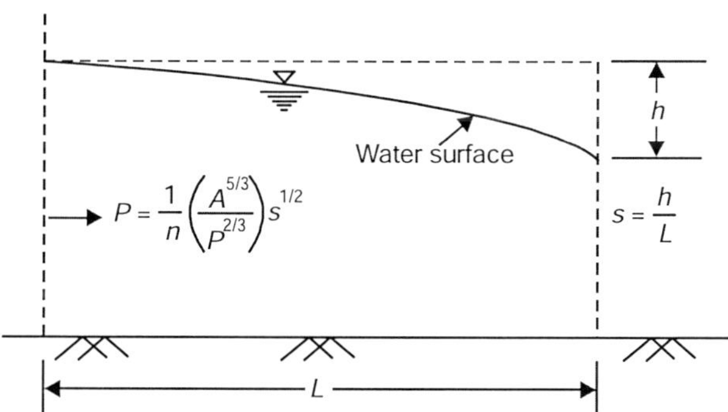

Figure 8.6 Slope area method.

Cross-sectional area of the reach is already measured during dry season below water level by echo sounding. It is also required to find Manning n of the river. Then discharge is:

$$Q = AV = A\frac{1}{n}\left(\frac{A}{P}\right)^{2/3} s^{1/2} = \frac{1}{n}\left(\frac{A^{5/3}}{P^{2/3}}\right)s^{1/2} \tag{8.9}$$

Here P is the wetted perimeter.

This method has limitations. The flow is assumed to be uniform in the river reach under consideration. Application of Manning n from table of Chow may also cause some error as n decreases with increase of depth. But for practical purpose, measuring Q by this method is applicable.

8.3.6 Rotameter: Variable Area Method

Rotameter is a variable area meter placed vertically in a flow path as shown in Figure 8.7. It consists of a transparent tube of increasing cross-section with a float and a calibrated scale on it. The float is heavier than water. Hydraulic principle is the combined action of drag, buoyancy and gravity. As force increases with velocity, the float rises to a higher level increasing the flow rate. The meter cross-section increases upwards to accommodate higher flow rate, which is obtained directly from the calibrated discharge scale. It is an excellent method to use in laboratory, pumps and pipes.

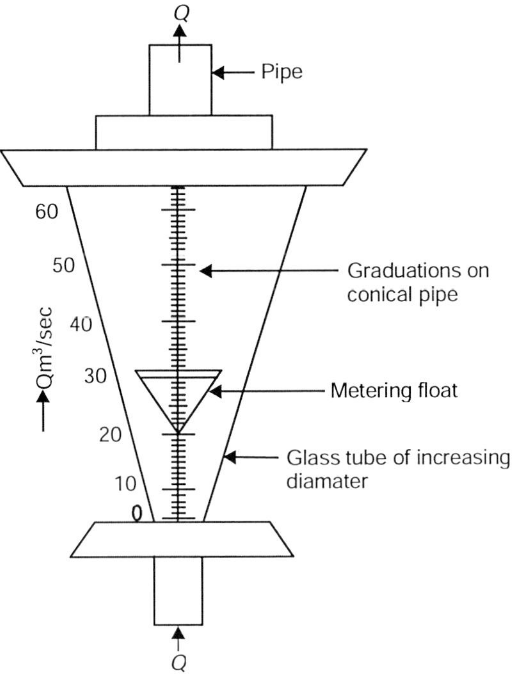

Figure 8.7 Rotameter.

8.3.7 Rating Curve Method

This is the most common and easy method of recording stage or depth of water level in a river. Various water level recorders are staff gauges [shown in Figure 8.8(a)], float gauge and self recording gauges. The gauges are fitted on a solid foundation or in piers and abutments (If the bridge is there in the gauging site). Graduations on the staff should be bold and big enough to read from a distance.

From the recorded stage on the gauge, discharge Q is obtained from the previously calibrated rating curve [Figure 8.8(b)] of the gauge side. As shown in Figure 8.8(b), when the gauge reading is 75 m, corresponding value of Q is equal to AE, which is equal to 25000 m^3/sec. This rating curve has lot of application in river engineering. Statistical method is also used to extrapolate the rating curve to measure high discharge. It is developed due to unique relationship between Q and H, which may be expressed as:/

$$Q = b(H - a)^n \tag{8.10}$$

where the value of constants a, b and exponent n can be obtained by plotting in log-log plot with the combination of least square method.

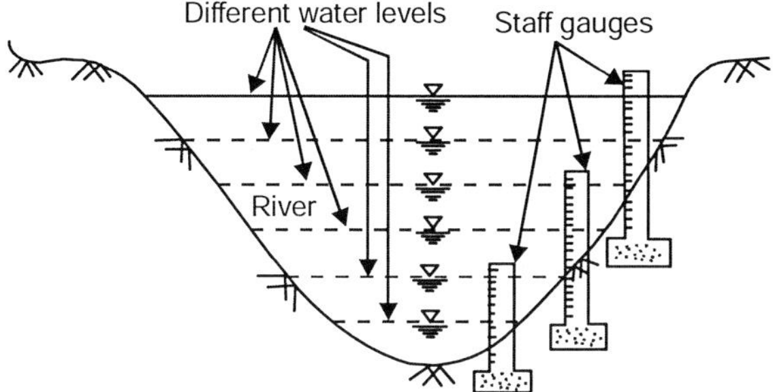

(a) Different staff gauges at different water levels in deep river

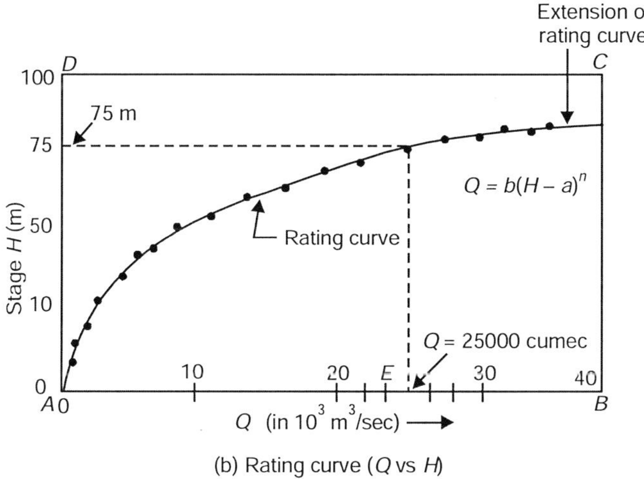

(b) Rating curve (*Q* vs *H*)

Figure 8.8 Rating curve method.

8.3.8 Obstruction Meter Method

The conventional obstruction meters are venturimeter, orifice meter and nozzle meter. Hydraulic principle of obstruction is to accelerate locally the flow which reflects in a measurable difference of pressure head h between inlet and obstruction [as shown in Figures 8.9(a) and 8.9(b)] and then apply Bernoulli equation at inlet and throat (obstruction). The difference of pressure is measured by piezometer or U-tube differential manometer. Application of Bernoulli's equation in almost horizontal datum leads to the following equation:

$$Q = C\frac{A_1 A_2}{\sqrt{A_1^2 - A_2^2}} = \sqrt{2gh} \tag{8.11}$$

Here A_1 and A_2 are areas of the meters at inlet and at throat and C is the constant of the meter, value of which lies around 0.97. But it is better to calibrate the constant of the meter before use.

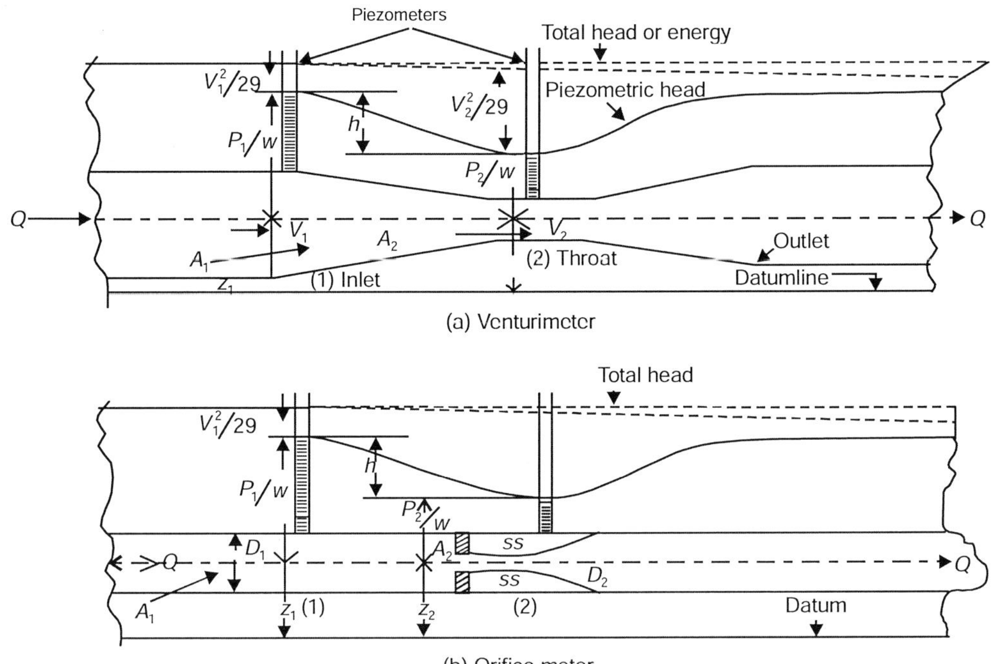

Figure 8.9 Obstruction meters.

8.3.9 Contraction Meter Method

The contraction meters are weirs or notches, spillways of dams, culverts and sluices. The use of above contraction meters in hydraulics laboratory and fields are well known although some limitations are there to use them in the field. Notches are used to measure the discharge from large tank used in sprinkler and drip irrigation. Weirs are used to measure the discharge in canals for irrigation and streams. In large natural rivers, they have some limitations due to large head, debris, and sediment load and backwater influence. Other structures like spillway of dam, culvert, sluices whose primary functions are other than flow measurement, can conveniently be utilized in the field for discharge measurement.

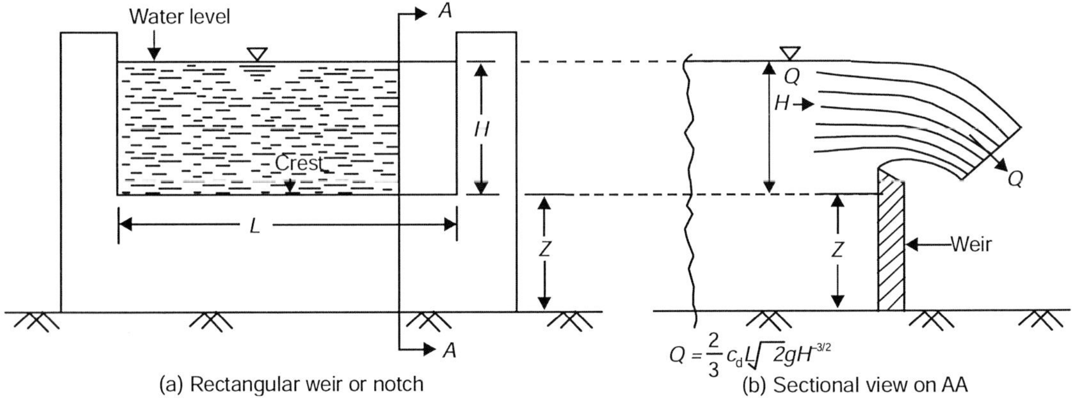

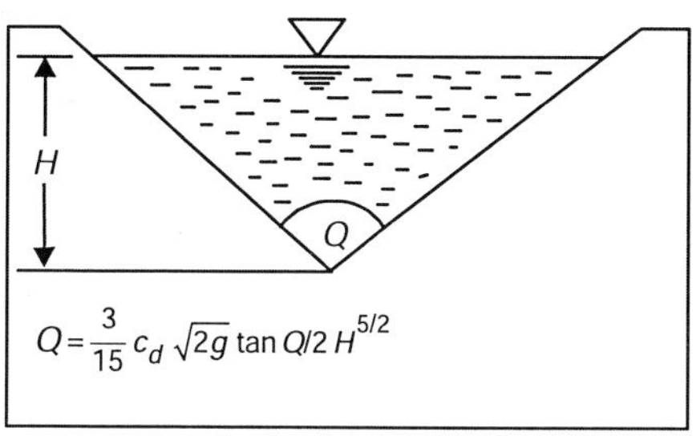

(c) Triangular weir or notch

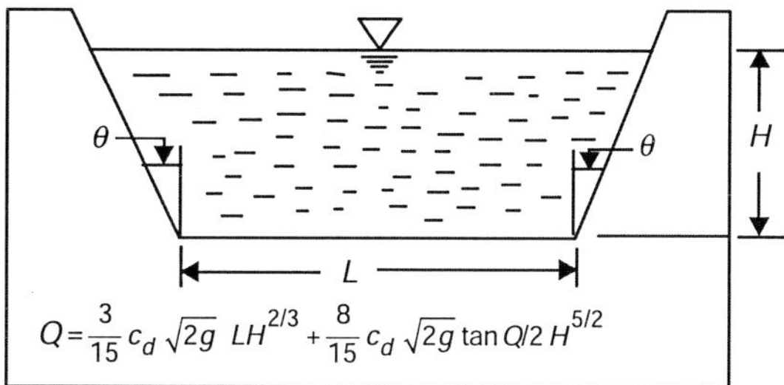

(d) Trapezoidal notch or weir.

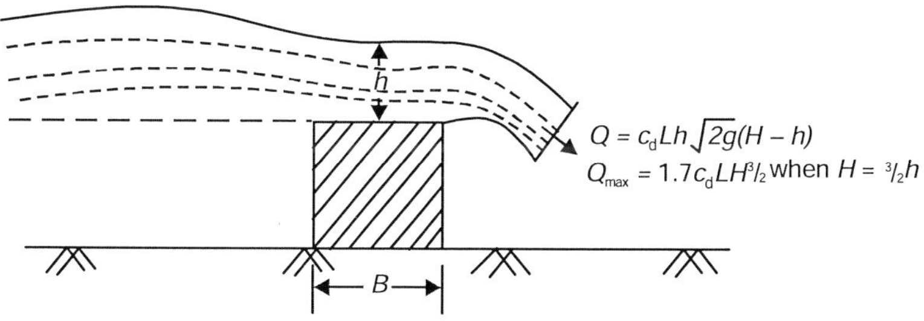

(e) Broad crested weir

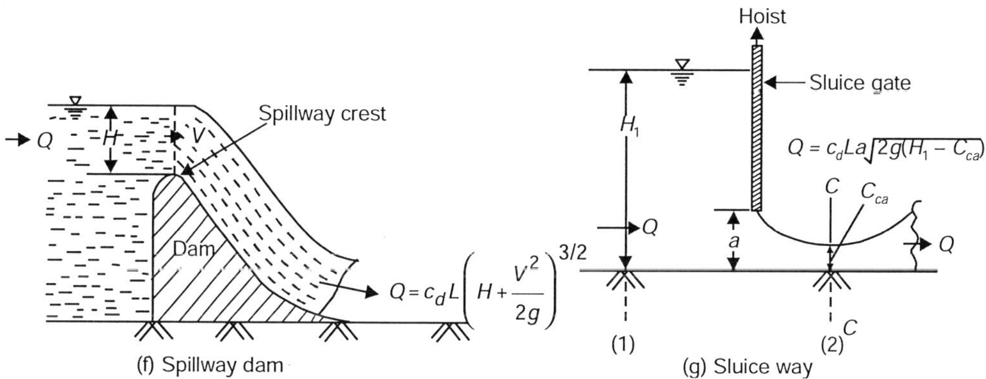

(f) Spillway dam

(g) Sluice way

Figure 8.10 Contraction meters.

The various discharges equations used in contraction meters are:

1. Rectangular weir or notch

$$Q=\frac{2}{3}c_d L\sqrt{2g}H^{3/2} \tag{8.12}$$

2. Triangular notch or weir

$$Q=\frac{8}{15}c_d\sqrt{2g}\tan\theta/2H^{5/2} \tag{8.13}$$

3. Trapezoidal crested weir

$$Q = c_d\sqrt{2g}H^{3/2}\left[\frac{2}{3}L + \frac{8}{15}H + \tan\theta/2\right] \tag{8.14}$$

4. Broad crested weir

$$Q = c_d Lh\sqrt{2g(H-h)} \tag{8.15}$$

$$Q_{max} = 1.7c_d LH^{3/2}, \text{when } h = \frac{2}{3}H \tag{8.16}$$

5. Spillway dam

$$Q = c_d L\left(H + \frac{V^2}{2g}\right)^{3/2} \tag{8.17}$$

6. Sluice way

$$Q = c_d\, La\sqrt{2g(H_1 - C_c a)} \tag{8.18}$$

8.3.10 Channel Transition Meter Method

Flow meters like venturiflume, critical flow flume and Parshall flumes are some of the transition meters as they are formed in channel transition and control. The hydraulic principle in the development is based on critical depth and specific force. Although they are very good for laboratory channels, can satisfactorily be applied to artificial channels made for irrigation, water supply and power generation. The general equations of discharge are essentially of the same pattern in all the flumes.

1. Venturiflume

$$Q = c_d \frac{Aa}{\sqrt{A^2 - a^2}}\sqrt{2g(H-h)} \tag{8.19}$$

2. Critical depth flume

$$Q = c_d bH^{3/2} \tag{8.20}$$

3. Parshall flume (Developed by R.L. Parshall in 1920)

$$Q = c_d bH^n \tag{8.21}$$

Some of the critical flow flumes are shown in Figure 8.11.

8.3.11 Moving Boat Method

This method has been developed by Smoot and Novak[4] (1969) in the US Geological Survey (USGS). It has some advantages than the standard current meter method. Let V_b be the velocity

[4] Smoot, G.F. and Novak, C.E., Measurement of Discharge by the Moving Boat Method, US Geological Survey, Tech Water Resource Inv., bk 3, Chap A11, 1969.

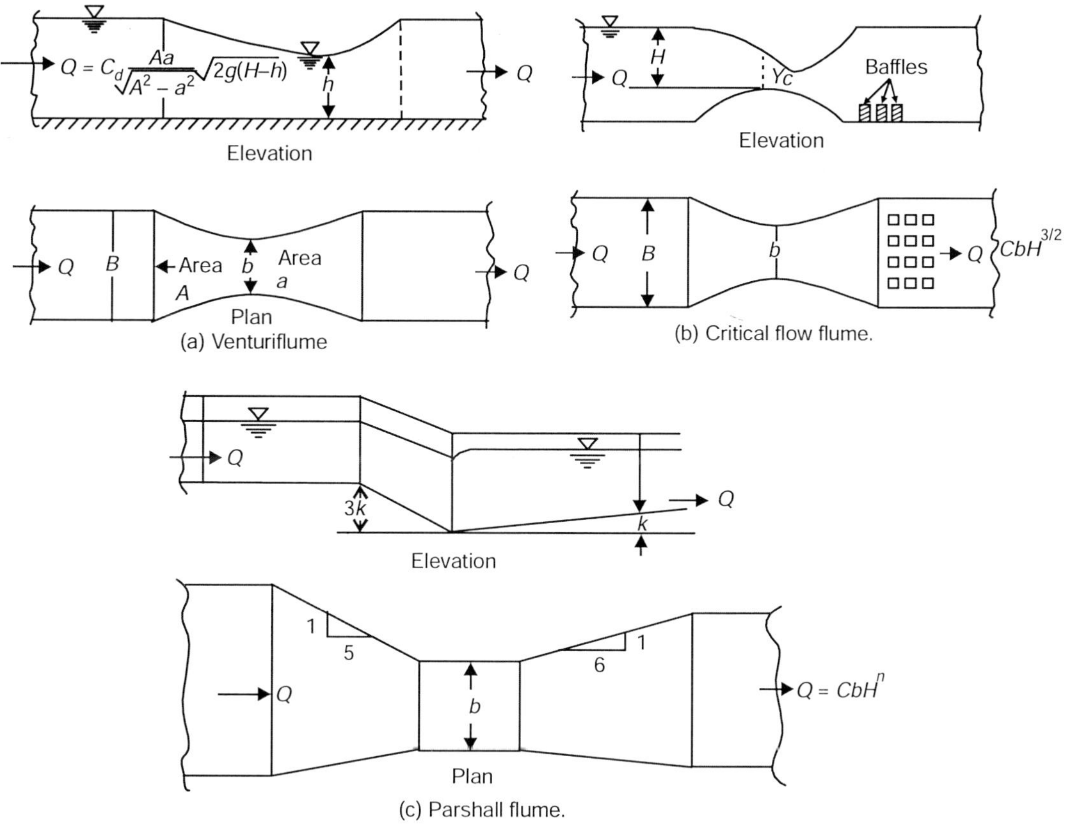

Figure 8.11 Some critical flow flumes.

of the boat with current meter at right angle to the velocity of flow V_f then the current meter will align itself in the direction of resultant velocity V_r.

From Figure 8.12, $V_b = V_r \cos\theta$ and $V_f = V_r \sin\theta$. Let Δt be the time of transit between two verticals placed at a distance W, then

$$V_b \Delta t = W \tag{8.22}$$

If the flow depth between two verticals are d_1 and d_2, then

$$\Delta Q = \left(\frac{d_1 + d_2}{2}\right) W V_f \tag{8.23}$$

Substituting W from Eq. (8.22) in Eq. (8.23),

$$\Delta Q = \left(\frac{d_1 + d_2}{2}\right) V_b \Delta t V_f \tag{8.24}$$

Again substituting V_b and V_f in Eq. (8.24),

$$\Delta Q = \left(\frac{d_1 + d_2}{2}\right) V_r \cos\theta.\ \Delta t V_r \sin\theta$$

$$\Delta Q = \left(\frac{d_1 + d_2}{2}\right) V_r^2 \sin\theta.\ \cos\theta\ \Delta t$$

$$\therefore \quad \text{Total } Q = \sum \Delta Q = \sum_{i=1}^{n} (d_i + d_{i+1}) V_r^2 \ \sin\theta \cos\theta\ \Delta t \tag{8.25}$$

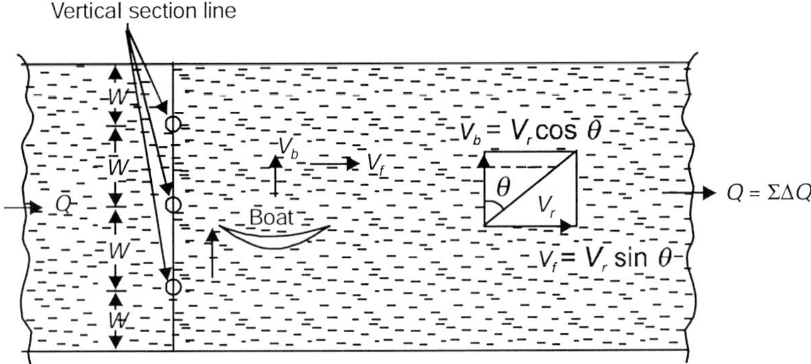

Figure 8.12 Moving boat method.

8.3.12 Hydraulic Model Method

When the river flows with very high depth, measurement of depth or stage becomes difficult. In such situation hydraulic model method is very useful. Normally low flood data are available or may be measured. A model in scale around 1.50 is constructed. Stage-discharge relationship of the model is established by experiments and its calibration is checked by low flood available data of the river. If those data come very close to the model rating curve, then extrapolating the model rating curve, high discharge of the river where gauging is extremely difficult, may be determined. It was successfully applied in Shiroto George in England.

8.3.13 Echo Sounder Method

In this method, electro acoustic instrument is used. The river, which is sufficiently deep, compared to its width with very high velocity and where conventional method of measuring its depth or stage is virtually impossible due to steep topography, this method, is suitable and useful. It is based on electrical principle of transmitting a pressure wave from water surface level to the riverbed by a transmitter or transducer for a short duration and receiving back the reflected pressure wave in the form of an echo on a receiver. The time of transmission and reception is plotted automatically. From this time, knowing the velocity of sound in fluid medium (i.e., approximately equal to 1470 m/sec in water), depth of flow is known.

8.3.14 Ultrasonic Flow Meter Method

This method has been developed by Swengel (1955) to measure the velocity of flow of a wide river by the use of two transducers fixed on river banks as shown in Figure 8.13. Ultrasonic signals can be received and sent by these two transducers. Velocity of flow at different depths of the river may be obtained by changing the position of the transducers. Then average velocity V of the river is determined. This sophisticated electronic method can perform the same velocity measurement at different depths in a much easier way than the laborious and tedious works required to perform by the current meter method.

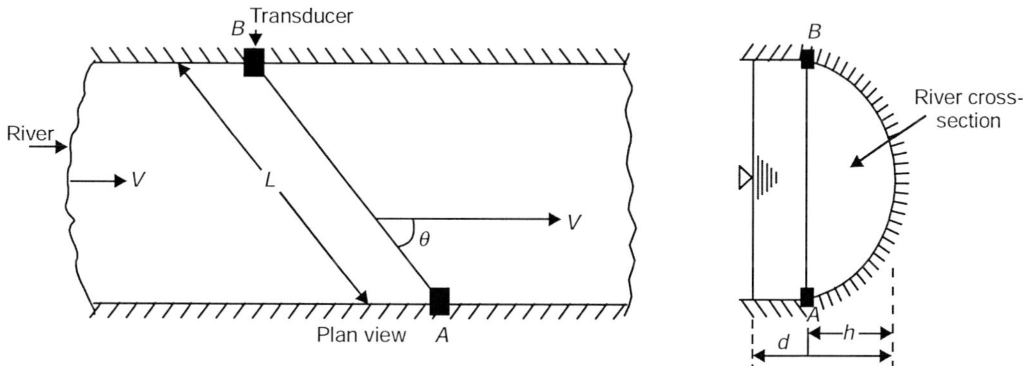

Figure 8.13 Ultrasonic flow meter method.

Let the two transducers are fixed at depth h from the bottom of the river. Assume L is the length of path from B to A, V_h is the component of flow velocity in the sound path, C is the velocity of sound in water, t_1 is the time taken by A to receive the ultrasonic signal sent from B to A and t_2 is the time required to receive back the signal from A to B. Then

$$t_1 = L/(C + V_h) \text{ and } t_2 = L/(C - V_h)$$

Thus

$$(1/t_1) - (1/t_2) = (C + V_h)/L - (C - V_h)/L = 2V_h/L = (2\ V \cos\theta\ )/L$$

Then

$$V = (L/2 \cos \theta)\ (1/t_1 + 1/t_2\ ) \tag{8.26}$$

From Eq. (8.26), average velocity at a depth h from bottom can be determined from a given L, θ, t_1 and t_2.

Similarly, fixing the transducers at different depths, average velocity at other depths can also be determined. The average velocity V of the river is then determined by taking the average of the velocity at different depths. Thus, it is seen that this method is essentially a area-velocity method to determine the discharge.

8.3.15 Inertia Pressure Method

This method, developed by N.R. Gibson, is based on valve closure and is used to measure average velocity of liquid in the pipe, which in turn gives the discharge Q. The device records the inertia pressure against time at a point immediately upstream of the valve. From the theory of water hammer,

$$\int p\,dt = \frac{WLV}{g}$$

or
$$V = \frac{g}{wL}\int p\,dt \tag{8.27}$$

RHS of the Eq. (8.27) is known; hence average velocity V of the pipe is calculated.

8.3.16 Tracer Technique Method

Tracer technique method is also known as **dilution technique** or **chemical method** based on continuity principle applied to tracer which is allowed to mix completely with flow (Figure 8.14). Tracers like common salt or sodium dichromate of known rate (q m^3/sec) and known concentration C_t is injected in the stream as shown in Figure 8.14. The solution will be diluted by turbulent action of stream. Concentration of tracer C_s at downstream section is determined.

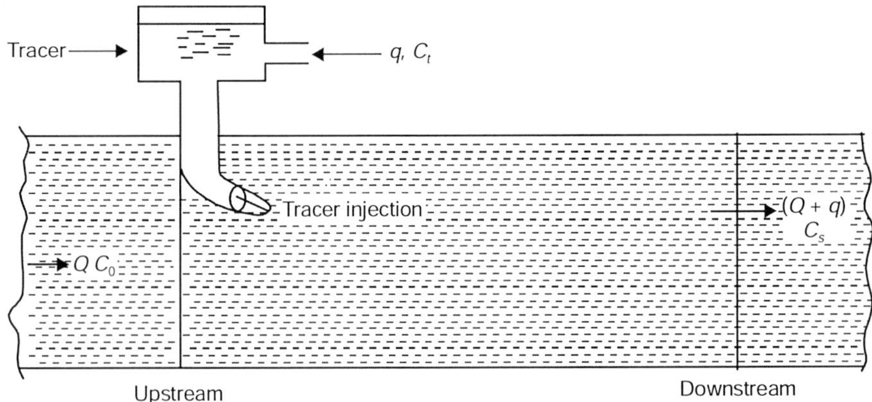

Figure 8.14 Tracer technique method.

Let Q is the flow stream, C_0 is the concentration of tracer initially present in water, C_s is the concentration measured at downstream after mixing and q is rate inflow of tracer. Then applying continuity,

$$QC_0 + qC_t = (Q + q)\,C_s$$

Solving for Q,

$$Q = \left(\frac{C_s - C_t}{C_0 - C_s}\right)q \tag{8.28}$$

The method is based on steady flow assumption in the reach from upstream to downstream. When flow is unsteady and spatially varied, there will be a change in storage volume affecting the method. However, this method has a major advantage of estimating Q in a absolute way. Method is specially useful and attractive for small turbulent stream.

8.3.17 Laser Doppler Anemometer (LDA) Method

It is the recent scientific method based on laser beam and Doppler effect. Hence it is called **laser Doppler anemometer** (LDA). The layout of this method is shown in Figure 8.15. In this LDA method, a low laser beam is focused on a fluid in motion through a transparent duct. A portion of the beam is transmitted through stream and some part of beam is scattered by solid particles present in water. The component of the beam-passing straight through the medium is colliminated by a lens outside the duct and reflected by a mirror to a beam splitter. The scattered beam condensed through another lens is also focused on the beam splitter. In other words, the two components of original laser beam now combine at beam splitter. The reflecting mirror position is so adjusted that the two components travel identical path lengths, whereas direct transmitted component of the beam maintains the same frequency of the laser beam emanating from the source and the frequency of scattered beam is slightly different. The difference arises due to well known Doppler effect, and is proportional to the velocity of scattering particles, i.e., the difference of frequency of the two beams combining at the splitter is a measure of velocity of scattering particles. Since the particle moves with same velocity as the water, Doppler frequency shift is also a measure of liquid velocity. The combined beams are sensed by a photo multiplier, the output of which is connected to a frequency (Spectrum) analyzer for measuring the Doppler frequency. Thus, velocity of flowing liquid in the duct is measured.

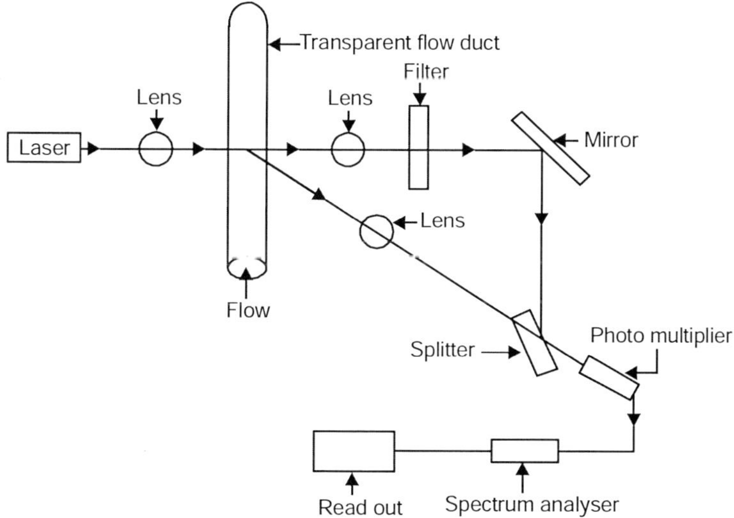

Figure 8.15 Laser Doppler anemometer (LDA) layout.

8.3.18 Hot Wire Anemometer Method

Iowa Institute of Hydraulic Research developed this anemometer to use in the laboratory to measure for both compressible fluid as well as incompressible water flow.

It is an instrument (Figure 8.16) used to measure velocity. It consists of platinum or tungsten wire of diameter 5×10^{-3} to 8×10^{-3} mm and about 6 mm long.

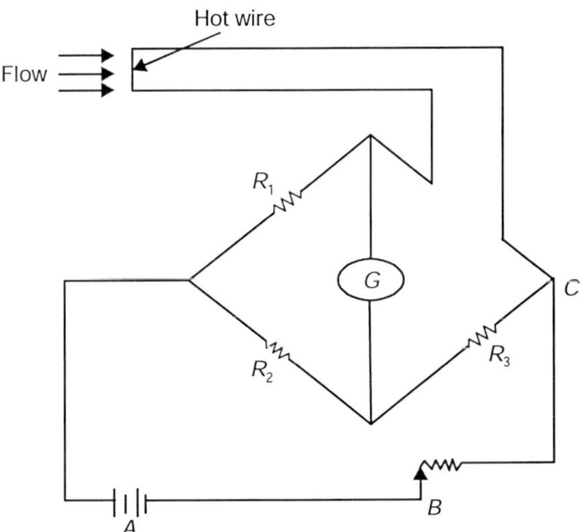

Figure 8.16 Hot wire anemometer.

The wire is mounted on prongs placed perpendicular to the flow. The temperature of the wire is raised relative to the temperature of liquid or fluid whose velocity is to be measured. Passage of liquid or fluid flow over the wire cools it which changes the resistance. This is reflected in a change of current or voltage across the wire, which can be seen in the galvanometer G shown in the figure. This change in current or voltage is calibrated against velocity. The instrumentation of this device is rather complicated in field, but suitable for laboratory.

8.3.19 Bubble Gauge Method

It is an automatic recorder of stage or depth of stream. It works on the principle of pressure. In this gauge (Figure 8.17), compressed air or gas is made to bleed out at very small rate through an orifice placed at bottom of the river. The gas is fed in the tubing system and is allowed to bubble freely. A pressure gauge measures the gas pressure P and depth H is obtained by

$$H = \frac{p}{w}$$

It consists of gas cylinder, servometer, transitor control gas purge system and a recorder as shown in Figure 8.17.

The bubble gauge is widely used in the world. It does not require the construction of stilling well or any other structure in the stream bed. It is also suitable for digital recording and hence facilitates telemetering and automatic long distance telecommunication.

8.3.20 Dynamometer Method

It is used on translation of momentum forces of stream into either deflection or stress. This is measured and calibrated against velocity or discharge. The deflection principle has been incorporated in the Keeler meter and successfully applied to the measurement of outflow from a lake in New Hampshire. Based on the same principle, hydrometric pendulum, developed at

Hydraulic Laboratory of Delft, has been utilized to measure velocity in extremely turbulent river of Nigeria.

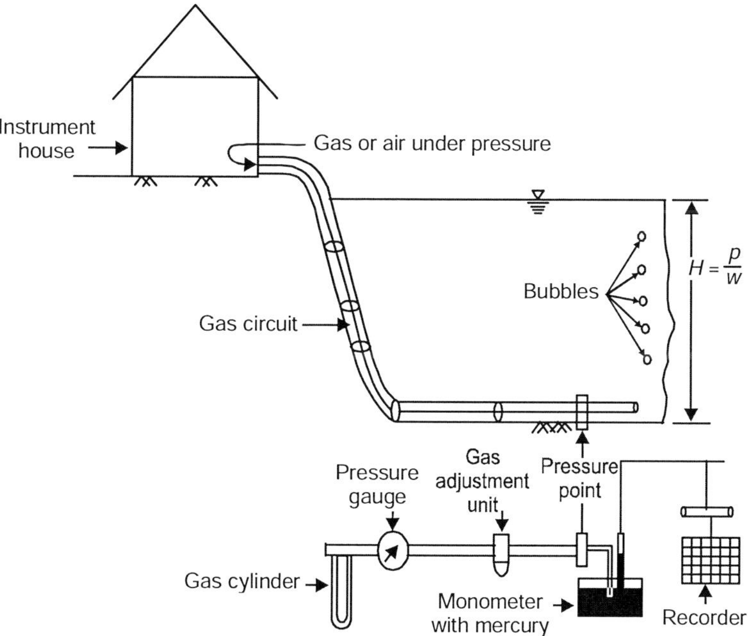

Figure 8.17 Bubble gauge typical installation and line dragram.

8.3.21 Electromagnetic Flow Meter Method

This method is based on Faraday's principle which states that an emf is induced in the conductor when it cuts a normal magnetic field. Large coils buried at bottom of the river carry a current I to produce a controlled vertical magnetic field. Electrodes provided at the sides of the river section measure the small voltage produced due to flow of water. If E is the signal output, d is the depth of flow, I is the current in the coil, then

$$Q - K_1\left(\frac{Ed}{I} + K_2\right)^n \tag{8.29}$$

where n, K_1 and K_2 are system constants. This method involves sophisticated and extensive instrumentation and has been successfully tried. It gives the total discharge when calibrated for system constants.

8.3.22 Thrupp's Ripple Method

It is developed by E.C. Thrupp. It is an approximate method, but rapid method of determining surface velocity.

This method is based on formation of ripples on the surface of water due to small obstacles when velocity exceeds 0.2 m/sec. As the velocity increases, the angle between the diverting lines of ripples becomes more acute. To afford a simple means of measuring the divergence rate, Thrupp used two 7.5 cm wire nails about 0.3 cm diameter at a definite distance apart.

If l is the distance (Figure 8.18) from the base line between the nails and point of intersection of the last ripples, the velocity of the stream given by him is:

$$V_s = 0.12 + 0.025l \tag{8.30}$$

When d = 15 cm,

$$V_s = 0.12 + 0.34\,l$$

When d = 10 cm. (8.31)

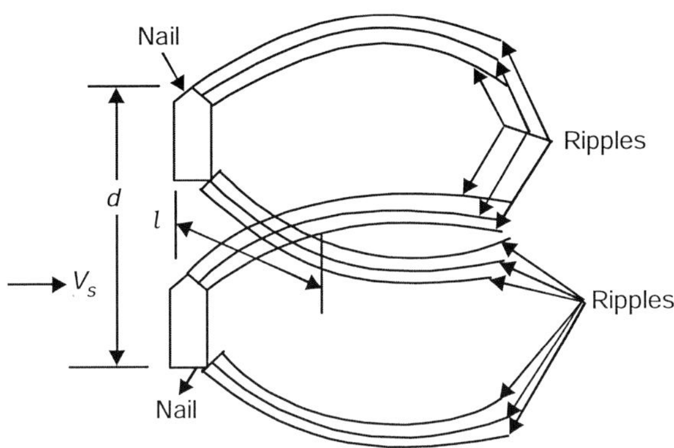

Figure 8.18 Thrupp's ripple method.

8.3.23 Salt Velocity Method

This method is based on the fact that when a salt solution is added to flowing stream, as shown in Figure 8.19, electrical conductivity of water increases. A length L of the stream is selected and salt solution is injected at upstream of reach L as shown in the figure. At upstream/downstream detection section, an ammeter is connected to source current and an insulated electrode is also connected at same section.

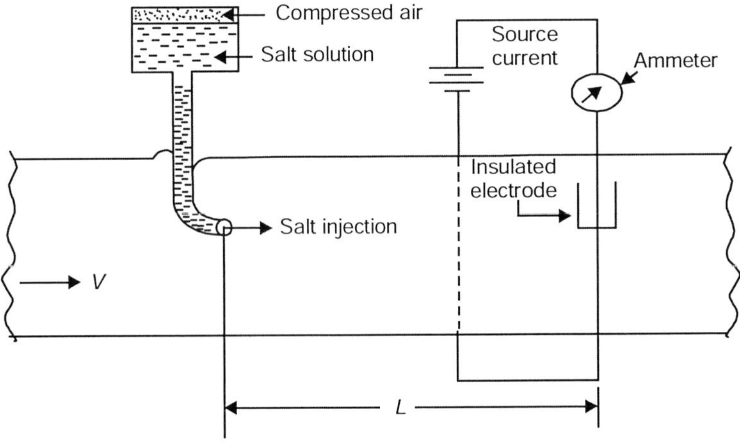

Figure 8.19 Salt velocity method.

When ordinary water without salt injection passes the section, a very little current is registered by the ammeter. When salt injected water reach the downstream section of L, salted water increases the conductivity and sudden increase of current is registered by the ammeter, which is automatically recorded on a chart or noted by an observer. Difference of time from salt injection to sudden increase of reading in ammeter is recorded. The distance L divided by this time gives the velocity of the stream. It is useful in turbulent river in which complete mixing of salt solution is possible.

8.3.24 Brink Depth Method

It is a simple and easy method. It is frequently used in irrigation canal fall to measure the discharge. The discharge which is to be measured is allowed to fall freely in a free overfall. The depth at free overfall is brink depth or end depth y_e. Various investigators like Rouse[5] (1936), Boer and Graf[6] (1971), Subramanya and Kumar[7] (1993), Das and Goswami[8] (2000, 2001) have worked on this problem. Almost similar discharge equations have been derived. Relationship between critical depth y_c and brink depth y_e is first established. Then discharge equation is expressed in terms of y_e. In most of the analysis by the investigators, relationship between y_c and y_e is obtained as

$$y_c = Cy_e \tag{8.32}$$

and

$$\text{Value of } Q = KBy_e^{3/2} \tag{8.33}$$

where k varies from 4.7 to 5.19.

The value of K should be calibrated for the free overfall to get a good assessment of Q.

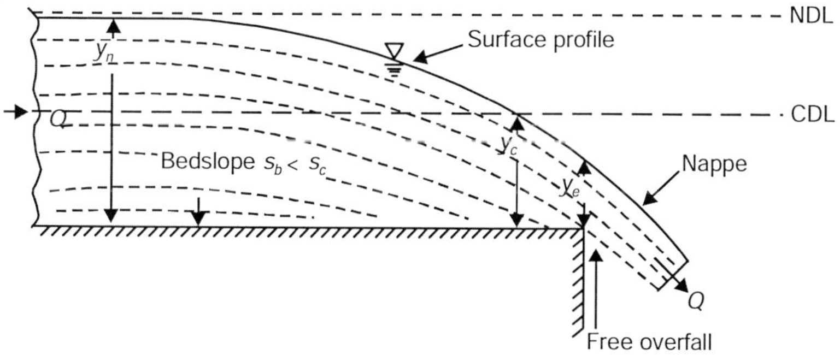

Figure 8.20 Brink depth method.

[5] Rouse, H., Discharge Characteristics of Free Overfall, *Jour. Civ. Engg*, ASCE, vol. 16, No. 4, pp. 257–60, April 1936.

[6] Boer, S.W. and Graf, W.H., Free Overfall as a Flow Measuring Devices, *Jour. Irrig. Engg. and Drainage*, Div. Proc., ASCE, vol. 97, March 1971.

[7] Subramanya, K. and Kumar, N., End Depth in Horizontal Circular Free Overall, *Jour.*, ICE (India), vol. 73, pp. 185–87, March 1993.

[8] Das, M.M. and Goswami, M.D., Brink Depth Technique of Free Overfall in Field Channel, *Jour. Agri. Engg.*, vol. 37(1), pp. 55–60, January-March, 2000.

8.4 CONCLUSION

Twenty-four methods of stream flow measurement (direct or indirect) are discussed in detail. Recent trend of discharge measurement has shown that investigators are inclined to adopt the electronic method of measurement. Time is perhaps not very far when the investigators will develop computer package to connect the flow system and discharge will immediately be assessed by the computer.

Whatever may be the method of stream flow measurement, correct assessment of it is very important in any design of water resource engineering project.

EXERCISES

8.1 A stream is assumed to be trapezoidal in cross-section in bed width of 12 m and side slope 2 horizontal: 1 vetical in a reach of 1 km. During flood time, high water levels recorded at both ends of the reach are tabulated in Table 8.1.

Table 8.1 Data of Question 8.1

Section	*Elevation of bed* (m)	*Water surface* (m)
Upstream	100.20	102.70
Downstream	98.60	101.30

If Manning's n = 0.03, estimate the discharge in the stream.

(***Ans***: 30.18 m^3/sec)

8.2 In a stream gauging operation, observations recorded with a current meter are as follows: Estimate the discharge in the river if the current meter rating is $V = 0.8N + 0.05$, V is m/sec and N is the revolutions/sec.

Distance from the bank (m) secs		*Depth of flow* D(m)	*Depth of meter* (m)	*Revolutions* N/sec	*Time*
0.6		1.0	0.6	15	50
at 0.6*D*					
1.2		4.0	3.2	30	55
0.8*D*					
0.2*D*			0.08	48	53
2.0		5.5	4.4	40	
46	0.8*D*				
			1.1	60	
64	0.2*D*				
3.0		6.5	5.2	45	
48	0.8*D*				
			1.3	67	
52	0.2*D*				
3.8		4.5	3.6	33	
55	0.8*D*				

			0.9	51
50	0.2*D*			
4.5		2.5	2.0	26
48	0.8*D*			
			0.5	44
55	0.2*D*			
5.0		1.0	0.6	20
47	0.6*D*			
5.6		0		
other bank				

(***Ans***: 14.97 m^3/sec)

8.3 Draw the rating from the stage-discharge relationship from Table 8.2. Assume the value of zero discharge to be 20.50 m^3/sec. Determine the stage of the river corresponding to a discharge of 2600 m^3/sec.

Table 8.2 Data of Question 8.3

Stage (m)	21.95	22.45	22.80	23.00	23.40	23.75	23.65	24.05	24.55	24.85	25.40	25.15	25.55	25.90
Q(m^3/sec)	100	220	290	400	490	500	640	780	1010	1220	1300	1420	1550	1760

(***Ans***: 26.90 m^3/sec)

SUGGESTED FURTHER READINGS

Ackers, P., White, W.R., Perkins, J.A. and Harrison, A.J.M., *Weirs and Flumes in Flow Measurement*, John Wiley, New York, USA, 1978.

Addison, H., *Applied Hydraulics*, 4th ed., Wiley, New York, 1954.

Addison, H., *Hydraulic Measurement,* Chapman and Hull, London, 1946.

Allen, C.H. and Taylor, E.A., The Salt Velocity Method of Water Measurement, Trans., *ASCE*, vol. 45, 1923.

Balloffet, A., Critical Flow Meters (Venturiflume), Paper 743, *Proc. ASCE*, vol. 81, pp. 1–31, July 1955.

Bos, M.G., Discharge Measurement Structures, Int. Inst. for Land Reclamation and Improvement, Wageningen, Netherlands, Pub. No. 20, 1976.

Chow, V.T, Maidment, D.R. and Mays, L.W., *Applied Hydrology*, McGraw-Hill, New York, 1988.

———, *Open Channel Hydraulics*, McGraw-Hill, N.Y., 1959.

———, (Ed.), *Handbook of Applied Hydrology*, McGraw-Hill, New York, 1964.

Cone, V.M., The Venturiflume, *Jour. Agri. Resear.*, vol. 9, No. 4, pp. 115–129, April 23, 1917.

Corbeat, D.M. and others, Stream Gauging Procedures, U.S. Geo. Surv. Water Supply Paper, 888, 1943.

Corbet, D.M. and others, Stream Gauging Procedure, US Geolo. Surv, Water Supply Paper 888, 1945.

Crump, E.S., Modeling Irrigation Channels, Punjab Irrigation Branch Publications, Paper No. 26 and 30A, Lahore, India, 1922, 1933.

Das, Ghanasyam, *Hydrology and Soil Conservation Engineering*, Prentice-Hall of India, New Delhi, 2000.

Das, M.M. and Goswami, M.D. Irrigation Measurement in Free Overfall Field Channels, *Jour. IWRS*, vol. 21, Roorke, July 2001.

De Marchi, G., New Experimental Researches on Standing Wave Flume Venturiflume, Ministry of Agri., Paris, France, 1937.

Dugal, K.N. and Soni, J.P., *Elements of Water Resources Engineering*, New Age International, New Delhi, 1996.

Engel, F.V.A.E., The Venturiflume, *The Engineer*, vol. 158, pp. 104–107, August 1934.

Faraday, M., Experimental Resources, Phil, Trans. Roy Society, 1832.

Frazier, A.H., William Gunn Price and Price Current Meter, *US Nat'l Mus. Bull.,* vol. 252, pp. 37–68, 1967.

Garg, S.K., *Hydrology and Water Resources*, Khanna Publishers, New Delhi, 2000.

Herschy, R.H. and Loosemor, W.R., Ultrasonic Method of River Flow Measurement, Symp. of river Gauging by Ultrasonic and Electromagnetic Methods, Univ. of Reading, 1974.

Herschy, R.H., New Methods of River Gauging, Chapter 5 in J.C. Rodd (Ed.), *Facts of Hydrology*, Wiley, New York, 1976.

Herschy, R.W. (Ed.), *Hydromerty,* Wiley Interscience, John Wiley, Chichester, 1978.

Inglis, C.C., Notes on Standing Wave Flumes and Flume Meter Baffle falls, Public Works Depart., Govt. of Bombay, Tech paper No. 15, India, 1928.

Jameson, A.H., The Development of Venturiflume, Water and Water Engg., London, vol. 32, No. 375, pp. 105–107, March 30, 1930.

Jameson, A.H., The Venturiflume and Effect of Contractions in Open Channels, Trans. Instn. of Water Engrs., vol. 30, pp. 19–24, 1925.

Kohler, M.A., Design of Hydrology Networks, WMO Tech. Note, No. 25, 1958.

Langbein, W.B., Stream Gauging Networks, Publ. 38, Int. Assoc. of Hydrology, Sci. General Assembly, 1954.

_____, Peak Discharge from Daily Records, USG-Survey, *Water Resour. Bull.*, 1944.

Linford, A., Venturiflume Flow Meters, Civ. Engg. and Public Works Review, London, vol. 36, No. 424, pp. 582–587, October 1941.

Linsley, R.L., Kohler, M.A. and Paulhus, J.L.H., *Hydrology for Engineers,* 3rd ed. McGraw-Hill, 1982.

Mutreja, K.N., *Applied Hydrology*, Tata McGraw-Hill, New Delhi, 1986.

Parshall, R.L., The Improved Venturiflume, Trans. ASCE, vol. 80, pp. 841–801, 1926.

———, The Parshall Measuring Flume, Colorado Agri. Expt. Station Bull., No. 423, March 1923.

Raghunath, H.N., *Hydrology: Principle, Analysis, Design,* New Age International, New Delhi, 1996.

Schaefer, H., Report on Hydrology, *Hydrology Sci. Bull.* vol. 17, pp. 145–166, July 1972.

Schuster, J.C., Measuring Water Velocity by Ultrasonic Flow Meter, *J. Hy. Div.* ASCE, vol. 101, pp. 1503–1517, 1975.

Sittner, W.T., Extension of Rating Curve by Field Survey, *Jour. Hy. Div.*, ASCE, vol. 89, March, 1963.

Subramanya, K., *Engineering Hydrology*, Tata McGraw-Hill, New Delhi, 1984.

US Bureau of Reclamation, Water Measurement Manual, 1953.

Wurbs, R.A. and James, W.P., *Water Resource Engineering*, Prentice-Hall of India, New Delhi, 2002.

Chapter 9

Estimation of Flood

9.1 INTRODUCTION

Flood is defined as abnormal high stage of flow which overtops the natural or artificial river banks in any reach and causes immense loss of crops, property, human lives and lines of communication. It is a natural event that results from excess runoff generated from a drainage basin due to severe combination of critical hydrologic and meteorological conditions over the region. In view of heavy losses due to occurrence of flood, its proper measurement or estimation is very much essential for its control and design of different hydraulic structures.

9.2 METHODS OF ESTIMATION

Some of the existing methods of estimating flood are:

1. Envelope curves
2. Empirical formulae
3. Physical indication of past floods
4. Rational method
5. Probable maximum precipitation (PMP) chart
6. Rating curve
7. Unit hydrograph method
8. Flood frequency analysis

Out of all these methods, flood frequency analysis, based on probability and statistics is favourable. It gives quite a reliable value. A lot of research works have already been done by quite a number of investigators.

9.3 ENVELOPE CURVES

Kanwarsain and Karpov (1967), after collecting large data from rivers of India, presented two envelope curves shown in Figure 9.1. The two curves prepared by them are for southern and north-central rivers of India.

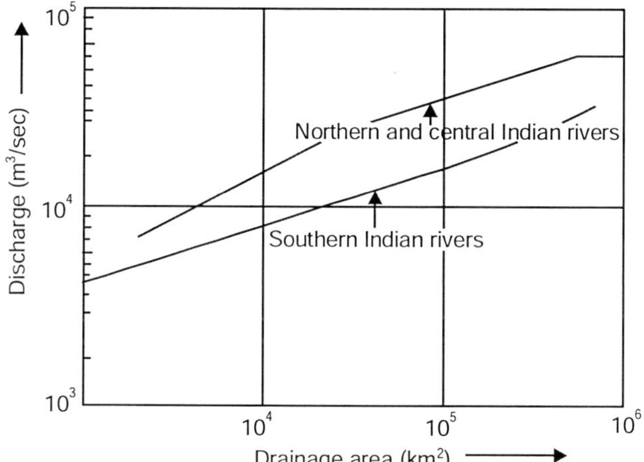

Figure 9.1 Envelope curves of Indian rivers.

The envelope curves have shown that the floods in northern and central Indian rivers are more than the southern rivers.

9.3.1 Limitations of Envelope Curves

The inherent drawbacks of these curves lie in the fact that it is based on the past recorded floods. But there lies a probability that still flood of higher magnitude may occur in future. No return period can be assigned to the flood estimated from the curve. Results from envelope curves can not be fully relied upon.

9.4 EMPIRICAL FORMULAE

Quite a lot of empirical formulae relating mainly to Q against area A of catchments with constants and indices have been developed around the world in both MKS and FPS unit.

Dickens' formula (1865): It is the oldest formula developed in Indian catchments. The formula is:

$$Q = CA^{3/4} \tag{9.1}$$

where Q is the maximum flood (m^3/sec), A is the catchment area (km^2) and C is the Dickens' constant, varying from 6 to 28. For north-Indian plains, north-Indian hilly regions, central India and coastal Andhra and Orissa, C values are 6, 11–14, 14–28, 22–28 respectively.

Ryve's formula (1884): The formula is:

$$Q = CA^{2/3} \tag{9.2}$$

Here unit of Q and A are same as Dickens' formula.

Ryve recommended value of C is 6.8 for areas within 80 km from east coast, 8.5 for areas within 80–160 km from the east coast and 10.2 for limited areas near hills. Further experiments in different situations show that the actual value of C is on higher side than given by Ryve.

Fuller's formula (1914): Fuller[1] from some catchments of the United States gave the following formula:

$$Q = CA^{0.8}\,(1 + \log_e T) \tag{9.3}$$

Fuller connected both area A (km^2) and frequency T (maximum 24-hr flood with a frequency T of years), Q (m^3/sec) and C varies from 0.18 to 1.88.

Inglis formula: It is based on data of catchment in Western Ghats in Maharashtra, and is given by

$$Q = \frac{124A}{\sqrt{A + 10.4}} \tag{9.4}$$

where Q is in m^3/sec and A is in km^2.

Ali Nabag Jung Bahadur formula: It is developed for Hyderabad Deccan catchment, and is given as:

$$Q = CA^{\left(0.993 - \frac{1}{14}\log_e A\right)} \tag{9.5}$$

where Q is in m^3/sec, A is in km^2 and C takes a value from 48 to 60.

Creager's formula: Creager developed this formula from the study of flood peaks in the USA, which is:

$$Q = CA^{0.894A - 0.048} \tag{9.6}$$

where Q is in cumecs, C = 40 to 130, lower values for ordinary floods and higher value for intense and acute flood and A is in sq. miles = 0.39 km^2.

Jarvis formula: It states that

$$Q = C\sqrt{A} \tag{9.7}$$

C varies from 1.77 to 177 as maximum.

Myer's formula: It is given by:

$$Q = 177P\sqrt{A} \tag{9.8}$$

P varies from 0.002 to unity.

Craig's formula: It is given by

$$Q = 440\ RBVI \log_e\left(\frac{8L^2}{B}\right) \tag{9.9}$$

[1] Fuller, W.E., Flood Flows, Trans. ASCE, vol. 77, pp. 564–617, 1914 and discussions by Allen Hazen, p. 623.

where Q in in ft^3/sec, L is greatest length of catchment in miles, B is average width of catchment in miles, V is velocity of runoff in ft/sec, R is coefficient of runoff and I is average rainfall in inches/hr.

Rhind's formula: It is given as:

$$Q = C\frac{SRA^P}{L} \tag{9.10}$$

where Q is in ft^3/sec, S is the average fall of last 3 miles of the river above the basin outlet in feet/mile, R is greatest average rainfall recorded in inches/hr, P is variable index, L is variable greatest length catchment in miles and C is a coefficient which depends on the ratio B/L.

Chmier's formula: It is given by

$$Q = 640\ IKA^{3/4} \tag{9.11}$$

where Q is in ft^3/sec, I is average rainfall in inches/hr, K is runoff coefficient and A is catchment area in (miles)^2.

Iszkowski's formula: It is obtained as:

$$Q = 0.03171\ CMHA \tag{9.12}$$

where Q is in m^3/sec, H is annual rainfall in metre, C is coefficient of runoff, A is catchment area in km^2, and M is a coefficient relating maximum to average discharge.

Dredge or Burge's formula: It is given by

$$Q = 19.6\frac{A}{L^{2/3}} \tag{9.13}$$

Here L is the length of drainage basin in km and A is in km^2. If B is average breadth of the catchment, then

$$A = BL$$

$$\therefore \quad Q = 19.6\,BL^{1/3} \tag{9.13a}$$

Pettis' formula: It is developed by Pettis in Northern United States.

$$Q = C(PB)^{5/4} \tag{9.14}$$

where P is probable 100-year maximum one day rain in cm, B is average width of basin in km and C is constant equal to 1.5 for humid area and 0.2 for desert area.

Boston Society of Civil Engineering's formula: The Committee of Boston Society of Civil Engineering's made a study of floods. Based on New England flood of 1927, they found that

$$q = 0.0056\frac{D}{t} \tag{9.15}$$

Here q is peak flow in (m^3/sec) square km, D is total depth of flood runoff on the basin in cm and t is the total flood period in hours or the base of the flood hydrograph.

Baird and McIllwraith formula: Based on the maximum floods throughout the world, they gave the following formula:

$$Q = \frac{3025A}{(278 + A)^{0} 78} \tag{9.16}$$

Here Q is in m^3/sec and A is in km^2.

Burkli Ziegler's formula: It is developed in the USA, and is given as:

$$Q = 412A^{3/4} \tag{9.17}$$

Here A is in km^2.

9.4.1 Limitations of Empirical Formulae

It has been observed from above empirical formulae, mainly catchment area A is involved, with constant C. This constant C varies from catchment to catchment in different countries. The major limitation of using these formulae is the subjective decision about the value of C to be adopted. However, in absence of flood data, they may be used by giving an allowance 33% increase.

9.5 PHYSICAL INDICATION OF PAST FLOODS

Ancient movements situated near the banks of the river always bear past flood marks. The old man in the villages near the river bank may be contacted to know the highest flood level attained during his life time. Thus, the highest flood mark may be attained. Then the cross-section of river valley is plotted. Suitable value of Manning's n is selected based on bed roughness. Slope of the river bed and hydraulic radius may be computed to calculate the velocity. Then the flood discharge is obtained from the formula $Q = AV$. Thus, approximate peak flood may be obtained from such physical past marks available near the banks when other data is not available.

9.5.1 Limitations of the Method

1. The method gives a rough estimate.
2. River cross-section does remain constant with time due to sedimentation erosion, and scouring and weed growth.
3. Alignment of river may change considerably from the time of past records.
4. The Manning's n value also changes with the change of river bed source.

9.6 RATIONAL METHOD

This method is based on the application of the formula:

$$Q = CiA \tag{9.18}$$

Here C is the runoff coefficient dependent on runoff qualities of the catchment. This method requires the estimation of three parameters—runoff coefficient C, time of concentration t_c and rainfall intensity i. The time of concentration t_c is the time taken by raindrops from the farthest point to reach the outlet. When a continuous rainfall occurs in a catchment, runoff curve from beginning of storm continuously rises. When all the runoff reaches the outlet, runoff becomes constant as shown in Figure 9.2. Time at which it begins to be horizontal is the time of concentration t_c.

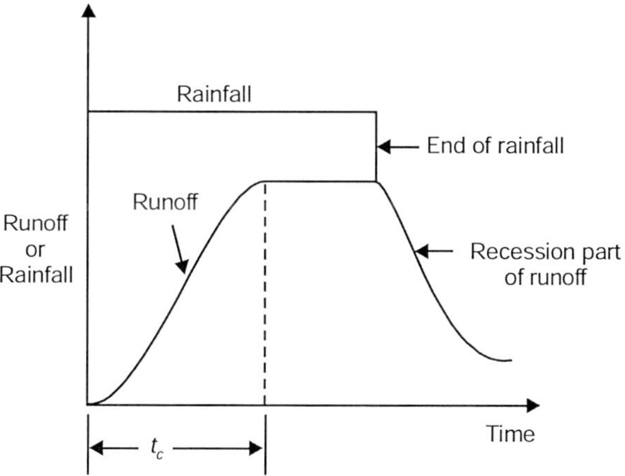

Figure 9.2 Runoff hydrograph due to uniform rainfall of long duration to obtain time of concentration t_c.

Normally rainfall intensity $i(= i_c)$ is obtained from the intensity duration frequency curve for the catchment corresponding to t_c and recurrence interval T. If this curve is not available for the catchment, and if P is the maximum precipitation occurs during a storm period t_r, then is calculated from the equation:

$$i = \frac{P}{t_r}\left(\frac{t_r + 1}{t_c + 1}\right) \tag{9.19}$$

Even if the time of concentration t_c of the catchment is not known, then

$$i \cong \frac{P}{t_r}$$

The value of the runoff coefficient C is given in Table 9.1. The values given may serve as rough tool as it may change depending on the previous few days' conditions of the catchment before the rainfall. Usually if the catchment is already wet before the rain comes, value of C increases as loss due to absorption will be less.

Table 9.1 Values of Runoff Coefficient C

Topography or terrain	*Value of C*
Flat cultivation land Sandy land	0.2
Hilly areas Forests Clay and loamy soil	0.5
Build up urban areas (impervious)	0.8
Flat residential areas	0.4
Moderately step residential area	0.6

If Q in Eq. (9.18) is in (m^3/sec), i is in mm/hr, A is in km^2, then

$$Q = CAi\left[(1000 \times 1000) \times \left(\frac{1}{10 \times 100 \times 60 \times 60}\right)\right]$$

$$\because \quad \text{km}^2 = (1000 \times 1000)\ \text{m}^2 \tag{9.20}$$

and

$$\text{mm/hr} = \left(\frac{1}{10 \times 100 \times 60 \times 60}\right)$$

$$Q = \frac{1}{3.6} CiA \ \ \text{m}^3/\text{sec}$$

9.6.1 Limitations of Rational Method

1. This method is also dependent on runoff coefficient C which changes with surface conditions and topography of the terrain.
2. The value of C again dependent on time of concentration t_c which is again evaluated by empirical equation.
3. Assumption of rainfall duration equal or greater than time of concentration for maximum flow is not always correct.
4. Assumptions in the rational method hold good to paved areas with gutters and drainages only.

9.7 PMP CHART

In the design of major hydraulic structures such as spillways in large dam, hydraulic engineers would like to maintain the failure probability as low as possible. In design and analyses of such structures, probable maximum precipitation (PMP) that may reasonably be expected in a basin is adopted. This PMP is defined as the greatest or extreme rainfall of given duration that is physically possible over a basin. From operational point of view, PMP can be defined as that rainfall over a basin which will produce a flood flow with virtually no risk of being exceeded. PMP chart is developed by the Meteorological Department of India. Those prepared PMP charts for some areas like Punjab, Haryana, Delhi and Rajasthan by the following two methods:

1. Meteorological method
2. Statistical study of rainfall data

Statistically,

$$PMP = \overline{P} + K\sigma \tag{9.21}$$

Here $\overline{P}$ is the average rainfall, σ is the standard deviation and K is a frequency factor which depends on statistical distribution of series, number of years of record and return period. The value usually hits in the neighourhood of 15.

9.7.1 Limitations of the Method

1. The design flood determined by PMP is only applicable to large and important project.
2. It gives overestimation for medium and small projects.
3. Orographic infuences interfere in the correct assessment of PMP.
4. It is also based on assumption of maximum air moisture with maximum air inflow for maximum precipitation, which may not happen all the time.

9.8 RATING CURVE

Rating curve method for the determination of flood Q has already been discussed in Chapter 8 (Section 8.10). Therefore, details have not been discussed here. In case of very high flood, it may have to estimate from the extended the part of the rating curve.

9.8.1 Limitations of Rating Curve

1. Variable backwater effects induce different gauge readings.
2. Weed growth, dredging, vegetations near the gauge site affects the rating curve.
3. Aggradations and degradations of the alluvial stream affect the stage of the stream.
4. Water surface slope changes with stages of flow leading to the discrepancy of the discharge measurement.
5. During high stage of flow, river becomes inaccessible and, therefore, extension of rating curve is necessary. Some error in the extension part is likely to occur in true values of rating curve.

9.9 UNIT HYDROGRAPH METHOD

Details of unit hydrograph (UH), synthetic unit hydrograph (SUH), and instantaneous unit hydrograph (IUH) have already been presented in Chapter 7. The use of UH to determine flood hydrograph and its peak value from the peak of UH are also discussed. If this linear model of the catchment is already known, flood of any storm can be obtained immediately.

To find the maximum possible flood in the catchment by the UH, following points are considered:

1. Selection of major storms
2. Maximization of selected storms by few percentages
3. Plotting the depth-area-duration (DAD) curves and analysis
4. Moisture adjustment of the catchment
5. Use of infiltration index to get the rainfall excess

9.9.1 Limitations of Method

This UH method has also some limitations. As per theory of hydrograph, catchment area preferably should not be more than 5000 km^2 to assume uniform rainfall distribution in the catchment. If the catchment area is much larger; one UH cannot be used for the whole basin. Whole basin has to be divided into some sub-basins and UH$_s$ for every sub-basin are to be computed which makes the estimation quite complex and difficult. The UH method is suitable for smaller catchment where the rainfall intensity and duration are uniform.

9.10 FLOOD FREQUENCY ANALYSIS

It is the best method out of all the earlier methods discussed so far provided sufficient data of flood is available. Floods are extremely complex natural events. It is very difficult to model flood analytically as it is the outcome of many component parameters. Therefore, estimation of peak flood is a complex problem leading to many approaches discussed earlier. Flood frequency analysis is based on the theory of statistics and probability (See Appendix A). It is supposed to be the best method provided flood data are available for a long period of years. Flood occurred in the river are treated as statistical events and may be considered to constitute discrete function of random variables.

9.10.1 Frequency or Probability of Flood *P*

It denotes the likelihood of flood being equalled or exceeded. A flood has 50% frequency means that this flood has 50 changes out of 100 chances of being equalled or exceeded. This frequency or probability is denoted by P.

Return period or recurrence interval T_r: It is defined as the number of years in which a flood can be expected once or a flood of given magnitude will be equalled or exceeded only once. The probability P and return period of flood of T_r are related by

$$P = \frac{1}{T_r} \tag{9.22}$$

9.10.2 Plotting Positions

This return period T_r is often used in lieu of probability to describe a design flood. To plot a series of peak flows as a cumulative frequency curves it is necessary to decide on a probability or return period to associate with each peak. There are various formulae for defining this value known as **plotting position** as per Benson[2] (1962), Weibull[3] (1939). To explain this plotting position, different methods or equations, annual floods of a river for N years are known. Arrange the magnitude of flood in decreasing order of magnitude and serial number to each floods is given. The highest flood will be at the top with serial number 1, second highest at serial number 2 and the lowest flood at the bottom with serial number m, then return period T_r of that flood is determined by the following methods:

1. Weibull method (1939)

$$T_r = \frac{N+1}{m} \tag{9.23}$$

where $$P = \frac{1}{T_r} = \frac{m}{N+1}$$

2. California method (1923)

$$T_r = \frac{N}{m} \tag{9.24}$$

3. Hazen method (1930)

$$T_r = \frac{2N}{2m-1} \tag{9.24a}$$

4. Gumbel method (1943)

$$T_r = \frac{N}{m+C-1} \tag{9.25}$$

C is known as Gumbel's correction given in Table 9.2

5. Gringorton's method (1963)

$$T_r = \frac{N+0.12}{m-0.44} \tag{9.26}$$

6. Chegodyev method

$$T_r = \frac{N+0.4}{m-0.3} \tag{9.27}$$

Table 9.2 Gumbel C values

C	1	0.88	0.845	0.78	0.73	0.66	0.59	0.52	0.4	0.38	0.28
m/N	1	0.8	0.7	0.6	0.5	0.4	0.3	0.2	0.1	0.08	0.04

[2] Benson, M.A., Plotting Position and Economics of Engineering Planning, *J.H.Y. Div. ASCE*, vol. 88. pp. 57–71, 1962.
——, Evolution of Methods of Evaluating the Occurrence of Floods, US Geolog. Surv. Water Supply Paper, 1580-A, 1962.

[3] Weibull, W.A., A Statistical Theory of the Strength of Materials, The Royal Swedish Institute for Engineering Research, Proc. No. 51, pp. 5–45, 1939.

9.10.3 Determination of Design Flood of any Frequency or Probability

1. Calculate return period T_r of all floods (Q_1 to $Q_{n\text{-}}$) by any method. Weibull method is most popular.
2. Find probability P (or Frequency) by equation $\frac{1}{T_r}$.
3. Plot flood Q vs T_r or Q vs P in probability graph paper (Figure 9.3).
4. Draw best fit line AB by the method of least squares or by extrapolation.
5. The line AB may be extended as shown by dotted line to determine the return period of flood of higher magnitude.
6. From the curve or line ABC, once the magnitude of flood is known, its return period or probability of occurrence can be determined.

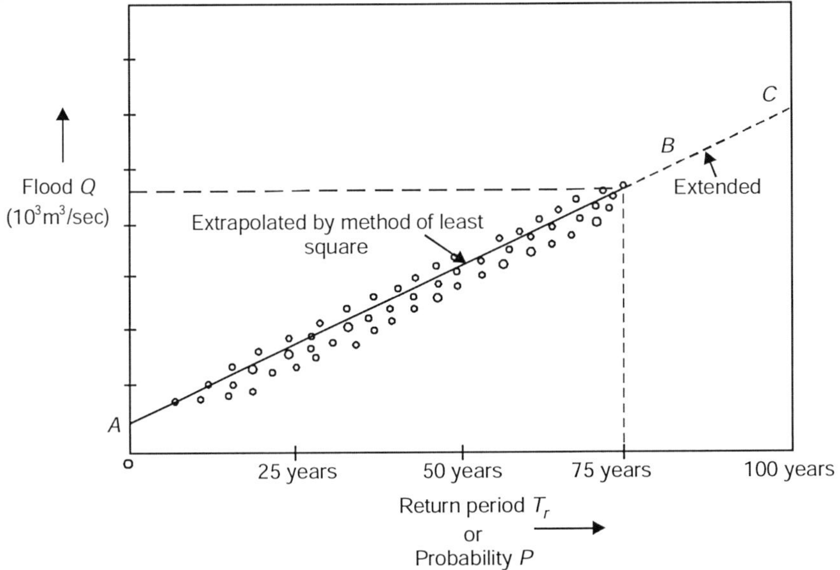

Figure 9.3 Plot of flood Q against return period T_r or probability P.

9.10.4 Limitations of Flood Frequency Analysis

This is the simplest method. It is acceptable only if one is interested for shorter return period. Fitting the curve at higher values of T_r is not much reliable. Therefore, better methods are available which are discussed in the forthcoming sections.

9.11 GUMBEL EXTREME VALUE DISTRIBUTIONS

E.J. Gumbel[4] in 1941 was the first to consider that the annual flood peaks are extreme value of

[4] Gumbel, E.J., The Return Period of Flood Flows, *The Annals Mathematical Statistics*, vol. 12, No. 2, pp. 163–190, June 1941.

floods in each of the annual series of recorded flood. Hence floods should follow the extreme value distribution.

Let $Q_1, Q_2, Q_3, \ldots, Q_n$ are annual extreme values of flood in a particular river of catchment area. In Gumbel method, the exceedence probability P of a given flow Q_t having a return period of T years being equalled or exceeded is given by:

$$P = 1 - e^{-e^{-y}} \tag{9.28}$$

where y is called **reduced variate**, e is the base of Naperian logarithm.

$$\because \qquad P = \frac{1}{T_r}$$

$$\therefore \qquad \frac{1}{T_r} = 1 - e^{-e^{-y}} \tag{9.29}$$

When size of the sample is infinite,

$$y = a(Q_t - Q_f) \tag{9.30}$$

where

$$a = \frac{1}{0.78\sigma} \tag{9.31}$$

Here σ is standard deviation which is given by

$$\sigma = \sqrt{\frac{\Sigma\left(Q - \bar{Q}\right)^2}{N-1}} \tag{9.32}$$

where

$$\bar{Q} = \frac{Q_1 + Q_2 + Q_3 + \cdots + Q_n}{N} \tag{9.33}$$

and

$$Q_f = \bar{Q} - 0.45\sigma \tag{9.34}$$

Using Eqs. (9.30), (9.31) and (9.34), we get

$$Q_t = \bar{Q} + \left(0.78y - 0.45\right)\sigma \tag{9.35}$$

However, if the flood records available are for limited period with limited samples as suggested by Chow,

$$Q_t = \bar{Q} + K\sigma \tag{9.36}$$

Here K is frequency factor obtained from Table 9.3.

Example 9.1

The maximum annual flows in a river Puthimari, a tributary of the river Brahmaputra for the period 1975 to 2004 are given as follows:

Year	*Flow* (10 m^3/sec)	*Year*	*Flow* (10 m^3/sec)	*Year*	*Flow* (10 m^3/sec)
1975	2.7	1985	2.4	1995	3.8
1976	4.0	1986	2.7	1996	7.8
1977	2.4	1987	5.0	1997	2.4
1978	4.6	1988	2.4	1998	3.0
1979	4.3	1989	1.7	1999	4.9
1980	2.0	1990	2.7	2000	9.2
1981	3.8	1991	5.0	2001	2.4
1982	2.6	1992	5.3	2002	4.8
1983	2.5	1993	2.3	2003	5.7
1984	4.4	1994	2.4	2004	3.8

Compute 100 years average return period of flood using Gumbel method. The frequency factor K for 100-year return period for Gumbel distribution is given as 3.653 (from Table 9.3).

Solution:

Step 1 Find the average discharge.

$$\bar{Q} = \frac{\Sigma Q}{N} = \frac{(2.7 + 4.0 + 2.4 + 4.6 + 4.3) + \cdots + 4.8 + 5.7 + 3.8}{30}$$

$$= \left(\frac{113.2 \times 10^3}{30}\right) = 3773.33 \text{ m}^3/\text{sec}$$

Step 2 Find:

$$\Sigma\left(Q - \bar{Q}\right)^2 = (2700 - 3773.33)^2 + (4000 - 3773.33)^2$$

$$+ \cdots + (5700 - 3773.33)^2 + (3800 - 3773.33)^2$$

$$= 86047177 \text{ m}^3/\text{sec}$$

Step 3 Find standard deviation:

$$\sigma = \sqrt{\frac{\Sigma\left(Q - \bar{Q}\right)^2}{N - 1}}$$

$$= \frac{86047177}{30 - 1}$$

$$= 1722.54 \text{ m}^3/\text{sec}$$

Step 4 Since data is not too large, Eq. (9.36) may be used to find Q_t, i.e., Q_{100}

$\therefore$

$$Q_{100} = \bar{Q} + K\sigma$$

$$= (3773.33 + 3.653 \times 1722.54)$$

or

$$Q_{100} = (10.065 \times 10^3) \text{ m}^3/\text{sec}$$

EXAMPLE 9.2

For a long record of annual peak flood of river Sankosh, a north bank tributary of the river Brahmaputra, the mean Q and standard deviation calculated by the Water Resource Department, Govt. of Assam are 143.9 m^3/sec and 56.65 m^3/sec respectively. Using Gumbel's approach, obtain the return period of flood for a flood of 350 m^3/sec of this river.

Solution: Probability of exceedence of a flood of magnitude Q_t (350 m^3/sec) is given by:

$$P(Q > Q_t) = -e^{-e^{-y}}$$

where $y = a(Q_t - Q_f)$ (assuming sample to be large)

and
$$a = \frac{1}{0.78\sigma} = \frac{1}{0.78 \times 56.65} = 0.022631$$

Now,
$$Q_t = 350 \text{ m}^3/\text{sec} \qquad \text{(Given)}$$

∴
$$Q_f = \bar{Q} - 0.45\sigma$$
$$= (143.9 - 0.45 \times 56.65)$$
$$= 118.4 \text{ m}^3/\text{sec}$$

Using Eq. (9.30),
$$y = a(Q_t - Q_f)$$
$$= 0.022631(350 - 118.40)$$
$$y = 5.24134$$

∴
$$P(Q > 350) = 1 - e^{-e^{-e^{-5.24134}}}$$
$$= 1 - 0.99472 = 0.00528$$

But
$$P = \frac{1}{T_r} = \frac{1}{0.00528}$$

∴
$$T_r = 189.4 \text{ years}$$

EXAMPLE 9.3

The following table gives the flood of a big river. Find flood of 500 years period by Gumbel method.

Sl. No	*Year*	*Flood in cumecs*	*Sl. No*	*Year*	*Flood in cumecs*
1	1990	10200	9	1998	11200
2	1991	3860	10	1999	6200
3	1992	3360	11	2000	9210
4	1993	5400	12	2001	3960
5	1994	8300	13	2002	3700
6	1995	7720	14	2003	2380
7	1996	2220	15	2004	6880
8	1997	1785			

Solution:

$$\bar{Q} = \frac{10200 + 3860 + 3320 + 5400 + \cdots + 3700 + 2380 + 6880}{15}$$

$$= 5755.66 \text{ cumecs}$$

Now, $$\sigma = \sqrt{\frac{\sum_{1}^{N}\left(Q - \bar{Q}\right)^2}{N-1}}$$

$$= \sqrt{\frac{(10200 - 5755.66)^2 + (3860 - 5766)^2 + \cdots + (6880 - 575566)^2}{14}}$$

$$= 3046.558 \text{ cumecs}$$

Then $$T_r = 500, P = \frac{1}{T_r} = \frac{1}{500} = 0.002$$

∴ $$P = 1 - e^{-e^{-y}}$$

or $$0.002 = 1 - e^{-e^{-y}}$$

Solving for y, $$y = 6.2136$$

We have $$y = (Q_{500} - Q_f) \quad \text{(i)}$$

Assuming sample to be large,

$$a = \frac{1}{0.78\sigma} = \frac{1}{0.78 \times 3046.558} = 0.0004208$$

and $$Q_f = \bar{Q} - 0.45\sigma = (5755.66 - 0.45 \times 3046.558)$$

$$Q_f = 4384.716 \text{ m}^3\text{/sec}$$

Substituting y, a and Q_f in Eq. (i),

$$6.2136 = 0.0004208\,[Q_{500} - 4384.716]$$

∴ $$Q_{500} = 10,382.39 \text{ cumecs}$$

9.11.1 Gumbel Method When Sample is Not Infinite

The discussion and problem already solved in Gumbel Method is based on infinite sample. In practical field, sample is not infinite. In such situations, solution is slightly different. It may be described as:

Probability of extreme value distribution of peak flood ≥ Q is given by

$$P = 1 - e^{-e^{-y}}$$

The reduced variate y is given by

$$y = -0.834 - 2.303\log\ \log\left(\frac{T_r}{T_r - 1}\right) \tag{9.37}$$

or

$$y = -0.834 - 2.303X_T$$

Here

$$X_T = \log\ \log\left(\frac{T_r}{T_r}\right) \tag{9.38}$$

The reduced variate y is linear with the variate Q (annual peak flood) itself and is given by

$$y = \left(\frac{Q - \overline{\overline{Q}}}{\sigma}\right)\sigma_n + \overline{y}_n \tag{9.39}$$

where σ_n (reduced standard deviation) and $\overline{y}_n$ (reduced mean to mode) are function of sample size N, and are given in Table 9.5.

∴

$$Q_T = \overline{Q} + \left(\frac{y + \overline{y}_n}{\sigma_n}\right)\sigma \tag{9.40}$$

or

$$Q_T = \overline{Q} + K\sigma \tag{9.41}$$

Here

$$K = \left(\frac{y + \overline{y}_n}{\sigma_n}\right)$$

and Q_T is annual flood which has return period of T.

EXAMPLE 9.4

The analysis of 30 years flood data at point of a stream yielded $\overline{Q}$ = 1200 m^2/sec and standard deviation σ = 650 m^3/sec. For what discharge would you design the structure at the point to provide 95% assurance that the structure would not fail in next 50 yrs. (Assume data to be finite in Gumbel method.)

Solution:

Assurance = 95%, Risk = 5%, n = 30, Life period = 50 yrs

Now,

$$\text{Risk } R = 1 - \left(1 - \frac{1}{T_r}\right)^n \qquad (T_r = \text{Return period})$$

∴

$$0.05 = 1 - \left(1 - \frac{1}{T_r}\right)^{30}$$

or

$$1 - \frac{1}{T_r} = (1 - 0.05)^{1/30} = 0.99829$$

or $$\frac{1}{T_r} = 1 - 0.99829 = 0.0017083$$

or $$T_r = 585.37 \text{ years}$$

Again $$p = \frac{1}{T_r} = 1 - e^{-e^{-y}}$$

or $$0.0017083 = 1 - e^{-e^{-y}}$$

or $$e^{-e^{-y}} = 1 - 0.0017083$$

or $$-e^{-y} \log_e(e) = \log_e(0.9982917) \qquad \because \log_e(e) = 1$$

or $$+e^{-y} = 0.00170976$$

or $$-y \log_e e = -6.37$$

or $$y = 6.37 \qquad \text{(i)}$$

Writing Eq. (i),

$$Q_T = \bar{Q} + \left(\frac{y - \bar{y}_n}{\sigma_n}\right)\sigma$$

The values of $\bar{y}_n$ and σ_n for n = 30 years from Table 10.5 are:

$$\bar{y}_n = 0.5465$$
$$\sigma = 1.1607$$

and $$\left.\begin{aligned}\bar{Q} &= 1200 \text{ m}^3/\text{sec}\\ \sigma &= 650 \text{ m}^3/\text{sec}\end{aligned}\right\} \text{(Given in the problem)}$$

$$Q_T = \left[1200 + \left(\frac{6.371 - 0.5465}{1.1607}\right) \times 650\right]$$

$$Q_T = 4461.76 \text{ m}^3/\text{sec}$$

Example 9.5

A flood of 300 cumecs has been exceeded 6 times during a period 25 years. What is the probability in percentage of this flood?

Solution:

Annual occurrence $P = m/n = 6/25 = 0.24$

Now, $T_r = 1/p = 1/0.24 = 4.166$ years

Hence Annual probability in p.c. $= (1 - e^{-P}) \times 100$

$= (1 - e^{-0.24}) \times 100$

$= 21.33\%$.

9.12 LOG PEARSON TYPE III DISTRIBUTION

It was K. Pearson[5] (1930) who developed this method. In this method, it is recommended to convert the data series to logarithms and then compute the following:

1. Compute logarithms of flow (log Q).
2. Estimate standard

$$\log \bar{Q} = \frac{\Sigma \log Q}{N} \tag{9.42}$$

3. Compute standard deviation

$$\sigma_{\log Q} = \sqrt{\frac{\sum \left(\log Q - \log \bar{Q}\right)^2}{N-1}} \tag{9.43}$$

4. Compute skew coefficient

$$C_s = \frac{n\Sigma\left(\log Q - \log \bar{Q}\right)^3}{(n-1)(n-2)(\sigma \log)^3} \tag{9.44}$$

5. Then

$$\log\ Q_t = (\log \bar{Q}) + K(\sigma \log Q)$$

where K is log Pearson frequency factor obtained from Table 9.4.

Table 9.3 Frequency Factors for K for Gumbel Extreme Value Distribution Method

Years of record (N)	*Return period, T_r year*								
	10 *yrs*	20 *yrs*	30 *yrs*	50 *yrs*	75 *yrs*	100 *yrs*	200 *yrs*	400 *yrs*	1000 *yrs*
15	1.703	2.410	2.823	3.321	3.721	4.005	–	–	6.265
20	1.625	2.302	2.690	3.179	3.563	3.836	4.490	5.150	6.006
25	1.575	2.235	2.614	3.088	3.463	3.729	–	–	5.848
30	1.541	2.188	2.560	3.026	3.393	3.653	4.28	4.91	5.727
35	1.516	2.152	2.520	2.979	3.341	3.598	–	–	–
40	1.495	2.126	2.489	2.943	3.301	3.554	4.160	4.780	5.727
45	1.478	2.104	2.464	2.913	3.268	3.520	–	–	–
50	1.466	2.086	2.443	2.889	3.241	3.491	–	4.56	5.576
55	1.455	2.071	2.426	2.869	3.219	3.467	–	–	–
60	1.446	2.059	2.411	2.852	3.200	3.446	–	–	–
65	1.437	2.048	2.398	2.837	3.183	3.429	–	–	–
70	1.430	2.038	2.387	2.824	3.169	3.413		–	5.359
75	1.423	2.029	2.377	2.812	3.155	3.400	–	–	–
80	1.417	2.020	2.368	2.802	3.145	3.387	–	–	
85	1.413	2.013	2.361	2.793	3.135	3.376	–	–	–
90	1.409	2.007	2.353	2.785	3.125	3.367	–	–	–
95	1.405	2.002	2.347	2.777	3.116	3.357	–	–	–
100	1.401	1.998	2.341	2.770	3.109	3.349	3.93	4.51	5.261

[5] Pearson, K., *Tables for Statisticians and Biometricians,* 3rd ed., Cambridge University Press, London, 1930.

Table 9.4 *K* Values for Log Pearson Type III Distribution

Skew coefficient C_s	*Return period* T_r *year*				
	10 *yrs*	25 *yrs*	50 *yrs*	100 *yrs*	200 *yrs*
3.5	1.180	2.278	3.152	4.051	4.970
2.5	1.250	2.162	3.048	3.845	4.652
2.0	1.302	2.219	2.912	3.605	4.298
1.8	1.318	2.193	2.848	3.499	4.147
1.6	1.329	2.163	2.780	3.499	4.990
1.4	1.337	2.128	2.706	3.371	3.828
1.2	1.340	2.087	2.626	3.149	3.661
1.0	1.340	2.043	2.542	3.022	3.489
0.9	1.339	2.018	2.498	2.957	3.401
0.8	1.336	1.993	2.453	2.891	3.312
0.7	1.333	1.967	2.407	2.824	3.223
0.6	1.328	1.939	2.359	2.755	3.132
0.5	1.323	1.910	2.311	2.686	3.041
0.4	1.317	1.880	2.611	2.615	2.949
0.3	1.309	1.849	2.211	2.544	2.856
0.2	1.301	1.818	2.159	2.472	2.763
0.1	1.292	1.785	2.107	2.400	2.670
0	1.282	1.751	2.054	2.326	2.576
–0.1	1.272	1.716	2.000	2.252	2.482
–0.2	1.258	1.680	1.945	2.178	2.388
–0.3	1.245	1.643	1.890	2.104	2.294
–0.4	1.231	1.606	1.834	2.029	2.201
–0.5	1.216	1.567	1.777	1.955	2.108
–0.6	1.200	1.528	1.720	1.880	2.016
–0.7	1.183	1.488	1.663	1.806	1.926
–0.8	1.166	1.448	1.606	1.733	1.837
–0.9	1.147	1.407	1.549	1.660	1.749
–1.0	1.128	1.366	1.492	1.588	1.664
–1.2	1.086	1.282	1.379	1.449	1.501
–1.4	1.041	1.198	1.270	1.318	1.351
–1.6	0.994	1.116	1.166	1.197	1.216
–1.8	0.945	1.035	1.069	1.087	1.097
–2.0	0.895	0.959	0.980	0.990	0.995
–2.5	0.771	0.791	0.798	0.799	0.800
–3.0	0.660	0.666	0.666	0.667	0.667

Table 9.5 Reduced Mean $\overline{y}_n$ and Reduced Standard Deviation σ_n as Function of N in Gumbel Distribution

Sample size N	$\overline{y}_n$	σ_n
10	0.4952	0.2457
15	0.5128	1.0206
20	0.5236	1.0628
25	0.5309	1.6915
30	0.5362	1.1124
35	0.5403	1.1283
40	0.5436	1.1413
45	0.5436	1.1518
50	0.5465	1.1608
55	0.5504	1.1681
60	0.5521	1.1747
65	0.5536	1.1803
75	0.5549	1.1898
85	0.5578	1.1973
90	0.5553	1.2038
100	0.5600	1.2065
200	0.5672	1.2359
500	0.5724	1.2588
1000	0.5745	1.2685

EXAMPLE 9.6

From the annual flood series of river Barnadi, a tributary of the Brahmaputra, following data are given. Estimate flood discharge for return period (i) 100 years (ii) 200 years by using log Pearson type III.

Table 9.6 Data of Example 9.6

Year	*Flow* (m^3/sec)	*Year*	*Flow* (m^3/sec)	*Year*	*Flow* (m^3/sec)
1978	2949	1987	4798	1996	6599
1979	3521	1988	4290	1997	3700
1980	2399	1989	4651	1998	4175
1981	4124	1990	5050	1999	2988
1982	3496	1991	6960	2000	2709
1983	2947	1992	4366	2001	3873
1984	5060	1993	3380	2002	4593
1985	4903	1994	7826	2003	6761
1986	3751	1995	3320	2004	1971

Solution:

In Column (2) all discharges from Column (1) are converted to log Q. Here N = 27.

$$\log \bar{Q} = \frac{\Sigma \log Q}{N}$$

$$= \frac{(3.469 + 3.546 + \cdots + 3.830 + 3.294)}{27}$$

$$\log \bar{Q} = 3.607$$

Now, find standard deviation of log Q, i.e.,

$$\sigma \log Q = \sqrt{\sum_{N-1}^{27} (\log Q - \log \bar{Q})^2}$$

$$\sigma \log Q = 0.1427$$

Compute skew coefficient by Eq. (9.39).

$$C_s = \frac{n \sum (\log Q - \log \bar{Q})^3}{(n-1)(n-2)(s \log Q)^2}$$

$$C_s = 0.043$$

Now, $\log Q_t = \log \bar{Q} + K \log Q$

When $C_s = 0.043$ and $T = 100$ yrs.

From Table 9.4, $K = 2.35$

$\therefore$ $\log Q_{100} = 3.607 + 2.35 \times 0.1427$
$= 3.942345$

$\therefore$ Q_{100} = antilog (3.942345) = 8756.8 m^3/sec

Similarly when T = 200 yrs, C_s = 0.043, K value from Table 9.4 is:
$K = 2.60$

$\therefore$ Q_{200} = antilog (3.97802) = 9506.48 m^3/sec

Table 9.7 Conversion of discharge of Table 9.6 to log Q

Discharge Q (1)	*log Q* (2)	*Discharge Q* (1)	*log Q* (2)
2947	3.469	4366	3.640
3521	3.546	3386	3.528
2399	3.38	7826	3.893
4124	3.625	3320	3.521
3496	3.543	6599	3.819
2947	3.469	3700	3.568
5060	3.704	4175	3.620
4903	3.690	2988	3.478
3751	3.574	2709	3.432
4798	3.368	3873	3.588
4290	3.632	4593	3.662
4652	3.667	6761	3.830
5050	3.703	1971	3.294
6960	3.838		

EXAMPLE 9.7

From the long records of floods, the following values are calculated:

$$\log \bar{Q} = 3.555$$

$$\sigma \log Q = 0.158$$

$$C_s = 0.7$$

Calculate 200 years flood magnitude using log Pearson type III distribution.

Solution:

We know

$$\log Q_{200} = (\log \bar{Q}) + K(\sigma \log Q)$$

When $C_s = 0.7$ and $T_r = 200$ years,

$K = 3.223$ (From the Table 9.4)

$\therefore$ $\log Q_{200} = 3.555 + 3.223 \times (0.158) = 4.064234$

$\therefore$ $Q_{200} = \text{antilog } (4.064234) = 11594.01879 \text{ m}^3\text{/sec}$

9.13 LOG NORMAL DISTRIBUTION METHOD

This method is based on log normal probability law, and assumes that floods are so distributed that their natural logs are normally distributed. This is skew distribution of unlimited range in both directions. This distribution defines a straight line on log normal probability paper.

The flood of any return period which follows the log normal probability law is computed from:

$$Q_t = \bar{Q} + K\sigma \tag{9.45}$$

where K is a log normal frequency factor. A function of skewness coefficient C_s is given by

$$C_s = 3C_v + C_v^3 \tag{9.46}$$

where C_v is the coefficient of variation and

$$C_v = \frac{\sigma}{\bar{Q}} \tag{9.47}$$

Here σ is the standard deviation.

Steps:

(i) Compute $\bar{Q}$ and σ from given flood series.

(ii) Find $C_v = \dfrac{\sigma}{\bar{Q}}$.

(iii) Find $C_s = 3C_v + C_v^3$.

(iv) Determine K from normal probability table.

(v) Find the required flood $Q_t = \bar{Q} + K\sigma$.

9.14 CHOW METHOD

Chow[6] (1951) made some modification of Gumbel method using a frequency equation like:

$$Q_T = a + bX_T \tag{9.48}$$

$$X_T = \log\log\frac{T}{T-1} \tag{9.49}$$

when a and b are the parameters estimated by method of moments from the observed data of flood. By least squares method, the following equations are used:

$$\Sigma Q = a_n + bX_T \tag{9.50}$$

$$\Sigma(QX_T) = a\Sigma X_T + b\Sigma X_T^2 \tag{9.51}$$

The only two unknowns can be solved from Eqs. (9.50) and (9.51). Arranging the flood in descending order, Weibull (1939) plotting position equation is taken, i.e.,

$$T_r = \frac{n+1}{m}$$

Substituting T from Eq. (9.23) in (Eq. 9.49), we get

$$X_T = \log\log\frac{n+1}{n+1-m} \tag{9.52}$$

For known annual discharges of n years, return period of T_r can be computed from Eq. (9.23). Then X_T for particular flood of mth order is computed, ΣQ, ΣX_T, ΣX_T^2 are also determined. Then the values of a and b are solved. Flood of return period T_r (mth order) is thus computed.

9.15 POWELL METHOD

Powell made a modification of Gumbel equation (9.44) (when sample is not indefinite) as:

$$K = \frac{\sqrt{6}}{\pi}\left(\gamma + \ell n \ell n \frac{T}{T-1}\right) \tag{9.53}$$

where γ is Euler's constant (0.5772)

and $\qquad \ell n = \log_e$

Simplifying, $\qquad K = -1.1 - 1.795X_T$

Then $\qquad Q_T = \bar{Q} + K\sigma \qquad (9.54)$

9.16 PARTIAL SERIES ANALYSIS

Normally in the frequency analysis of flood of river, the highest annual flood of every year is considered as one data point of the year. It has been seen that in certain year, few more floods

[6] Chow, V.T., A General Formula for Hydrologic Frequency Analysis, *Trans. Am Geophysical Union*, vol. 12, pp. 231–237, April 1951.

are much higher than the highest floods of some annual floods. And therefore, peak discharge frequency analysis may be based on two types of data sets: an annual series where only peak flow of the year is recorded as data points and other is partial series in which peak floods are included above a base flood magnitude where multiple higher floods in the same year above that base flood are included in the series. The base flood magnitude for partial duration series could be the channel capacity, level at which significant damages may occur. Such data series is called **partial duration series**.

Annual series analysis is generally adopted for determining events of larger recurrence events, i.e., more than 20 years, or so. On the other hand, partial duration series analysis is generally adopted in estimating events of frequency occurrence, having return period of less than 5 years or so, as may be the ease in the design of urban drains, coffer dams for river diversion etc.

Langbein[7] (1949) related the recurrence interval of annual series T_A and recurrence interval T_p of partial series by the following equation:

$$T_P = \frac{1}{\log_e A - \log_e(T_A - 1)} \tag{9.55}$$

Langbein equation shows that the difference between T_A and T_P is significant for $T_A < 10$ years and for $T_A > 10$ years, the difference is negligibly small. It is calculated and shown in Table 9.8.

Table 9.8 Return Period of Partial and Annual Series

T_A	1.16	1.58	2.00	2.54	5.52	10.5	50.5	100.5
T_p	0.5	1	1.45	2.0	5.0	10	50	100

i.e. return of period $T_A \cong T_P$ when $T_r \geq 10$ and $T_A \neq T_P$ When $T_r < 10$ Linsley et al (1958) are in the opinion that the design problem where return period is less than 5 years, partial series is preferable. For higher return period, annual series should be considered.

In partial series, since several occurrences in one year may be selected, the probability P_a is introduced as:

$$P_a = \frac{m}{y}$$

where m is the number of occurrences in y years and probability P in percentage of time is given by

$$P = (1 - e^{-Pa}) \times 100 \tag{9.56}$$

Recurrence interval $T_r = \dfrac{1}{P_a}$ $\left(\text{not } \dfrac{1}{P}\right)$ Example 9.6 is on annual series distribution.

[7] Langbein, W.H., Annual Floods and Partial Duration Series, *Trans Am. Geo Phys. Union.*, vol. 30, pp. 879–881, December 1949.

9.17 STOCHASTIC METHOD

Long back in 1713, J. Bernouilli[8] stated: "The science of conjecture or the stochastic science, is defined as the art of estimating as best one can, the probability of things so that in our judging and acting we may choose to follow the best, the safest, the surest, or most soul-searching way".

In statistics, the stochastic is synonymous with random, but in hydrology it has been used in a special way to refer to a time series which is partially random. The stochastic hydrology fills the gap between deterministic method and probability hydrology. Stochastic methods were first introduced into hydrology to cope with the problem of reservoir design. The basic approach to stochastic generation applies to stream flow, evaporation, precipitation and other hydrologic factors. Works on stochastic method of flood frequency have been done by Yevjevich (1963), Chow[9] (1964), Sharma (1975) and others.

One of the well known equations, based on annual flood data using Poisson probability law and theory of sums of random number of random variables, is:

$$Q_T = Q_{\min} + 2.3(\bar{Q} - Q_{\min}) \log\left(\frac{n_f}{n} T\right) \tag{9.57}$$

Here $T = \dfrac{n}{m}$, n_f is the number of recorded floods, counting only one for same flood peak occurring in different years.

9.18 REGIONAL FLOOD–FREQUENCY METHOD

This frequency method is applied when the available data to make frequency analysis is not adequate in the catchment under consideration. With such few years' record, frequency studies described earlier are of little use. The procedure designed to overcome this problem is the regional flood frequency analysis according to Darlymple[10] (1950). In this method, hydrologically homogenous region from the statistical point of view is considered. Available long term data of neighbouring catchments are tested for homogeneity and a group of stations which satisfy this homogeneity is selected. This group of stations constitutes a region. All the data of this region are analysed as a group to find the frequency characteristics of the region. To do this analysis, flood data selected of the region are made non-dimensional dividing by mean flood $\bar{Q}$ which corresponds to return period of 2.33 years.

From Eq. (9.35),

$$Q_T = \bar{Q} + (0.78y - 0.45)\sigma$$

When $$Q_T = \bar{Q},$$

$$(0.78y - 0.45)\sigma = 0$$

or $$0.78y = 0.45$$

$\therefore$ $$y = 0.577$$

[8] Bernouilli, J., Ars Conjectandi, p. 213, Basel, 1713.

[9] Chow, V.T. (Ed.), *A Handbook of Applied Hydrology*, McGraw-Hill, New York, 1964.

[10] Dalrymple, T., Regional Flood Frequency, High Res. Board Res. Rep. 11-B, pp. 4–20, 1950.

Now, $$T_{\text{mean}} = \frac{1}{P} = \frac{1}{1 - e^{-e^{-0.577}}}$$

$\therefore$ $$T_{\text{mean}} = 2.33 \text{ yrs}$$

The ratio $\left(\frac{Q}{\overline{Q}}\right)$ is plotted against corresponding return period T_r as shown in Figure 9.4 which is called **regional frequency curve**. From this curve, flood of ungauged catchment may be predicted.

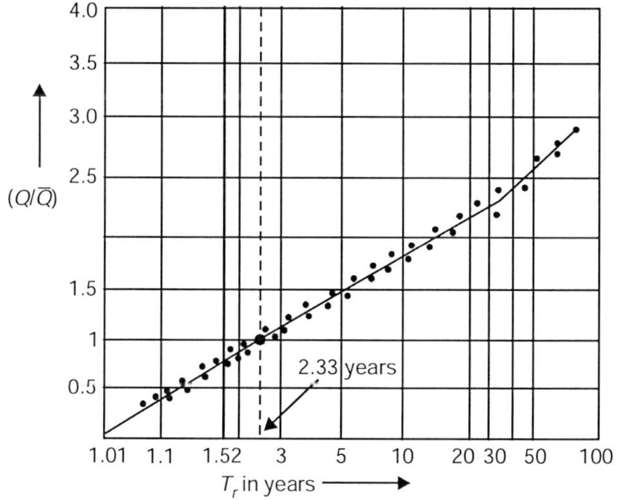

Figure 9.4 Regional frequency curve.

But this frequency analysis has the weakness in the assumptions that all streams in the region maintain the same variance $\left(\frac{Q}{\overline{Q}}\right)$. Linsley and Paulhus (1958) are of the opinion that, some streams may have same $\left(\frac{Q}{\overline{Q}}\right)$, but all streams will never prove that $\left(\frac{Q}{\overline{Q}}\right)$ are identical. Therefore, using this regional frequency approach, care must be taken to select streams carefully as nearly similar in hydrologic characteristics as possible. They should have similar vegetal cover, land use, topographic conditions, geologic characteristics, and rainfall and evapotranspiration regimes.

9.19 CONCLUSION

Investigators are of the opinion that methods other than frequency analysis are approximate. They are suitable for quick and approximate design with a margin of few percentages on higher side for safety. Flood frequency analysis specially the Gumbel extreme value analysis is supposed to be the best and is popularly used by the design engineers. Frequency analysis also has limitations if series of flood data is not long enough.

The flood of river becomes abnormally high causing lot of damages. Therefore, in Chapter 10, attention is given for flood control and hazard mitigation measures.

EXERCISES

9.1 The maximum recorded flood in the river Dhansiri, a tributary of the river Brahmaputra for the period from 1979 to 2005 is given as follows:

Year	*Flow* (m^3/sec)	*Year*	*Flow* (m^3/sec)	*Year*	*Flow* (m^3/sec)
1979	2949	1988	4798	1997	6599
1980	3521	1989	4290	1998	3700
1981	2399	1990	4651	1999	4175
1982	4124	1991	5050	2000	2988
1983	3496	1992	6960	2001	2709
1984	2947	1993	4366	2002	3873
1985	5060	1994	3380	2003	4593
1986	4903	1995	7826	2004	6761
1987	3751	1996	3320	2005	1971

Estimate flood discharge with recurrence interval of (i) 100 years (ii) 150 years by Gumbel method. Assume the sample to be large.

(***Ans***: Q_{100} = 8758.7 m^3/sec Q_{150} = 9213.6 m^3/sec)

9.2 The flood frequency computation of the river Chambal at Gandhi Sagar dam by using Gumbel's method yielded the following results:

Returns (yrs)	*Peak flow* (m^3/sec)
50	40809
51	46300

Estimate the flood magnitude in this river with a return period of 500 years.

(***Ans***: 58991.114 m^3/sec)

9.3 The annual flood peak in m^3/sec of the river Brahmaputra at Guwahati (Pandu station) from the year 1955 to 1977 are respectively 52010, 52454, 58352, 61977, 53448, 58606, 36193, 40480, 49386, 35009, 47734, 51100, 49476, 59470, 45542, 53389, 51744, 72748, 47442, 51338, 32950, 58188 and 45620.

Estimate the flood of 100 years return period by Gumbel method if the required frequency factor is 3.91.

9.4 The observed annual floods peaks for a river for the period 1969 to 2003 in m^3/sec are: 588, 432, 396, 420, 500, 336, 900, 300, 600, 504, 396, 578, 420, 528, 384, 698, 610, 408, 372, 408, 672, 720, 912, 624, 336, 324, 360, 696, 456, 636, 684, 756, 527, 312

Estimate (i) 50 yrs (ii) 100 yrs flood using (a) Gumbel (b) Log Pearson III type.

(***Ans***: (a) 1013.7 m^3/sec and 1115 m^3/sec, (b) 960 m^3/sec and 1050 m^3/sec)

9.5 The observed annual flood peaks in m^3/sec of a stream for the period 40 years from 1961 to 2000 are: 395, 619, 766, 422, 282, 990, 705, 528, 520, 436, 697, 624, 496, 589, 598, 359, 686, 726, 527, 310, 408, 721, 814, 459, 440, 632, 343, 634, 464, 373, 289, 371, 522, 342, 446, 366, 699, 560, 450, 610.

(i) Construct probability plot for annual flood on ordinary or semi log graph paper. From graph determine flood magnitude of 100 years return period.

(***Ans***: Q_{100} = 1060 m^3/sec)

(ii) Determine by Gumbel method of 100 yrs flood.

(***Ans***: $Q_{100} \cong$ 1038 m^3/sec)

(iii) Solve the Q_{100} flood by writing a computer program by both Gumbel and Log Pearson.

SUGGESTED FURTHER READINGS

Allen, T.G. and Cassidy, J., *Hydrology for Engineers and Planners*, Iowa State University Press, 1975.

Benson, M.A., Characteristics of Frequency Curves Based on a Theoretical 1000 Years Record, U.S. Geolog. Surv. Paper 1543-A, pp. 51–74, 1952.

———, Uniform Flood Frequency Estimating Methods for Federal Agencies, *Water Resource Res.,* vol. 4, pp. 891–898, October 1968.

Bobee, B., The Log-Pearson Type III Distribution and its Applications in Hydrology, *Water Resource Res.*, vol. 10, No. 3, 1980.

Dennis, P.L. and Berges, S.J., Gumbel's Extreme Value I Distribution: A New Look, *Jour. Hy. Div. Proc.*, ASCE, vol. 108, No. HY4, April 1982.

Eagleson, P.S., *Dynamic Hydrology,* McGraw-Hill, New York, 1970.

Fisher, R.H. and Tippett, L.H.C., Limiting Forms of Frequency Distribution of the Smallest and Largest Member of a Sample, *Proc. Camb. Philo. Soc.*, vol. 24, pp. 180–190, 1928.

Foster, H.A., Theoretical Frequency Curves and Their Application to Engineering Problems, Trans. ASCE, vol. 87, pp. 142–173, 1924.

Gringorten, I.I., A Plotting Rule for Extreme Probability Paper, *J. Geo Phys. Res.*, vol. 68, 1963.

Gumbel, E.J., On the Plotting of Flood Discharge, Trans. *Am. Geo Phys.,* Union, Pt. II, pp. 699–716, 1943.

Gumbel, E.J., *Statistics of Extremes*, Columbia University Press, New York, 1958.

Hazen, A., *Flood Flows,* Wiley, New York, 1930.

Kazmann, R.G., *Modern Hydrology*, 2nd ed., Harper and Row Publishers, New York, 1972.

Kimball, B.F., On Choice of Plotting Position on Probability Paper, *Jour. Am. Statistical Assoc.*, 55, 291, pp. 546–560, 1960.

Kulandaiswamy, V.C., Lecture Notes on Hydrology, ISTE Summer School, Guindy, 1971.

Langbein, W.B., Annual Floods and Partial Duration Series, *Trans Am. Geo Phys. Union*, vol. 30, pp. 879–881, December 1949.

Linsley, R.K., Kohler, M.A. and Paulhus, J.L.H., *Hydrology for Engineers*, McGraw-Hill, Auckland, 1982.

Paulhus, J.L.H. and Miller, J.F., Flood Frequencies Derived from Rainfall Data, *J. Hy. Div.*, ASCE, vol. 83, December 1957.

Pearson, K., On the Systematic Filling of Curves to Observations and Measurement, *Biometrika*, vol. 1, No. 3, pp. 285–303, 1902.

Rao, A.R. and Hamed, K.H., *Flood Frequency Analysis,* CRC Press, Boca Raton, FL, 2000.

Raudkivi, A.D., *Hydrology,* Pergamon Press, Oxford, 1979.

Sing, K.P. and Sinclair, R.A., Two Distribution Methods for Flood Frequency Analysis, *Jour. Hydro. Div. Proc.*, ASCE, vol. 108, No. HY1, January 1972.

Sing, V.P., *Elementary Hydrology*, Prentice Hall Inc., Upper Saddle River, N.J. 1992.

Stall, J.B. and Neil, J.C., Partial Duration Series for Low Flow Analysis, *J. Geo Phys Res.*, vol. 66, pp. 4219–4225, December 1961.

Stedinger, J.R., Fitting of Log-normal Distribution to Hydrological Data, *Water Resource Res.*, vol. 10, No. 3, 1980.

U.S. Army Corps of Engineers, Hydrologic Frequency Analysis, EM 1110-2-1415, Washington, DC, 1993.

Viessman, W. and Lewis, G.L., *Introduction to Hydrology*, 4th ed., Harper Collins, New York, 1996.

Yevjevick, V., *Probability and Statistics in Hydrology*, WRP, Fort Collins, Colorado, 1972.

Chapter 10

Flood Disaster Mitigation Measures and Damage Estimation

10.1 INTRODUCTION

Disaster around the world occurrs due to flood, wind and storms, land slide, waves and surges, droughts, feminine, earth quake and water related epidemic. Recently Kukami et al[1] (2006) have made an extensive survey on water related (flood) disasters that have taken place since 1960 to 2005. The findings of their survey reveal that water related disaster, i.e., flood disaster is the major cause of natural disaster in the world. The trend of flood disaster is increasing in time and continent wise. The flood disaster of Asia accounts for one third of the world's total water related disasters. When this flood disaster is considered country or region wise, northeastern part of India, where one of the world's biggest Himalayan Rivers the Brahmputra and its tributaries are flowing, is the worst sufferer. This increasing trend for Asia has some relationship with the facts that many regions particularly the northeastern region of India is under the highest influence of monsoon with annual rainfall intensity up to 1450 cm, cyclones/typhoons, plenty of sediments yields from the rivers, erosion, etc. Similarly, flood creates lot of devastation in other parts of the country. Therefore, flood disaster mitigation measures are considered as the topmost disaster management along its damage estimation.

10.2 CAUSES OF FLOOD

Flood is the abnormal high stage of flow that overtops the banks of river in any reach to cause immense damage houses, human and wild lives. It is a natural event that results from excessive rainfall due to severe combination of critical hydrometeorological conditions over the catchment.

In addition to this flood occurring commonly by overtopping banks during high rainfall, another flood of almost infinite magnitude occurs instantly when a very high dam with a large reservoir fails suddenly due to piping, erosion, foundation failure, earth quake, overtopping due

[1] Kukami, K., Inomota, H., Hapurrachichi, P. and Oki, R., Flood Forecasting and Warning System in Asia, Proc. Int'l Symposium on Managing Water Supply and Group Demand, Bangkok, Thailand, 16–20 October, 2006.

to sudden high flow to reservoir or sudden land slide in the reservoir. All over the world such flood, flash flood occurs instantly when the river embankment fails during very high stage of flow. This is called **embankment breaching** or **dike failure flood**. This flash flood (dam-break flood or dike failure flood) starts suddenly with tremendous velocity and depth, the disaster mitigation of which becomes virtually impossible. There are quite a lot of examples of this flash dam-break and dike failure flood all over the world, and devastation of this type of flood is considerably more.

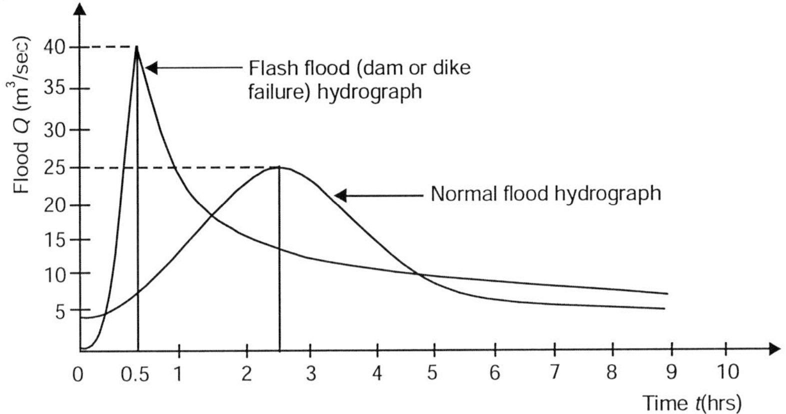

Figure 10.1 Normal flood and flash flood hydrographs.

In view of such flood disaster, mitigation measures are necessary although complete mitigation of this disaster is neither possible physically nor economically. Yet the following existing measures can mitigate the flood disaster to a major extent.

10.3 MITIGATION MEASURES

There are two types of measures to mitigate the flood disaster:

1. Structural mitigation measures
2. Non-structural mitigation measures

10.3.1 Structural Mitigation Measures

They are of following types:

Storage reservoirs: It is the most effective measures of flood disaster mitigation measure. The modern reservoirs are mostly multipurpose. In India, the use of reservoir for flood control, hydropower, irrigation had been started after 1951 when National Flood Control Policy had been adopted. The aim of reservoir is to store excess water during flood period and release it when flood subsides. Figure 10.2 shows the mitigation or attenuation of peak of the flood hydrograph by constructing a storage reservoir.

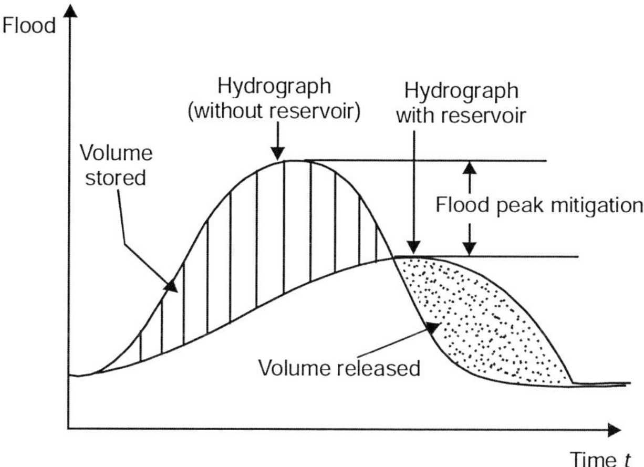

Figure 10.2 Hydrographs of flood without and with reservoir.

Some notable multipurpose Reservoirs in India are Bhakra-Nangal and Thien in Punjab, Rihand, Ramaganga, Yamuna I and II, Tehri in Uttar Pradesh, Ukai in Gujarat, Koyana I and II, Ujjani in Maharashtra, Rana Pratap Sagar, Beas, Sutlej I and II, Mahi Bajaj Sagar in Rajasthan, Kundah stage I and II in Tamilnadu, Hirakud, Balimela in Orissa, Srisailam, Nagarjun Sagar, Upper Seleru in Andhra Pradesh, Gandhi Sagar in Madhya Pradesh, Sharavati stage I and II, Kalindi, Tungabhadra in Karnataka, Sabarogiri and Idikki in Kerala and Salal in Jammu and Kasmir. An excellent example of flood disaster mitigation is the Tennessee Valley Authority (TVA) of the United States, which was established in 1933 during the then President F.D. Roosevelt with an initial aim of flood disaster mitigation project in one of the most undeveloped and flood affected region of the country comprising Virginia, North Carolina, Georgia, Tennessee, Kentucky, Alabama and Mississippi. Starting with flood hazard mitigation project, TVA has completed 36 hydroelectric, 22 multipurpose and 14 number single purpose reservoirs. This TVA has completely changed the quality of life in those states.

Confining river flow by embankments: Earthen embankments have been the principal methods of controlling flood as a short-term measure. In India, it has been adopted after the disastrous and devastating flood in 1954. The choice of embankment has been guided due to the following with reference to Figure 10(a) and 10(b).

1. They are easy to construct parallel to riverbanks by available earth.
2. They prevent overflowing the banks.
3. Initial cost of construction is less.
4. Maintenance is easy.
5. It increases the storage capacity of river as shown in Figure 10.3(b). These embankments have the following drawbacks and limitations:

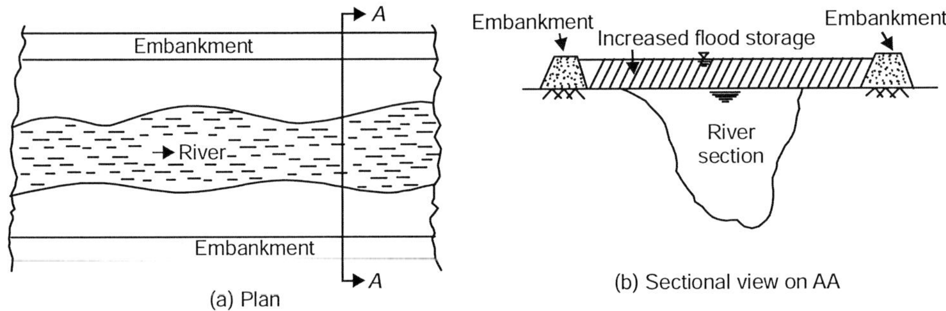

Figure 10.3 Earthen embankment for increased flood storage.

(i) Instead of flood disaster mitigation, sometimes it suddenly creates catastrophic flood exactly like dam-break flood when embankment fails during high stage. These situations have been occurring quite frequently around the world. In the north-eastern part of India in the valley of the Brahmaputra, it is a regular annual phenomenon. A survey for 5 years (1990–1995) conducted by Sarma (1999) reveals that average 70 failures of dike per year occur in this again.

(ii) During flood time, round the clock maintenance is very much essential against failure.

(iii) Due to increase of river capacity, velocity decreases, hence sediments flowing along with flood water are deposited in the bed thereby depth of riverbed goes on decreasing with time.

Unless the height of the embankment and maintenance are not proper, embankment may create flood disaster.

Channel improvement works: Work performed to increase the discharge or velocity of stream or to decrease the stage and duration of flood is called **channel improvement works.**

It includes

1. Widening and depending of the stream.
2. Straightening the meandering type river by cut off (shown in Figure 10.4) to reduce travel time of flood.

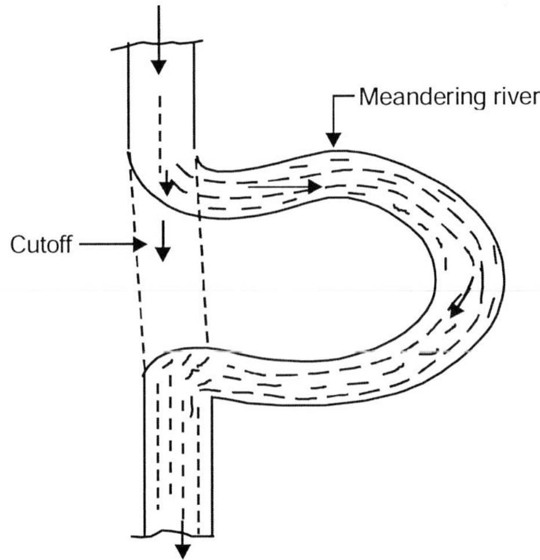

Figure 10.4 Cutoff decreases the river length.

3. Smoothing the river bed and sides, and removing roughness offered by sand bed, weed growth, etc.

Dredgers are used for channel improvement works. But the big river like the Brahmaputra, dredging is not fruitful and possible as it has plenty of sediment. Elimination of meanders has been done in the river Mississippi in the USA.

Diversion works: It includes the diversion works from places where flood disasters are likely to occur. A diversion channel with a regulator upstream of the important area is constructed (Figure 10.5). The important area is placed on convex side of the river where inundation and erosion are possible. The diversion channel decreases the stage of flood near the important area, and thereby possible flood disaster may be reduced.

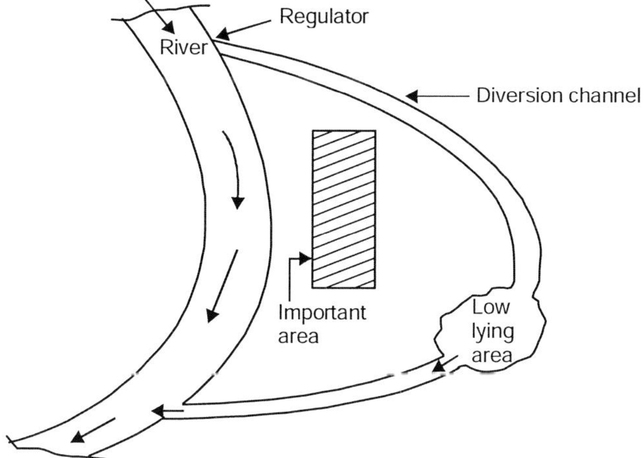

Figure 10.5 Diversion works by a channel for flood mitigation in an important area.

Flood wall: Flood wall (Figure 10.6) is constructed as local flood hazard prevention for some important area situated at low ground level. Walls are generally RCC or masonry retaining wall conducted almost parallel to the river. They are designed against water pressure, factor of safety against sliding with sufficient free board (FB). The ends of the flood walls end in a high ground level otherwise it has to construct on all sides of the important town or town area.

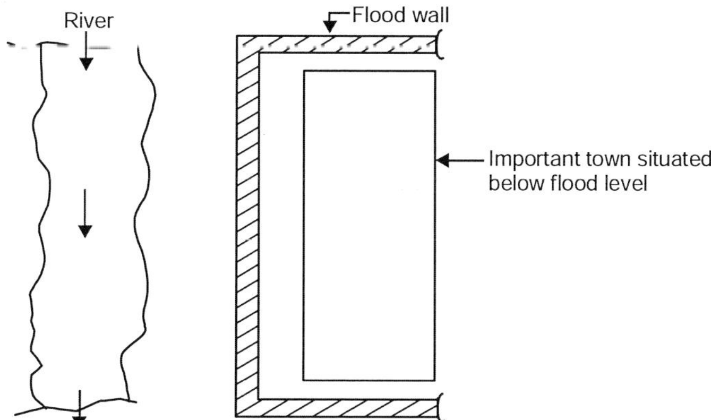

Figure 10.6 Flood wall to save a town from flood disaster.

Construction of high earthen platforms: Earthen platforms higher than high flood level (HFL) may be constructed in vulnerable flood prone areas to save human and wild life. Most of the national parks are situated near the river bank and flood normally occurs in such parks where precious wild life exists. Besides, some of villages are normally below the HFL. For the safety of those precious animals and human life, such earthen platforms (Figure 10.7) are the real need against the flood disaster.

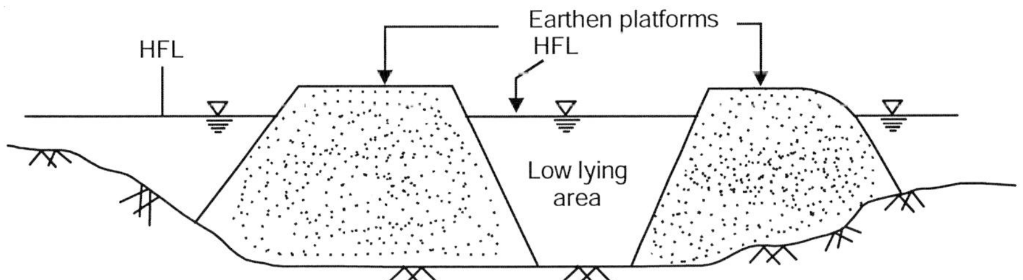

Figure 10.7 Earthen platform higher than HFL for temporary flood hazard mitigation.

Sluices: Sluices are provided and located in suitable places of embankments, flood walls and small rivers. Depending upon the upstream and downstream water levels, the sluice gates are operated (Figure 10.8). If water level downstream of the sluice gate is more, gate is closed [Figure 10.8(a)] and water from upstream is discharged to the river by pumping. If the level downstream is low, gate is opened for water to be discharged to the mainstream.

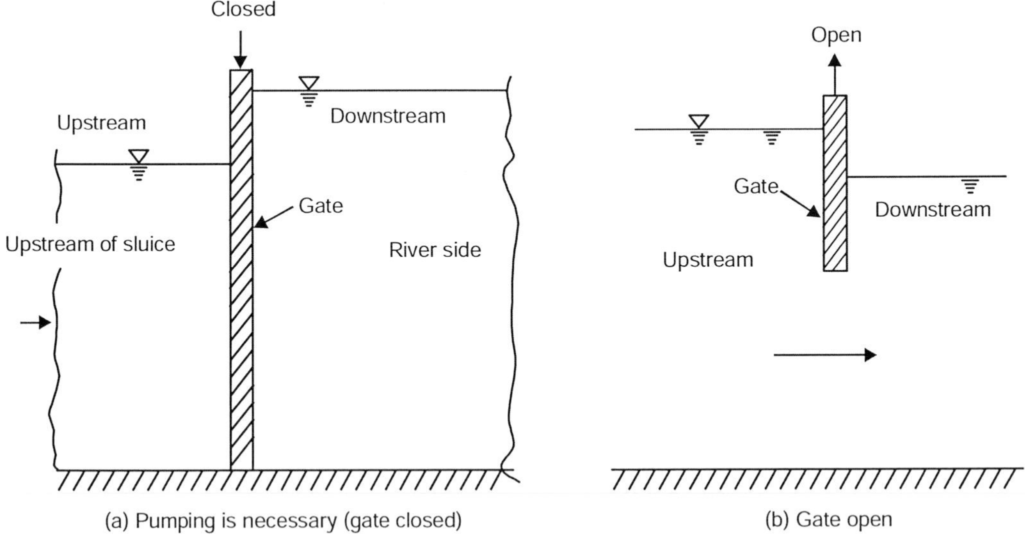

Figure 10.8 Sluice gate operation per flood mitigation.

Runoff reduction by watershed management: It is an indirect method applied to the watershed which has long-term effect on flood disaster mitigation. Watershed management is a primary part of soil conservation, i.e., soil is conserved from erosion and prevented and arrested to flow along with runoff. Erosion control and arresting the sediment to flow with water finally help in:

1. Mitigating flood in the watershed.
2. Prolonging the life of reservoir and more flood water can be stored.
3. Improving fertility of soil.
4. Travel time of flood becomes more and magnitude of peak flood becomes less, i.e., when soil erosion is prevented, vegetal cover increases, and it offers more resistance to runoff. Same volume of flood from the watershed takes more time to drain out from the watershed. Thus peak flood magnitude decreases, thereby chance of overtopping or like failure is much less. Different watershed management are: (a) Afforestation (b) Contour farming (c) Contour bunds or trenches (d) Check dam (e) Gullying (f) Spurs (g) Toe walls (h) Bank protection (i) Diversion drains (j) Jeo jute (k) Crib structures (l) Bench terracing (m) Strip cropping.

10.3.2 Non-structural Mitigation Measures

They are of the following types:

Flood plain zoning: Areas near the river are the most vulnerable for the flood hazard provided the areas are not a highland. Therefore, people should not be allowed to those flood prone areas for dwelling houses. These areas may be used as parks, recreation ground, etc. so that inundation of such areas may result in saving life and properties. Places below HFL should not be recommended for inhabitation. Thus, zoning of flood prone areas should be declared by the Government.

Flood forecasting: Flood forecasting or warning if could be conveyed earlier, disaster due to the probable flood can be minimized to a great extent. Temporary evacuation of persons and shifting the important property to safer places could be done before the flood arrives. Once the flood occurs, the normal activities of the society are badly disrupted with immense losses. Therefore, flood forecasting is a real necessity to minimize the flood havoc. In India, seven flood forecasting units have been operating at New Delhi, Guwahati, Jalpaiguri, Patna, Bhubaneswar, Surat and Lucknow. There are another 22 substations, and they have been functioning to give timely warning of floods in some major rivers. For forecasting the flood, historical data of the river like stream flow record, stage discharge relationship, rainfall runoff relationship of river are essential. Current information like hydrological conditions of the basin including amount and aerial distribution of rainfall and its duration, snow conditions, i.e., contribution of melting snow at the time of flood and the stage of river at some key points must be known. A communication network is essential between the control room with gauge stations, i.e., records of rainfall and stage. Sometimes telephonic communicating lines fail during heavy storms, so wireless network becomes essential.

Forecasting of flood magnitude is evaluated by the following methods:

1. Rainfall runoff relationship
2. Unit hydrograph (UH)

3. Synthetic UH (SUH)
4. Flood routing model both hydrological and hydraulic
5. Stage discharge relationship

Some of the techniques mentioned above have some limitations. For example, rainfall-runoff relationship is very much dependent on storm frequency, initial soil moisture conditions and storm durations.

Similarly UH Method may produce error due to its limitations of assumption of uniform rain distribution throughout the whole catchment.

Flood routing technique is scientific. It is taken to be best to predict flood stage or flood hydrograph at some downstream position.

Whatever may be the technique, flood forecasting is a better tool in the mitigation of flood hazard provided forecast can be made at least a day ahead to take necessary precautions.

Flood proofing: It essentially consists of a combination of structural change and emergency action. Structural change includes construction of the building wall with some water proofing material, closure of lower level windows, construction of the house with the arrangement of stilts, i.e., an arrangement with bracket to raise the house when required. In some indigenous process, dwelling houses in flood areas, floor of the house is constructed at some upper level, even moderate flood is not able to touch the floor of the houses. During flood period people arrange their shelter in their own house. Impact of water pressure below the floor level is almost negligible as only the columns of the house are exposed. Building should be constructed such that ground floor should be reserved for vehicles and other materials which can be moved out and the high value materials are on the floor above the flood level.

Weather modification: The scientific process of alteration of precipitation pattern to modify the flood has already been discussed in Chapter 2. Yet a short description of it as a flood disaster measure has been presented. A number of research works for determining the possibility of altering the weather in reliable and consistent manner are underway in various parts of the world. Attempts are also on to redistribute the precipitation both in respect of time and space and thereby reduction of flood hazard in a given area may be possible. As such, weather modification is more a concept than a reliable means of altering stream pattern. Efforts of producing rainfall by cloud seeding i.e., by injecting chemicals to cloud to force early rainfall at desired place have been in experimental stage. This method has not yet established whether overall gains would exceed the overall costs. Further redirecting the storm and inducing cloud to precipitate may reduce damage to property at certain areas, but it may aggravate the flood disaster in some other areas. It may also deprive certain areas from required rainfall for agriculture, industries, domestic water supply, etc. Weather modifications may need international regulations as its effort could extend far beyond the area where the method of modification is applied. This is still in experimental and research stage.

Mathematical modelling: Mathematical model has become a very powerful tool in this age by the use of modern high-speed computer in indifferent fields of sciences and technology. In the flood disaster mitigation measures, mathematical model by computer can predict flood intensity and area of inundation whether it is flood in river or a flash flood due to dam-break or dike failure.

Various investigators developed mathematical model in hydrologic and hydraulic flood routing in river to predict the maximum flood downstream from the known hydrograph at upstream. When high dam-reservoir project is undertaken, dam-break flood analysis and damage estimation are usually done before completion or starting of the project. This is normally done by developing model both in one-dimensional and two-dimensional flows. This model predicts the flood inundation if the dam suddenly fails. Accordingly downstream of the dam is rehabilitated before completion of project so that the dam-break flood disaster could be minimised if the dam fails in future. Again when natural flood upstream of a river is known, mathematical model of flood routing can predict the flood intensity at some downstream point of the river. Thus, mathematical modelling technique can alleviate the flood disaster in the river valley in case of both natural and dam-break flood situation.

10.4 FLOOD DAMAGE ESTIMATION

Damages in the following forms occur due to flood disaster:

1. Damage to crops
2. Damage to houses
3. Damage to human lives and live-stock
4. Damage to public utilities, roadways, rail road, etc.
5. Cost of relief measures

Important of the assessment of the damages is aimed to find a realistic assessment of benefit-cost ratio of the flood control scheme. The assessment of wealth of any flood control schemes involves the study of applied economics in the sense that annual savings due to this scheme must be evaluated against cost of construction and maintenance of the scheme. Condition of economic analysis requires that the marginal benefits of producing output must be equal to the marginal cost of producing it.

$$\frac{\text{Marginal benefit}}{\text{Marginal cost}} = \frac{M_B}{M_c} = 1$$

10.5 TYPES OF FLOOD DAMAGES

They are of two types:

Direct damage: Direct damage results due to physical contact of floodwater, for example, crop damage, private property, loss of life and cattle.

Damage to crops constitutes more than 60% of the total flood damages in India and hence greater care is needed for its assessment. Direct assessment of crop damage just after the flood becomes overestimated if replanting and resowing another crop after the flood is adopted. Replanting another crop after flood is quite common if flood damage takes place early in the season. It is likely that preliminary estimate is prepared quickly for the purpose of relief administration. Output of replanting is known only after sometime. Therefore, preliminary estimate becomes an overestimation. Thus, the correct estimate of flood damage of crops is the preliminary estimate prepared just after the flood less the return obtained from replanting after the flood.

Loss of human lives in flood may be due to capsizing boats, house collapse, etc. Loss of human lives is assessed only in number not in monetary values. But in case of loss of cattle, monetary value of such losses should be estimated.

Damages to Government properties such as building, public utilities, roadways bridges, canals, embankments and telephone and electric lines bear a significant part of flood damages. These damages are assessed by respective Government on the basis of estimated cost of repairs. State Government do not assess the damages done to Central Government property such as railways, post and telegraphs, defense installations, major central industries, etc. True assessment of public properties damages are obtained only when all the estimates of State and Central Government damages are combined in the form of total estimated damage.

Indirect damage: The flood damages discussed so far belong to direct damage arising out of direct contact of flood water. Indirect damage results to property or services which are not touched by flood water, but it is a loss or damage as a result of interrupted trade or diversion of rail or roadway traffic or other effect. Thus, indirect damage may be caused by cessation of normal economic activities due to flood even though the persons or assets affected may not come in contact of flood. For example, factories, shops and business centre in flood affected areas may be closed, resulting in the loss of their owners, stoppage of industrial production and temporary unemployment of workers.

Rail and road communication in affected areas may be interrupted resulting in reduced earning of railways and transport operations.

Daily wage earners and small peddler working in the affected areas may suffer loss in their daily wages or earnings.

Such losses may occur not only in the flooded areas but also outside due to forward and backward linkage. For a factory located elsewhere but depending on raw materials from a flooded area may suffer loss on cessation of supplies.

Similarly firms selling their output to flood affected areas may suffer loss.

Thus, the monetary loss of a farm or an industry which does not function due to flood can be considered to have suffered indirect flood damage, i.e., the damage caused due to interruption in normal course of everyday affairs. This indirect damage, although important, is difficult for objective estimation or assessment.

10.6 METHODS OF DATA COLLECTION FOR ASSESSMENT

Two standard methods are used for collection of data for flood damage estimation:

1. Complete enumeration
2. Sample surveys

A methodology based on sample survey has been recommended by National Council of Applied Economic Research (NCAER). In this method the states are divided into zones on the basis of their exposure of flood risks. The villages for the study would be randomly selected. The numbers of villages are large to obtain good estimate. Quite a lot of field works like preflood enquiry, inter flood enquiry, post flood enquiry are required. The sample survey technique requires expert statistical guidance.

National Commission of Flood Report Department of Irrigation, March 1980, recommended the methodology through complete enumeration. In this method, remote or aerial survey, record of damage of crops, and past records of yield of crops are required.

10.7 FLOOD DAMAGE SUSCEPTIBILITY REDUCTION BY EMERGENCY MEASURES BASED ON FLOOD WARNING

Damage or loss due to flood may be reduced by emergency measures when flood forecasting or warning has been made in appropriate time. These emergency measures can reduce the damage of different properties to a major extent. The classes of loss, degree to which loss may be reduced and types of emergency measure adopted is tabulated in Table 10.1. For example, in an agricultural land in row 1 of this table, if emergency action like early and rapid harvesting could be done, loss may be small. Even the kharif (summer) crops have been damaged, measures like massive rabi (winter) crops can compensate the loss and thus average loss becomes mostly small. In row 2, emergency action like early removal of stored crops, livestock, automobiles, furnishing, personal belongings to higher safe level like first floor of building or to some safer high lands, can reduce the loss to minimum. If such facilities are not available, loss may be high. Thus, losses may be reduced to medium if protection and rescheduling measures are taken.

Table 10.1 Flood Damage Reduction by Emergency Measures

Sl no.	*Classes of Loss*	*Degree to which loss may be reduced by emergency measures*	*Emergency measures*
1.	Agricultural	Mostly small.	Rescheduling by early or rapid harvesting
2.	Stored crops, livestock, residence, furnishings, personal belongings, automobiles, etc.	Some are large, some are small	By early removal
3.	Urban residential foundation, superstructures decorations, garage.	Mostly small	Protection and removal
4.	Communications like roadways, railroad, etc.	Mostly medium and small	Protection
5.	Bridges	Medium	Protection
6.	Relief expenditures	Medium	Rescheduling
7.	Public health, i.e., sickness, injury, etc.	Medium	Rescheduling

10.8 FORMAT OF FLOOD DAMAGE ESTIMATION

The format of flood damage estimation is represented in Tables 10.2 and 10.3. Table 10.2 shows the nature of flood damage estimation. Different damages can be estimated from known area affected, frequency of flood in a year, probable depth and duration of flood. If damages to crops are R_1 crores, R_2 crores to houses, R_3 crores to cattle, R_4 crores to public utilities, then total damages (in crores) is the sum of these four losses which is equal to R_5 crores.

Table 10.2 Nature of Flood Damage Estimation

Year	*Area affected in* h_{as}	*Frequ- ency of flood tion* (m)	*Probable depth of inunda- tion on* (days)	*Dura- tion of inunda- inunda- tion* (h_0)	*Damage to crops*		*Da mage to hours*			*Cattlelost*		*Damage public utilities of* Rs.	*Total damage in* crs.
					Area of of crops (crs. of Rs)	*Values*	*No in* crs	*Values lives*	*Human*	*No in* crs.	*Valuc*		
2005	Z	P	Y	D	K	R_1	N_1	R_2	M_1	N_3	R_5	R_4	$(R_1+R_2+R_3+R_4) = R_5$

Table 10.3 shows the evaluation of cost of relief and total cost R crores. If R_6, R_7 and R_8 are the relief costs due to gratuities, agricultural and land revenues respectively, then total relief cost is equal to the summation of these three, i.e., R_9 and the total annual cost of flood is $(R_5 + R_9) = R$ crores.

Table 10.3 Cost Relief Measures

Gratuitous relief (in crs)	*Agricultural & other loan* (crs of Rs)	*Remission of land revenues* (in crs of Rs)	*Total relief expenditure in* (in crs of Rs)	*Total cost of damage relief* (in crs)
R_6	R_7	R_8	$(R_6 + R_7 + R_8) = R_9$	$(R_5 + R_9) = R$

10.9 FLOOD DAMAGE ESTIMATION MODEL

The evaluation of socioeconomic impacts of floods can be done using different methodological procedures. The unit loss model is based on property by property assessment of potential damage. Some countries like UK and Australia have established detailed methodologies for estimation of tangible losses. In Australia and many other countries, damage estimation methodologies vary in different region within the country according to individual studies. Dutta et al[2] (2003) opined that the relationships between flood parameters and possible damage are derived on historical flood damage information, questionnaire survey based on laboratory experiences. Penning-Rowsell and Chitterton[3] (1977) gave detailed works for the residential and commercials services, industrial and agricultural sectors based on priority basis of damages for England and Wales. Another more rccent cxamplc carricd out for Japan is the work of Dutta,

[2] Dutta, D., Herath, S. and Musiake, K., A Mathematical Model for Flood Loss Estimation, *Jour. of Hydrology*, 277, pp. 24–49, 2003.

[3] Penning Rowsell, E.C. and Chitterton, J.B., The Benefits of Flood Alleviation, Saxon House, Teak field, West mead, U.K., 1977.

Herath and Musia Ke (2003). A mathematical model for the flood damages of Assam has been developed by Das Saikia[4] (2006) from the data of flood damages of Assam, India from 1953 to 2005. Three main components of flood damages, i.e., area inundated (A_{inun}) in hrs, crop area affected (A_{crop}) in hectares and population affected ($P_{population}$) in million have been selected for regression analysis of the available data from the Revenue Department. A regression equation of total damages (D_{total}) in crores Rs in 2006's rate has been developed in the form of:

$$D_{total} = C + C_1 A_{inun} + C_2 A_{crop} + C_{3population}$$

Here C, C_1, C_2 and C_3 are some constants.

Other parameters and their units are already defined. The values of C, C_1, C_2 and C_3 are obtained in 2006's rate are $C = 186.6$, $C_1 = 186.6$, $C_2 = 150.2$, and $C_3 = -2.1$.

Saikia-Das[5] (2007) has used the above statistical model to estimate flood damage of dam-break flood of the proposed high dam Dibang, in the river Dibang, a tributary of the Brahmaputra. It has been estimated that the total flood damage will be Rs. 391.2 crores (in 2006's rate) provided the dam fails and downstream of the dam is not rehabilitated after the construction of the dam. It shows that the modern trend of flood damage estimation is popular by developing mathematical model based on statistical approach.

10.10 CONCLUSION

Flood damage mitigation measures are discussed in detail. Along with flood damages mitigation measures, flood damage estimation is also analysed. Flood is chronic problem everywhere. Therefore, its mitigation measures and damage estimation have become very important all over the world.

EXERCISES

10.1 Describe different structural flood mitigation measures with sketches with reference to north-eastern region of the country.

10.2 Explain the non-structural flood mitigation methods.

10.3 What are the types of flood damage? Write an essay on flood damage estimation.

---

[4] Saikia-Das M., and Sarma, A.K., Dam-Break Flood Simulation for Analysis of Risk Management and Damage Estimation, 1st India Disaster Management Congress, New Delhi, Vigyan Bhavan, November 28–30, 2006.

[5] Saikia-Das, M., Simulation Dam-Break Hydraulics in National Flood Plain Topography, Ph.D. Thesis Submitted to the Indian Institute of Technology (IIT), Guwahati, July 2007.

SUGGESTED FURTHER READINGS

Chow, V.T. (Ed.), *A Handbook of Applied Hydrology*, McGraw-Hill, New York, 1964.

Das, M.M., Flood Disaster Mitigation Measures with Special Reference to North East India, Presented and Published in National Building Congress (NBC) Congress, India, held at Agartala, Tripura, February 16–19, 2007.

Grant, E.L., Ireson, W.G. and Leavenworth, R.S., *Principles of Engineering Economics,* 6th ed., Ronald, New York, 1976.

James, L.D. and Lee, R.R., *Economics of Water Resources Planning*, McGraw-Hill, New York, 1971.

Raghunath, H.N., *Hydrology: Principle, Analysis Design*, New Age International, Delhi, 1996.

Saikia-Das and Sarma, A.K., Numerical Model for Simulating Flow and River Bed Profiles in a Natural River under Dam Failure Conditions, Proc. 15th, IASTED Conference on Applied Simulation and Modelling, vol. 522, pp. 227–232, Rhodes, Greece, 26–28 June 2006.

Saikia-Das, M. and Sarma, A.K., Dam Flood Forecasting with Simple Numerical Model, 3rd Asia Pacific Hydrology and Water Management Association (APHW), Bangkok, Thailand October 16–18, 2006.

Saikia-Das, M. and Sarma, A.K., Numerical Simulation Model for Computation of Dam-Break Flood in Natural Flood Plain Topography, *Jour. Dam Engineering*, vol. 117, No. 1, pp. 31–50, June 2006.

Saikia-Das, M., Analysis for Adopting Logical Channel Section For 1-D Dam-Break Analysis, *APRN Jour. Engg. and Appl. Science,* vol. 1, No. 2, August 2006.

Sarma, A.K. and Das, M.M., Numerical Simulation of Flood Wave from Dike Failure, *Jour. Water and Maritime*, London, Issue 1, 2003.

Varshney, R.S., *Engineering Hydrology*, 2nd ed., Nem Chand & Bros, Roorkee, India, 1979.

Wilson, E.M., *Engineering Hydrology*, ELBS Macmillan, London, 1980.

Wurbs, R.A. and James, W.P., *Water Resources Engineering*, Prentice-Hall of India, New Delhi, 2002.

Chapter 11

River Engineering and River Training Works

11.1 INTRODUCTION

Rivers are well defined drainage channels formed to drain off the runoff of the watershed due to rainfall or melting of snow at high altitude. The formation and development of this well defined drainage take years together. Thus, it is clear that river finds its origin at high altitude and follows the natural slope of the ground and finally joins the sea.

If this natural channel is allowed to flow freely, it may cause devastations and troubles to human beings. River water carries lot of sediments and amount of sediment will increase when river current causes erosion at the banks. Thus, the process of erosion of banks and deposition of sediments are being continued by the river if it is not trained. River follows its own path which may be curved or meandering. Therefore, river training works are necessary to direct the river to flow in a desired path as far as possible, so that problems of flood, erosion, deposition, meandering may be mitigated to a major extent.

11.2 IMPORTANCE OF RIVER

The rivers have played a very important role in human civilisation and development. It has been observed that all the big cities, towns, important places have been developed near the bank of rivers since the days of prehistoric civilisation. River is a source of cheap energy. They provide us water for domestic and industrial use, irrigation, water power, fishery, recreation and for other purposes.

11.3 STAGES OF RIVER

Each and every river from its source to the sea undergoes following stages:

Mountain stage: This stage is also called **rocky stage** or **incised river stage**. It is the first stage of river flowing through steep valleys in the hill. The bed of the river is composed of rocks with steep slope forming rapids along its course. Many abrupt falls may exist in this stage. The bed and banks are less susceptible to erosion. Water is usually very clear.

Sub-mountain stage: This stage is also called **boulder stage**. The river here leaves mountain and enters the sub-mountainous tract. The slope becomes a bit flatter than the first stage, but yet it maintains steep slope. The river flows through a wide shallow bed, interlaced channels and develops a straighter course. During flood period, boulders, gravels, shingles are transported downstream, but as the flood subsides they are piled up in heaps and the flow is bifurcated. The river acquires a tendency to increase its bed width. Thus, shifting braided and interlaced channels are formed in this stage of river.

Trough or alluvial stage: This stage of river is also known as **river in flood plain** or **alluvial river stage**. During this stage, slope becomes flatter; river deposits its excess sediment. Velocity becomes less. During flood time, it inundates the areas near the river causing a considerable damage.

Another main characteristics of the river in this stage is meandering. When river flows in the pattern shown in Figure 11.1(a), meandering of alluvial stage is said to develop. Sediments begin to deposit themselves on convex bank and in concave bank materials begin to erode as shown in Figure 11.1(b) It happens when the river slightly deviates from its straight path, unbalance created goes on multiplying with constant erosion from the concave side and deposition on convex side.

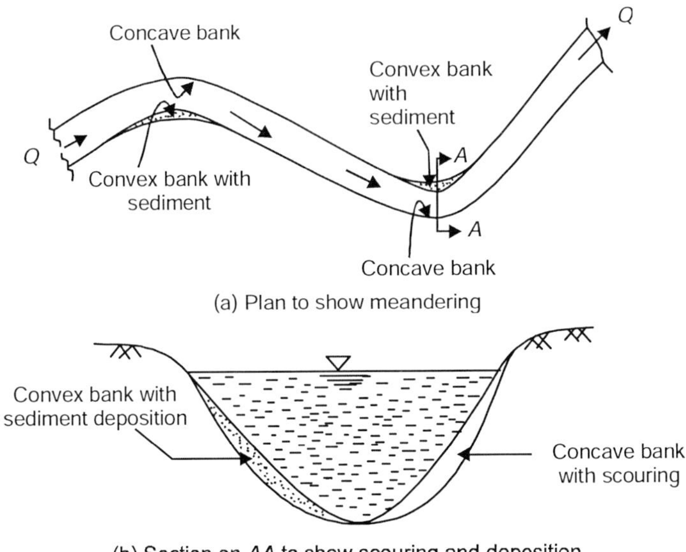

(b) Section on *AA* to show scouring and deposition.

Figure 11.1 Silting, scouring and meandering in alluvial river.

This process continues resulting in formation of a large meander if it is allowed to flow freely.

Delta and tidal stage: This is a stage when river is about to end its course by joining or spilling over to the sea. The slope in this last stage becomes flat, so velocity is very low and channels get silted. Water during flood spills over the banks to form a new channel and thus the river may spilt up into number of channels or branches. The interior portions of branches takes the form of Δ (delta) and hence this stage is called **delta** (Figure 11.2). In delta portion, river receives water from tidal wave of the sea or ocean. During flood tide, ocean water enters the river and recedes during ebb tide. The river in which periodic change of water level occurs due to tides is called **tidal stage**. Thus, delta and tidal stage may come under the same stage.

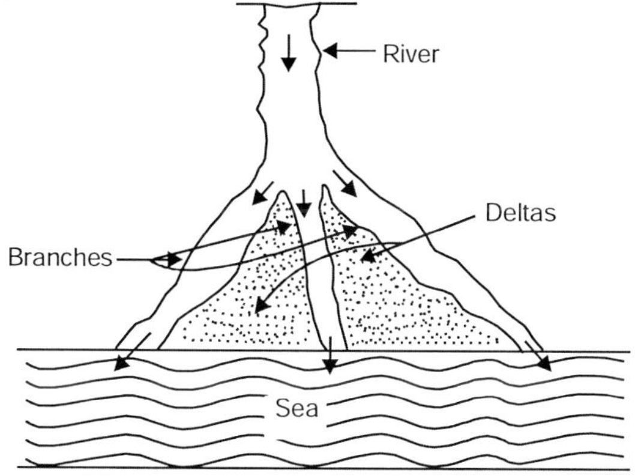

Figure 11.2 Delta formation.

11.4 CLASSIFICATION OR TYPE OF RIVER

Rivers are classified as follows:

Aggrading river: It is also called **accreting river** which is in the process of building up its own bed and certain slope due to excess sediment in the river. Sudden influx in the river from tributary, and due to construction of dam, weir and barrage downstream change of river regime takes place. This results in silting up the bed to increase its slope as shown in Figure 11.3(a). This process is called **building up of slope**. This type of river usually has straight and wide reaches with shoals in the middle, which shift with floods, dividing the flow into number of braided channels.

Degrading river: If the river bed is constantly getting scoured to reduce and dissipate available excess slope as shown in Figure 11.3(b), then the river is called **degrading river**. Degradation occurs below dam or weir or barrage or above a cutoff.

Stable river: When there is no silting or scouring, the river does not change its alignment, slope and its regime, and the river is then called **stable channel**. The river is able to carry the sediment it receives on its wary. This type of river has no appreciable change of alignment with time. In real field, such river is almost rare.

Flashy river: A river is said to be flashy if flood occurs and recedes quickly. In this type of river, the hydrograph has a very quick rising peak and falling limb with smaller base period. The natural slope of the ground or the river slope is steep and therefore, flood due to rainfall occurs and vanishes within a very short period. In this context, flash flood hydrograph shown in Figure 10.1 is referred.

Virgin river: The river which dries up before joining the sea due to excessive evaporation and high percolation is called **virgin river**. This is most common in desert area like Rajasthan in India.

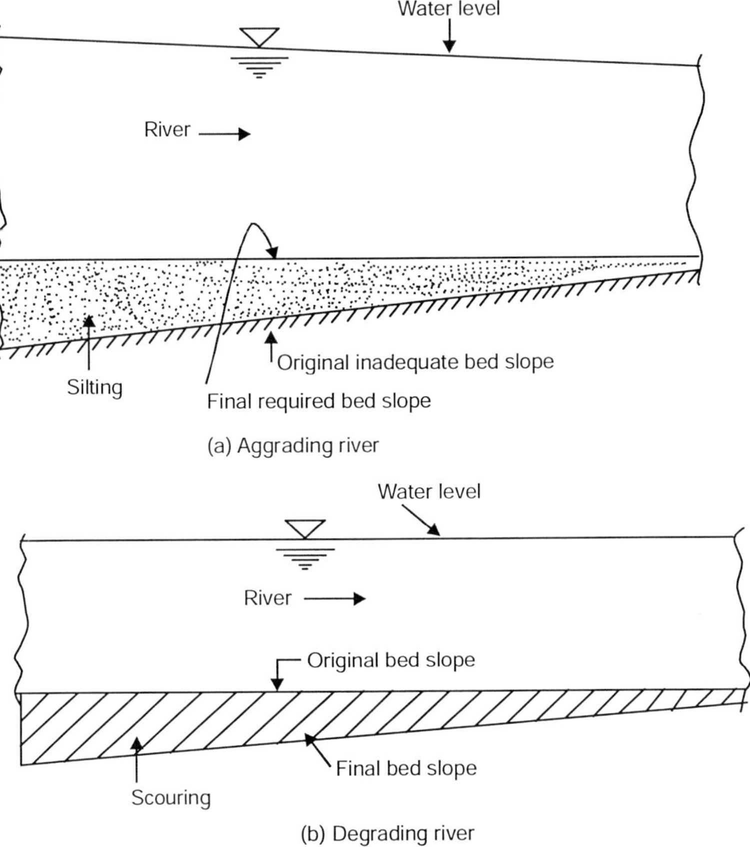

Figure 11.3 Aggrading and degrading rivers.

Himalayan river: The rivers, which originate from the Himalaya are termed as the **Himalayan Rivers**, and are applicable only in India. Supply of water in those rivers is due to melting of snow in summer and spring, and the rain during summer time. Thus, water in those rivers is available throughout the whole year, i.e., they are perennial. These rivers like the Brahmaputra and its tributaries during summer time always have been creating flood in the plain of Assam valley. Other Himalayan rivers are the Indus, Ganges, Jhelum, Chenab, Sutlej, Gomti and Kosi.

Non-Himalayan river: The rivers of India whose origin is not the Himalaya are called **non-Himalayan river**. The origins of these rivers are some other mountains of India. The rivers of South and Central India fall under this classification. Their origins are from the mountains like Aravali, Vindhya and Satpura. Non-Himalayan rivers are Mahanadi, Cauvery, Chambal, Godavari, Narmada and Tapti. All of them are mainly rain bed. They are much stable than the Himalayan rivers. Few of them tend to dry up during the period of no rain.

Braided river: In the alluvial plain when the sediment load is more in the river, in some reaches, the slope becomes small. Velocity decreases and thus allowing alluvial islands to form in the bed of the river. The flow has taken place through two or more channels. Braided pattern has been developed when local deposition of heavy sediment takes place and flow is unable to carry them downstream along with water. A sketch of braided river is shown in Figure 11.4.

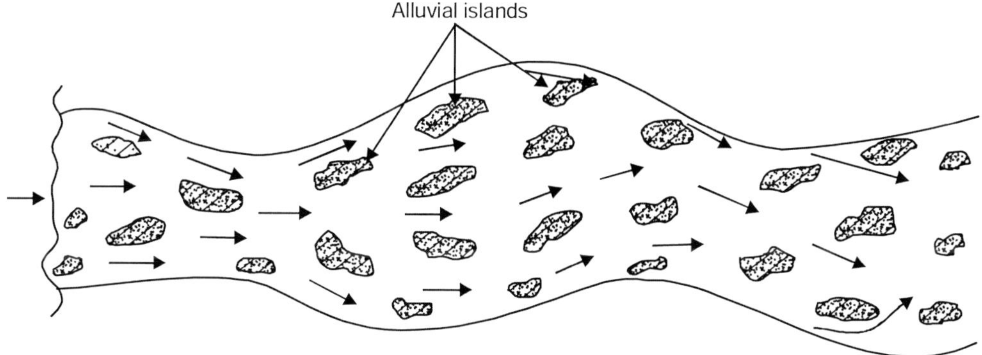

Figure 11.4 Braided reach of river.

11.5 MEANDERING RIVER: INVESTIGATION AND CAUSES

A river is said to be meandering type when it departs from its straight course and follows a sinuous winding path as shown in Figure 11.5. In Figure 11.5, M_L is the meander length and M_D is the transverse distance between apex point of one curve and apex point of the reverse (or meander belt).

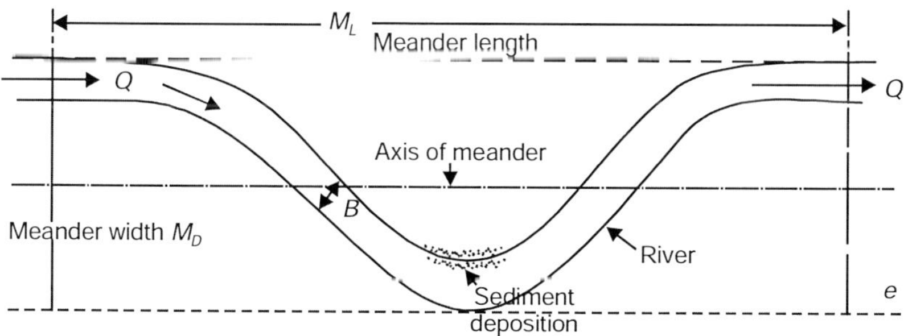

Figure 11.5 Meandering of river.

The ratio (M_L/M_D) is called **meander ratio**. The ratio between the curved length and straight air distance is called **degree of sinuosity**.

11.5.1 Investigation on Meandering

A lots of research and investigation works have been performed by many investigators. Meandering of river still remains a topic of interest for the investigators. Mathematical and statistical models, experimental works have been done on the topic around the world as many of the rivers are of meandering types which have caused lot of problems during flood periods. The science of meandering is because of turbulence and instability. The path of meandering has been a subject of scientific investigation, because understanding of the morphological feature and its major controlling factors is of fundamental importance in river control and regulation. Ferguson[1] (1973) summarized mathematical model on regular meander paths as sine wave, circular arcs, Fergus's spiral, Von Schelling's curve or sine generated curve. The last three models are generally similar. Sinha and Parker[2] (1996) presented a set of models for the longitudinal profile of rivers based on horizontal wavelike propagation, aggradations of the river profile balancing subsidence etc. Ikeda et al[3] (1981) analysed longitudinal velocity using a dynamic equation and bank erosion using kinematic equation to give mechanistic founding of sine generated curve. Works on this meandering have been investigated by Chang[4] (1984), Callender[5] (1978), Odgaard[6] (1986), Zhang[7] (1987) and others.

More recently Goswami[8] (2002), Goswami and Das[9] (2002) analysed this problem to develop a mathematical model with a view to identify the zone of influence of soil-perturbation along the river banks in flood prone region, phenomenalised by flood and soil characteristics. Meandering behaviour is investigated taking a case study of a river Puthimari, a tributary of the Brahmaputra, in lower Assam to identify different meander sections and establish a prediction criteria for magnitude of bank scouring. Also an analysis on environmental influence due to land degradation in river site is carried out.

11.5.2 Causes of Meandering

The causes of meandering are:

1. When bed load is heavy, river cannot follow a straight path. The heavy sediment deposited in the bed causes unevenness of the bed, and therefore, cross currents are developed leading to deposition on one side building shoals and deep channel on the other bank.

[1] Ferguson, R.I., Regular Meander Path Models, *Water Resource. Res.*, vol. 9, 1005, pp. 1079–1086, August 1973.

[2] Sinha, S.K. and Parker, G., Causes of Concavity in Longitudinal Profile of River, *Water Resource. Res.*, vol. 32, pp. 1417–148, May 1996.

[3] Ikeda, S., Parker, G. and Sawai, K., Bend Theory of River Meanders, Part I, Linear Development, *Jour. of Fluid Mechanics*, vol. 112, pp. 363–377, 1981.

[4] Chang, H.H., Analysis of River Meanders, *Jour. Hy. Engg*. ASCE, vol. 110, No. 1, pp. 37–50, January 1984. ———, Regular Meander Path Model *Jour. Hy. Engg*, ASCE, vol. 110, No. 10, pp. 1398–1410, October 1984.

[5] Callender, R.A., River Meandering, *Annual Review of Fluid Mechanics*, vol. 10, Annual Reviews Inc., Polo, Alto, Calif., pp. 129–158, 1978.

[6] Odgaard, A.J., Meander Flow Model I: Development, *Jour. Hy. Engg.*, ASCE, vol. 112, pp. 1117–1137, December 1986.

[7] Zhang, A. and Kahawita, R., Non-linear Model for Aggradations in Alluvial Channels, *Jour. Hy. Engg*, ASCE, vol. 113, No. 3, pp. 353–369, March 1987.

[8] Goswami, M.D., Sinusoidal Behaviour in Meandering Alluvial Channels, Ph.D. Thesis submitted to Gauhati University, January 2002.

[9] Goswami, M.D. and Das, M.M., Sinusoidal Channel Development in Alluvial Flood Plain, *Jour. IWRS*, vol. 22, No. 2, Roorkee, April 2002.

2. Meandering also results from local bank erosion and consequent local over loading and deposition by the river of the heaving sediments which move along the bed. Whatever may be the causes of meandering, but when a meander is fully developed, it has a definite pattern of curvature, length and breadth. This pattern changes when there is a change in discharge or change in material transported by the stream.

11.5.3 Factors Controlling the Process of Meandering

The basic factors that control the meandering are:

Valley slope: This is the average slope of the terrain traversed by the stream and is measured towards the outlet from the origin. A change in the valley slope always produces the change in the meander pattern.

Stream load: Composition of stream load as well as the rate of movement of this load affects the meander pattern.

Discharge: A close relationship between the rate of flow and rate of bed load movements affects the meandering.

Bed and bank resistance: Boundary resistance is characterised by the nature, size of the material comprising the alluvium, more significantly its resistance to erosion. Grain size, specific gravity, cohesion, roughness are important factors. If it is composed of gravel and boulder, meander pattern is not observed, river flows with interlaced channels. If the boundary soil has less resistance to erosion and is composed smaller size soil, meandering tendency is more.

11.5.4 Empirical Equations

Inglis analysed data of some Indian rivers collected by Jefferson and gave the following empirical relationship between M_L, M_D, B and Q (Figure 11.5) in Table 11.1.

Table 11.1 Empirical Relationship Between M_L, M_D B and Q

M_L, M_D, B (m) and Q (m³/sec)	*Rivers in flood plain*	*Incised river*
M_L	$6B$	$11.45B$
M_D	$17.4B$	$27.30B$
M_D	$2.86M_L$	$2.2M_L$
M_L	$53.25\sqrt{Q}$	$46\sqrt{Q}$
M_D	$153.75\sqrt{Q}$	$102.25\sqrt{Q}$
B	$8.87\sqrt{Q}$	$4.54\sqrt{Q}$

11.6 SEDIMENT TRANSPORT IN ALLUVIAL RIVERS

The bed and sides of rivers in alluvium consist of non-cohesive soil. As such problems of sediment transportation and deposition are always there due to erosion and scouring of banks and bed. Increase or decrease in discharge will increase or decrease the depth, water level, cross-section, bed and side slopes, bed forms and roughness coefficient. Sediment is the loose non-cohesive material which is transported or kept in suspension or deposited by action of flowing water in the river based on its flow conditions. Analysis of sediment transport and deposition in rivers in very much complicated and very difficult to obtain exact analytical solution and therefore, experimental works and empirical formulae are preferred by the investigators.

11.6.1 Sediments and Their Movement

The types of sediment load that move along with the water of alluvial river are bed load, suspended load and wash load. Bed load is heavier which moves or gets deposited on the bed, suspended load moves in suspension in middle layer of flow and wash load, i.e., finer part is carried in the upper layer. Bed load part depending on flow condition moves by sliding, rolling or by saltation. Only on certain condition, initiation of movement begins. At low depth, bed load does not move, gets deposited on the bed. A slight imbalance of the flow of the river from its straight path causes erosion of bank, sediment amount increases and the river begins to deposit the sediment on the concave side thereby it begins to meander in the alluvial stage. Although initiation of motion of the sediment, i.e., incipient motion conditions in the river is created, these conditions do not exist all the time of flow and deposition of the sediment takes place. Once the sediment starts depositing, the process continues and meander belt in the river is formed. If the incipient motion condition could be maintained, process of deposition, erosion and formation of meandering could be mitigated. A complete reliable analytical theory of flow condition is not available in literature and therefore, experimental works are preferred.

According to Garde and Ranga-Raju[10] (1985), experimental data on incipient motion condition have been analysed and used by different investigators to develop some equations, experience curves and tables. But all these developments are based on one of the following approaches:

1. Competent velocity approach – which causes the particle to move.
2. Lift force approach – in which lift force generated from bed, keeps the particles above for movement.
3. Critical Tractive force – based on drag force exerted by flowing water on bed to cause the particle to move.

Out of these three approaches, critical reactive force is best and preferred. Before discussing this logical approach in detail, it is better to discuss briefly some of the old theories of sediment movement.

11.6.2 Some Theories of Sediment Movement

These theories will be discussed in brief. [For details, refer Chapter 4 of Das, M.M., *Open Channel Flow,* Prentice-Hall of India, Delhi, 2008.]

[10] Garde, R.J. and Ranga-Raju, K.G., *Mechanics of Sediment Transpiration and Alluvial Stream Problem*, 2nd ed., Wiley Eastern, New Delhi, 1985.

Brahms[11] (1754) and Aiery[12] (1845) theory: Brahms had initiated the theory and finally Aiery gave the following equation of sediment movement when drag force exceeds frictional resistance:

$$V_{cr} = \frac{4}{3}\frac{dg}{\alpha_s c_d}(s-1)\tan\phi \qquad (11.1)$$

where V_{cr} is the critical velocity, d is particle size diameter, α_s is the shape coefficient, c_d is the drag coefficient, s is the specific gravity of sediment and ϕ is the angle of friction. Finally they gave

$$V_{cr} = kW^{1/6} \qquad (11.2)$$

where W is the weight of the sediment and k is a proportionality constant.

Jaffrey's lift concept[13] (1925): Based on difference of velocity, i.e., pressure by Bernoulli's equation, lift force L_D is produced to help the sediment to move with water, i.e.,

$$L_D = \left[\pi\rho_w r V_s^2\left(\frac{1}{3}+\frac{1}{9}\pi^2\right)\right] \qquad (11.3)$$

where r_w is the density of water, r is the radius of cylindrical shape of grain and V_s is the free upstream velocity.

White's approach[14] (1940): He gave for steady viscous inflow as:

$$\lambda_{cr} = \alpha\left[\frac{\pi}{6}\eta d\left(w_s - w_w\right)\tan\phi\right] \qquad (11.4)$$

where λ_{cr} is critical reactive force, α is an experimental coefficient and η is a packing coefficient ($\simeq$ 0.6).

For turbulent flow,

$$\lambda_{cr} = \left[\frac{\pi}{6}\eta d\left(w_s - w_w\right)\tan\phi\right] \qquad (11.5)$$

He further gave a simple equation for turbulent flow:

$$\lambda_{cr} = 80.1d$$

where λ_{cr} is in kg/m^2 and d is diameter of grain in metre.

Meyer-Peter, Muller[15] equation (1948): They gave the following formula:

$$G = 5.1844B\left[\left(\frac{10.844Q_B}{Q}\right)\left(\frac{D_{90}^{1/6}}{n_p}\right)ds - 0.637D_m\right] \qquad (11.6)$$

---

[11] Brahms, A., *Elements of Dam and Hydraulic Engineering*, Aurich, Germany, 1754 and 1757.

[12] Aiery, Sir G.D., Tides and waves, *Encyclopedia Metropolitana*, pp. 241–396, London, 1845.

[13] Jaffrey, H., Flow of Water in an Inclined Channel of Rectangular Section London, *Philosophical Magazine and Jour. of Science*, Ser. 6, vol. 49, pp. 793–807, Edinburgh and Dublin, 1925.

[14] White, C.M., Equilibrium of Grains on the Bed of Streams, *Proc. Roy. Soc.* of London, Series A, vol. 174, 1940.

[15] Meyer-Peter, E. and Muller, R., Formulas of Bed load transport, *Proc. 2nd IAHR Congress*, Stockholm, 1948.

where G is bed load in t/day, B is width of river in m, Q_B is water just flowing over the bed load in m^3/sec, D_{90} is the particular size in mm of stream of bed such that 90 per cent of the particle is finer than this size, n_p is Manning's n for the river bed, D_m is effective bed material size (weight mean diameter) in mm, d is mean depth of water in m and s is hydraulic gradient.

Laursen[16] (1958) and Garde and Albertson[17] (1958) formula: Studies by Laursen and Garde and Albertson (1958) gave the following formula:

$$q_s = c_4 \frac{v^4 n^3}{d_p^{1.5} d} \tag{11.7}$$

where q_s is sediment transport rate per unit width of the river, c_4 is a constant of river, v is water velocity, n is Manning's roughness of the river bed, d_p is sediment diameter and d is the depth of water.

The settling velocity V_s of suspended particles in still water is approximated by Stoke's law:

$$V_s = \frac{(w_s - w)D^2}{18\mu} \tag{11.8}$$

where w_s and w are specific weights of sand grain and water respectively, D is the diameter of the particle, μ is the dynamic viscosity of water.

In turbulent flow, the gravitational settling of particles is encountered by upward transport in turbulent eddies. The system is in equilibrium if gravity movement and turbulent transport are in balance and amount of suspended material remains constant.

The general two dimensional non-equilibrium equations for suspended sediment are:

$$\frac{V\partial c_s}{\partial x} = V_s \frac{\partial c_s}{\partial y} + \frac{\partial \varepsilon_x}{\partial x}\frac{\partial c_s}{\partial x} + \frac{\partial \varepsilon_y}{\partial y}\frac{\partial c_s}{\partial y} + \varepsilon_x \frac{\partial^2 c_s}{\partial x^2} + \varepsilon_y \frac{\partial^2 c_s}{\partial y^2} \tag{11.9}$$

where c_s is sediment concentration for a particular size of sediment, V_s is the settling velocity of the sediment, ε is the mixing coefficient, V is the velocity of flow in river, x and y are longitudinal and vertical axes respectively. Till now, no exact solution of this equation exists, and therefore, investigators have been concentrating on experimental works and empirical equations.

Du Boys[18] formula (1879): Bed load transport for last many years has been based on the classical solution of Du Boys (1879):

$$G_i = \gamma \frac{\lambda_0}{w}(\lambda_0 - \lambda_c) \tag{11.10}$$

[16] Laursen, E.M., The total Sediment Load in Streams, *Jour. of Hy. Div.*, ASCE, No. 1530, February 1958.

[17] Garde, R.J. and Albertson, M.L., Discussion of Paper by Laursen (Ref 22), *Jour. Hy. Div.*, ASCE, 1856, November 1958.

[18] Du Boys, P., The Rhone and stream with movable beds, Annuals des ponts at chausses, section, vol. 18, pp. 141–195, 1879.

where G_i is the bed load transport per unit width of the river, γ is an empirical coefficient depending on the size and shape of the sediment particles, w is the specific weight of water, λ_0 is the shear stress at stream bed, λ_c is the critical or magnitude of shear stress at which transport begins.

Graf[19] (1971) proposed numerous variation on the original equation of Du Boys, all using the concept of critical tractive force to initiate motion.

The successful application of Eq. (11.10) lies on the proper selection of γ values. Straub[20] (1935) performing experiments in laboratory gave the following values of λ_c and γ in Table 11.2.

Table 11.2 Values of γ and λ_c in Du Boys Equation by Straub

Particle diameter in mm	γ *in* m^6/kg^{2}s	*In* λ_c kg/m^2
1/8	0.0032	0.078
1/4	0.0017	0.083
1/2	0.0011	0.107
1	0.0007	0.156
2	0.0004	0.249
4	0.0002	0.439

Einstein bed load function (1950): A widely used approach is that of Einstein who defines the intensity of bed load (ϕ) transport as:

$$\phi = \frac{G_i}{w}\sqrt{\frac{\rho}{(\rho_s - \rho)}\frac{1}{gd^3}} \tag{11.11}$$

and

$$\text{Flow intensity } \psi = \left(\frac{\rho_s - \rho}{\rho}\right)\frac{d}{gR} \tag{11.12}$$

where G_i, ρ_s, ρ, d are already defined and R is the hydraulic radius.

Einstein bed load function ϕ is plotted in log-log paper against ψ and compared with data of Meyer-Peter et al (1934) and Gilbert (1914) for two different sizes of sediment. An empirical relationship between ϕ and ψ from Figure 11.6 allows solution for G_i in Eq. (11.11). Graf (1971) has illustrated a detailed computation of Einstein procedure by computer solution.

Shields' tractive force theory[21] (1936): It appears to be the most popular theory of sediment transport in stream. According to Shields, the tractive force exerted by flowing water on the sediment to cause its motion is equal to boundary shear stress λ_0. He expressed the

[19] Graf, W.H., *The Hydraulics of Sediment Transport,* McGraw-Hill, New York, 1971.

[20] Straub, L.G., H.R. Doc. No. 238, 73rd Cong. 2nd sess., 1935.

[21] Shields, A., (Application of Similarity Principles on Turbulence Research to Bed Load Movement (available in translation by W.P. Qtt and J.C. Uchelen, soil conservation service co-operative laboratory, California Institute of Tech., Pasadena) No. 26, Berlin, 1936.

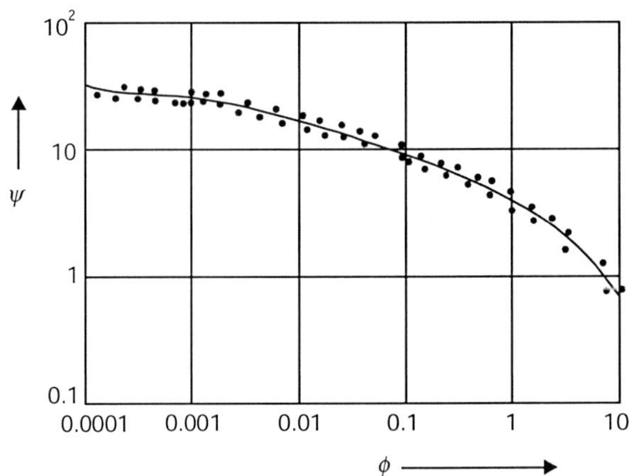

Figure 11.6 Einstein function ϕ vs ψ.

magnitude of this tractive force by a no dimension member or ratio $\left[\dfrac{\lambda_0}{(w_s - w)d}\right]$ which represents one significant parameter in bed load transport. The term $\left[\dfrac{\lambda_0}{(w_s - w)d}\right]$ is called **entrainment function** and magnitude of this term depends on whether grain size is outside or inside of laminar sub layer. If ∂_b is the thickness of laminar sub layer and d is the diameter of the grain size, $\partial_b/d < 1$, then sediment is completely submerged in laminar sublayer. The tractive force on the grain depends on laminar flow or viscous flow. If $\partial_b/d > 1$, the grain size remains outside the laminar sublayer, then tractive force depends on turbulent flow.

Thus, $$\text{Tractive force} \propto \frac{d}{\partial_b}$$

i.e. $$\frac{\lambda_0}{(w_s - w)d} \propto \frac{d}{\partial b} \tag{11.13}$$

Equation of laminar sublayer in boundary layer theory, Prandtl got from Nikuradse's experimental data to be:

$$\partial_b = 11.6\frac{\upsilon}{v_*} \tag{11.14}$$

where $$V_* = \sqrt{\frac{\lambda_0}{\rho}} \tag{11.15}$$

Hence from Eq. (14.14), $$\partial_b \propto \frac{\upsilon}{v_*}$$

$$\frac{1}{\partial_b} \propto \frac{v_*}{\upsilon}$$

or
$$\frac{d}{\partial_b} \propto \frac{v \cdot d}{\upsilon} \tag{11.16}$$

But $\frac{v \cdot d}{\upsilon} = R_*$ (Roughness or shear Reynolds number)

Replacing $\frac{d}{\partial b}$ by $\frac{\lambda_0}{(w_s - w)d}$ from Eq. (11.13),

$$\frac{\lambda_0}{(w_s - w)d} \propto \frac{v_* d}{\upsilon}$$

$\therefore$
$$\frac{\lambda_s}{(w_s - w)d} = f(R_*) \tag{11.17}$$

If Eq. (11.17) is plotted in log-log papers (Figure 11.7), the well known Shields' curve in sediment transport theory is obtained. If the flow condition is above Shields' curve, it is a mobile bed, i.e., sediment is transported and if below the curve, sediment has no motion.

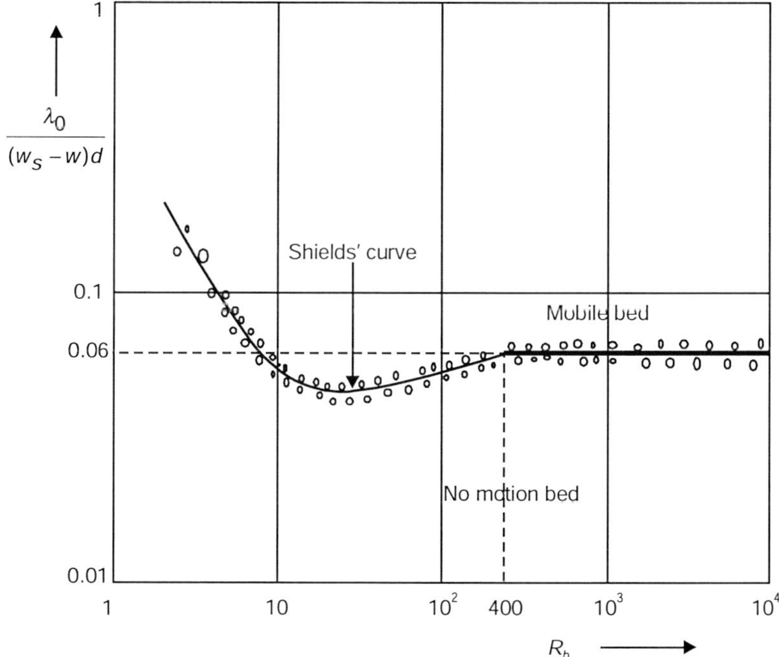

Figure 11.7 Shields' curve for incipient motion condition.

Characteristics of the Shields' curve

(i) It slowly rises when $R_* > 11.6$.

(ii) Curve becomes independent of R_* from $R_* \cong 400$ and value of $\frac{\lambda_0}{(w_s - w)d}$ at that condition ≈ 0.06 and the curve becomes horizontal after that value.

(iii) Above the curve, bed is in motion with sediment, and below the curve, there is no motion of sediment.

Relationship of grain diameter d, hydraulic saddles R and slope s_b

When the bed is mobile (i.e., sediment moves),

$$\frac{\lambda_0}{(w_s - w)d} \geq 0.06$$

or
$$\lambda_0 \geq 0.06(w_s - w)d$$

But
$$\lambda_0 = wRs_b \qquad \text{(for uniform flow)}$$

and if
$$\text{specific gravity sediment } \frac{w_s}{w} \simeq 2.65$$

or
$$wRs_b \simeq 0.06\ (w_s - w)d$$

or
$$Rs_b \simeq 0.06\ (w_s/w - 1)d$$

or
$$Rs_b \simeq 0.06\ (2.65 - 1)d$$

$$d \simeq \frac{1}{0.099} Rs_b \tag{11.18}$$

$$d \simeq 10 Rs_b$$

EXAMPLE 11.1

In a wide river, average velocity is 0.85 m/sec, depth of flow is 0.75 m, sediment size is 1 mm, specific gravity of sediment is 2.65, s_b = 1/4000, $\nu = 10^{-6}$ m^2/sec. Examine possibility of sediment movement by Shields' theory.

Solution:

Assume the stream to be wide, i.e., $P = B$ or $R \cong y$

∴
$$\lambda_0 = wRs_b = wys_b$$

$$= \left(9.81 \times 1000 \times 0.75 \times \frac{1}{4000}\right)$$

or
$$\lambda_0 = 1.84 \text{ N/m}^2$$

∴ Shields entrainment function
$$= \frac{\lambda_0}{(w_s - w)d} = \frac{1.84}{9.81 \times 1000(2.65 - 1) \times (1/1000)}$$

$$= 0.11367$$

and $$R_* = \frac{v*d}{\upsilon} = \frac{\sqrt{gys_b} \times d}{\upsilon}$$

$$= \frac{\sqrt{9.81 \times 0.75 \times (1/4000) \times \frac{1}{1000}}}{10^{-6}} = 42.887$$

If point (0.11367, 42.887) is plotted in Shields' diagram, the point lies above the curve. So there is a motion of bed load.

Further works on Shields' theory have been done by Mittal and Swamee[22] (1976), Yalin and Karahan[23] (1979). Readers are referred to Das M.M., *Open Channel Flow,* Prentice-Hall of India, New Delhi, 2008. Further works on sediment transportation and alluvial stream problems, resistance relationship in alluvial channel flow, regime criteria for alluvial streams are given by Garde and Ranga Raju[24] (1963) and (1966).

11.7 RIVER CONTROL WORKS

The previous sections of this chapter have been confined mainly to types of river and its sediment movement. The rivers are natural streams and if these natural streams are allowed to flow freely, they cause lot of damage in the form of flood and erosion. Therefore, in this part, discussion will be presented on different river control or river training works which may (i) direct and guide the river to flow in desired direction (ii) control and regulate the river bed and bank (iii) increase the low water depth (iv) control the erosion of banks (v) control the deposition of silt and sediment on river bed. Thus, the very essence of river control or training works is to stabilize the river along a desired alignment.

11.8 OBJECTIVE OF RIVER TRAINING

The river training works may serve the following objectives:

1. To pass safely and quickly the high flood discharge through the river reach.
2. To prevent the river from changing its course.
3. To control bank erosion to make the river course stable.
4. To provide the minimum required water depth for allowing uninterrupted navigation.
5. To prevent flooding the surrounding area.
6. To make provisions to transport the high bed and suspended sediments.
7. To prevent outflanking of structures like bridges, weirs and aqueducts.
8. To deflect the flow of river from the attacking banks.

---

[22] Mittal, M.K. and Swamee, P.K., An Explicit Equation for Critical Shear Stress in Alluvial Streams, CBIP, *Jour. Irrigation and Power*, New Delhi, April 1976.

[23] Yalin, M.S. and Karahan, E., Inception of Sediment Transport, *Jour. Hy. Div. Proc.*, ASCE, vol. 105, 1979.

[24] Garde, R.J. and Ranga-Raju, K.G., Resistance Relationship of Alluvial Channel, *Jour. Hy. Div. Proc.*, ASCE, vol. 92, 1966.

———, Regime Criteria for Alluvial Streams, *Jour. Hy. Div., Proc*, ASCE, November 1963.

9. To increase discharge capacity of river by deepening and straightening.
10. To provide escapes or diversion of water from the main stream to an auxiliary channel when the main stream is not capable of dispose the water safely.
11. Construction of small storage reservoirs to store safely the excess flow.

11.9 CLASSIFICATION OF RIVER TRAINING

Based on the purpose for which the objectives of river training programmes are adopted, river training works may be classified into the following three types:

High water training: It is also called **training of discharge**. It trains the river for sufficient and efficient cross-sectional area throughout the whole length for expeditious passage of maximum flood. It concerns with alignment and height of embankment for a given flood discharge thereby making the area of the river almost flood proof.

Low water training: It is the training work required at low depth of the river. Here river is trained to provide required depth for navigation during the period of low flow. It is done by closing other small interlaced channel, removing alluvial island, if any, contracting the width of the channel. They are done by bandalling, groynes, dredging, etc.

Mean water training: This is the training work to dispose the bed and suspending sediments to maintain a stable and efficient river section throughout. The maximum accretion capacity of river occurs when river reaches its mean flow. It is the important water training work as sediment once deposited creates lot of problem for the other two watering works.

11.10 METHODS OF RIVER TRAINING WORKS

11.10.1 Embankments (or Levees or Dikes)

After the disastrous flood of 1954 in the whole counting (India), National flood control policy was formed which recommended embankments as a short term flood control/river control works. These are earthen embankment constructed almost parallel to the banks of the river. These two embankments increase the flow area of the river (Figure 11.8a).

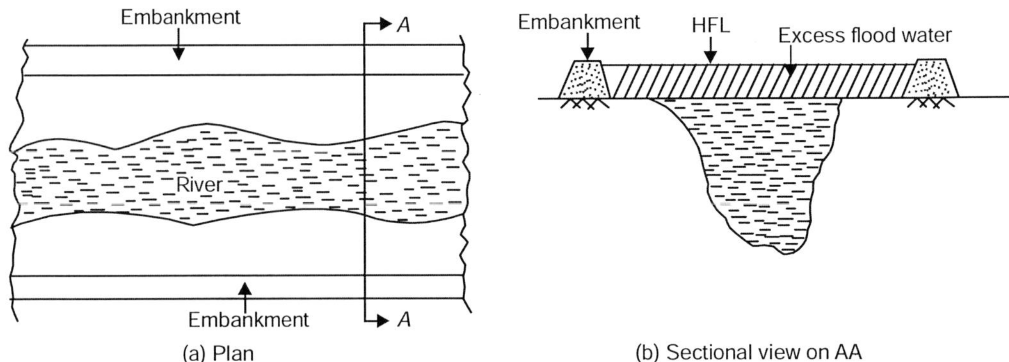

Figure 11.8 Embankments (a) Plan (b) Sectional view.

Embankment can store excess flood water shown by shaded line shown in Figure 11.8(b), hence embankment helps both river training works and flood control measures. A section of embankment with usual dimensions is shown in Figure 11.9.

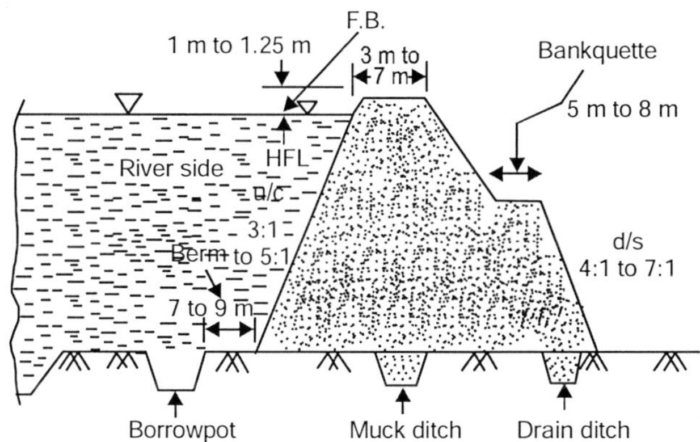

Figure 11.9 A general section of embankment (height is dependent on HFL).

Causes of failure of embankments: The embankments may fail due to

1. Caving of river bank may result in under cutting river side of embankment
2. Piping, i.e., seepage below foundation at flood time causes sand boiling downstream
3. Overtopping during flood
4. Foundation failure by infiltration
5. Cross-section leakage due to shrinkage of soil and rat holes

Advantages:

1. They are cheaper.
2. They are easy to construct.
3. Easy means to prevent overflowing the banks.
4. Maintenance is easy.

Disadvantages:

1. It raises the flood level.
2. If failed, catastrophic dike failure flood occurs.
3. Heavy maintenance during rainy days is essential.
4. Embankments decreases the velocity, hence sediments are deposited on the bed. River depth goes on decreasing with time.

11.10.2 Spurs (or groynes)

Spurs are structures constructed transverse to the river flow extending from the banks into the river to serve the following purposes:

1. To protect the river bank.

2. To train the river to flow along a desired direction by attracting or repelling the flow.
3. To create still pond along a particular bank with an aim of silting up the area.
4. To contract the wide river for improving the navigation depth.

Classification of spur: The Classification of spur is based on the following:

1. Based on material of construction:
   (a) Permeable
   (b) Impermeable

Permeable spur. It is constructed by wood, bamboo, mattress, etc. held in position by driving piles in bed. The framework is built by piles and wood or bamboo around them. The intervening spaces are filled by branches of trees or bamboos. These spurs are temporary and considerable amount of water flow takes place through them. Hence they are called **permeable spurs.** Permeable spur is also constructed with stone filled with wire crates and stone filled ballicrates.

Impermeable spurs. They are made of concrete or masonry and earthen embankment with pitching. Flow does not take place through the body of the spurs. Hence they are called impermeable spurs.

2. Based on height:
   (a) Submersible
   (b) Non-submersible

When the height of the spur is submerged in water, it is called **submersible spur,** and if the spur is not submerged, it is non-submersible spur.

3. Based on functions:
   (a) Attracting spur
   (b) Deflecting spur
   (c) Repelling spur

Attracting spur. Spur pointing downstream is shown in Figure 11.10(a) attracts river water at its nose and thus scour holes are formed. It is usually used in meandering river. The purpose is to attract river flow towards the bank from where it extends so that an important area placed approximately at $0.4L_M$ (L_M is meander length) can be saved.

Deflecting spur. It is simply placed perpendicular to the bank or flow. It deflects the water from the bank from where it emanates. It is shown in Figure 11.10(b).

Repelling spur. This spur points upstream at an angle approximately 60° to 80° to the bank. The purpose is to repel water from this bank and create still pocket upstream of the spur. This pocket is filled with sediment just to deflect river from this bank which is vulnerable for erosion [see Figure 11.10(c)].

4. Based on special type:
   (a) Denehy's T-head spur
   (b) Hockey spur

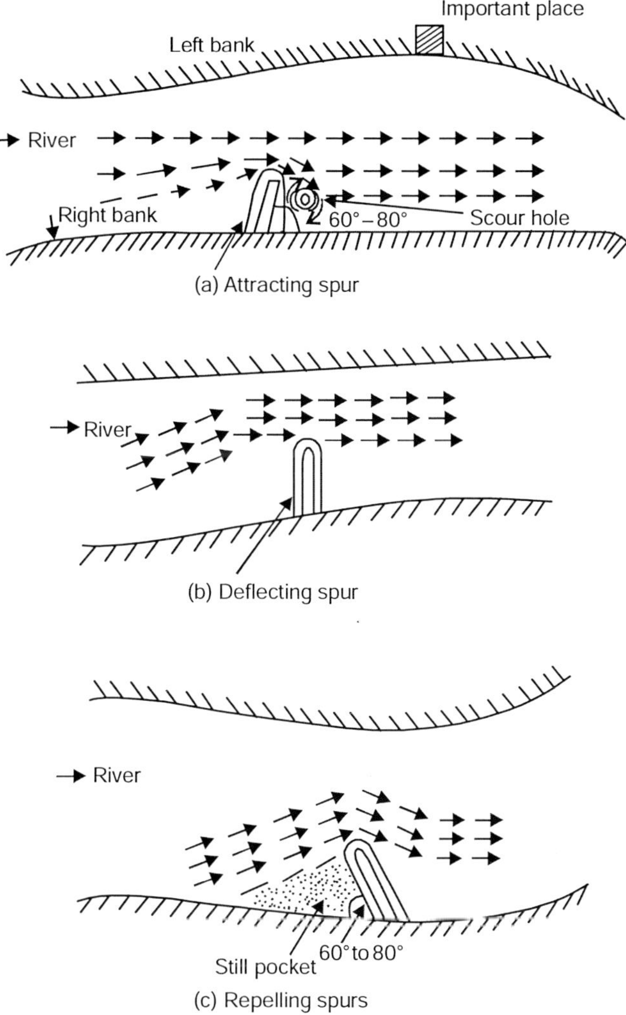

Figure 11.10 Different spurs based on function.

T-head spur. They are originally proposed by Denehy. The shape is like the inverted "**T**" as shown in Figure 11.11(a). The aim is to deposit silt on concave bank from which they extend, thus erosion on this concave bank is alleviated as shown.

Hockey spurs. In usual hockey spur [Figure 11.11(b)], scour is equal on both upstream and downstream. In inverted hockey spur [Figure 11.11(c)], silting takes place downstream.

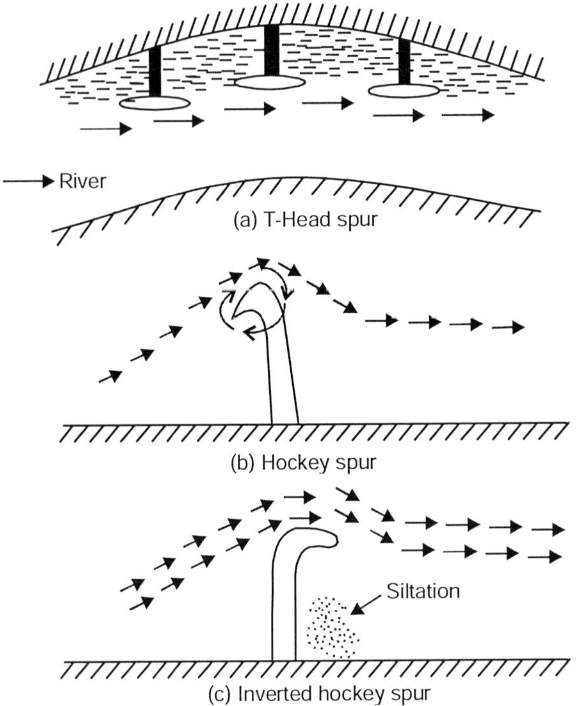

Figure 11.11 T-shaped and Hockey spur.

11.10.3 Artificial Cutoff

A hair pin bend or a horse shoe bend on river caused by meandering may be dispensed with an artificial cutoff as shown in Figure 11.12. The river then flows straight and bend gets silted up in course of time.

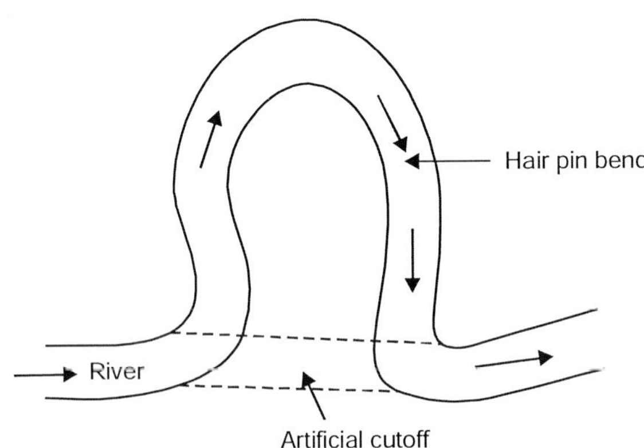

Figure 11.12 Cutoff.

Advantages:

1. It reduces the distance of navigation.
2. It also reduces the travel time of flood. Flood in cutoff flows with higher velocity, so stage is reduced.
3. It reclaims useful land and property near the bend.

11.10.4 Guide Banks (Bell Bunds)

This type of river training work was introduced by J.R. Bell and designed later by Francis I.E. (1903). Therefore, this bund is called sometimes **Bell bunds**. Guide banks are constructed to guide the flow near the bridges, weirs and barrages so as to confine it in a reasonable width of river which are generally wide.

It becomes extremely expensive to construct bridges spanning the whole width of the river. To economise the construction works, training works are constructed to confine the flow in a reasonable and shorter waterway at the site of hydraulic structures. This is done with the help of guide banks as shown in Figure 11.13(a) and (b).Guide banks consists of a heavily built embankment [Figure 11.13(c)]. As seen from Figure 11.13(a) and (b), the shape of guide banks are bell month on both sides of the constricted channel. Materials of construction are locally available earth with protection on the river side along with its bed.

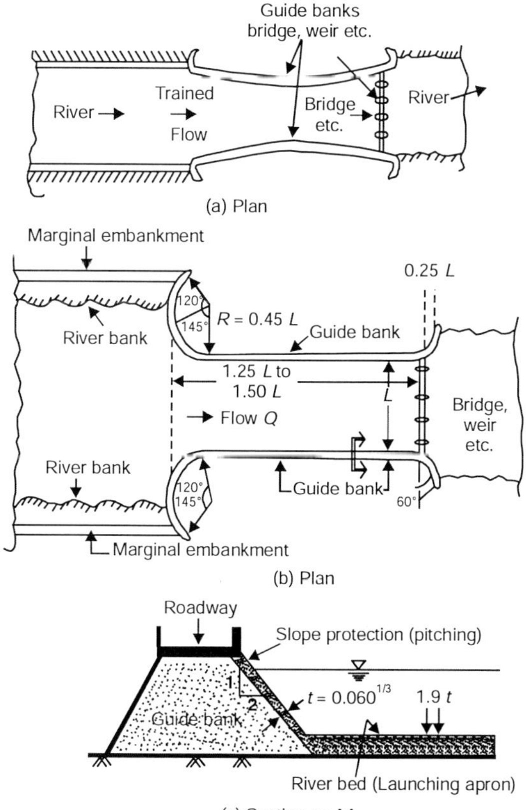

Figure 11.13 Plan and sectional views (a), (b) and (c) of guide bank.

Types of guide banks:

1. Parallel type
2. Divergent type
3. Convergent type

Design proportions: Design proportions of guide bank are shown in Figure 11.14.

$$L - 4.75\sqrt{Q} \qquad \text{(by lacey formula)}$$

$$= 5\sqrt{Q}$$

Length of upstream (from Hydraulic structure) portion = $1.25L$ to $1.5L$
Length of downstream

$$\text{portion} = \frac{1}{4} L \text{ to } \frac{1}{10} L \simeq 0.25L$$

$$R \text{ (radius)} \cong 0.45L \text{ (upstream)}$$

$$\text{Angle upstream} = 120° \text{ to } 145°$$

$$\text{Angle downstream} = 45° \text{ to } 60°$$

$$R_1 \text{ (downstream)} = R/2$$

Thickness of pitching [Figure 11.13(c)], $t = 0.06Q^{1/3}$

$$\text{Launching apron} = 1.9t$$

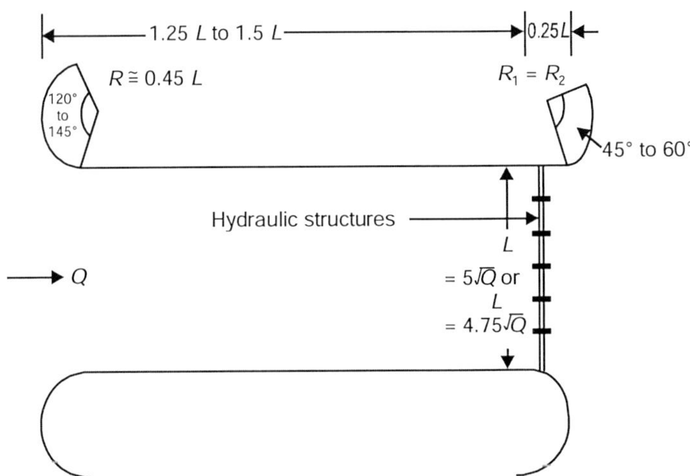

Figure 11.14 Design proportions of guide banks.

11.10.5 Banks Protections

It is the work to save the banks of the river mainly from erosion, sliding of slope, undermining, sloughing, piping and drawdown.

Methods adopted:

1. Protection of slope by vegetal cover by turfing or sodding. Effective cover is the low growth shrubs and willows. This method is effective if the current is not very strong.
2. Pavement of the bank may be constructed with such material which can resist quick erosion. Temporary covering of the pavement should be given by rush wood or mattresses weighed by stone. It is used against strong current.
3. In case current is very strong, protection is provided by stone revetment or various types of mattresses such as willow, lumbar or asphalt or articulated concrete.

11.10.6 Pitched Island

A pitched island is an artificially created island. It is a river bed protected by revetment or stone pitching on all sides. These islands obstruct the flow, and turbulence is created. The islands are placed in series to form deep channel and to divert the flow away from the bank in danger. For wide and shallow rivers, pitched island is not effective. Plan and sectional views are shown in Figures 11.15(a) and (11.15(b).

11.10.7 Pitched Bank and Launching Apron

Protection of banks is a part and parcel of river training works. Banks are pitched by stone and in the bed of the river upto some distance from the bank launching apron is provided so that the pitched bank cannot be eroded by action of water. However, such arrangement cannot be made for the whole river due to constraints of fund and time. In most of the vulnerable cross-sections, such arrangement may be provided (Figure 11.16).

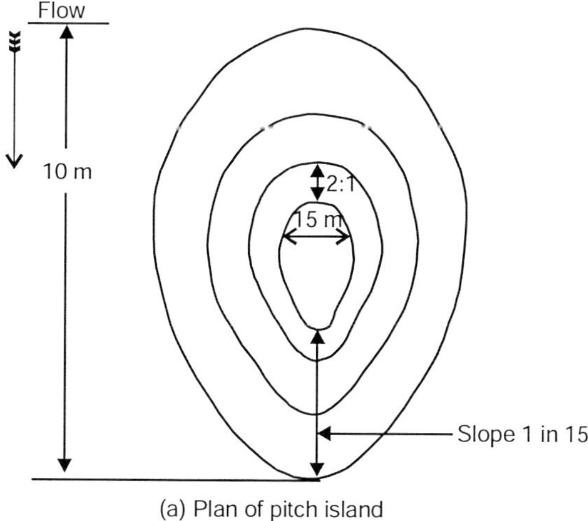

(a) Plan of pitch island

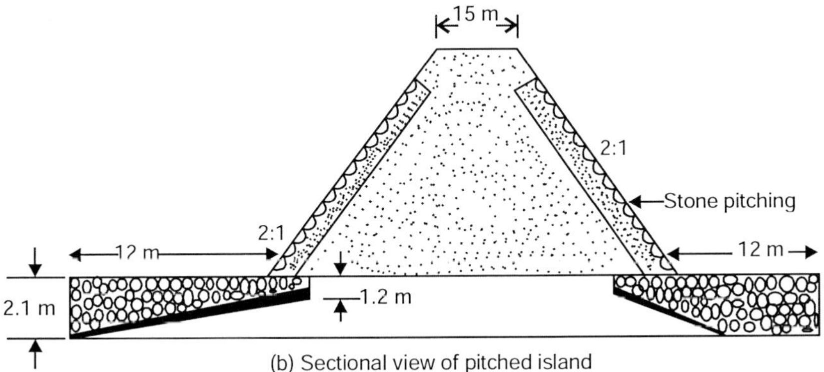

(b) Sectional view of pitched island

Figure 11.15 Pitched island.

11.10.8 Miscellaneous Methods

Bandalling, deepening river by dredging, bottom paneling, submerged sill or dikes, closing dikes, Nanal plantation are some of the other methods of river training.

The bandals are vertical mats or screen of bamboos. The process mainly consists of placing the rows of bandalling at the entrance of secondary channel. The flow is obstructed and caused to flow towards the main channel. Bandalling thus helps in diverting the channel and encourages the deposition of sand bed load outside the main stream and there by it helps in deepening the main channel for navigation.

Bottom paneling are submerged screens placed in the streams and obstructs flow in secondary channel. They are placed 10° to 45° with current to divert the bottom current out of channel and thereby favours the deposition of sediment load. Thus, bottom paneling when placed in secondary channels closed the channels after few operations. This method is suitable for protection of banks and checking erosion.

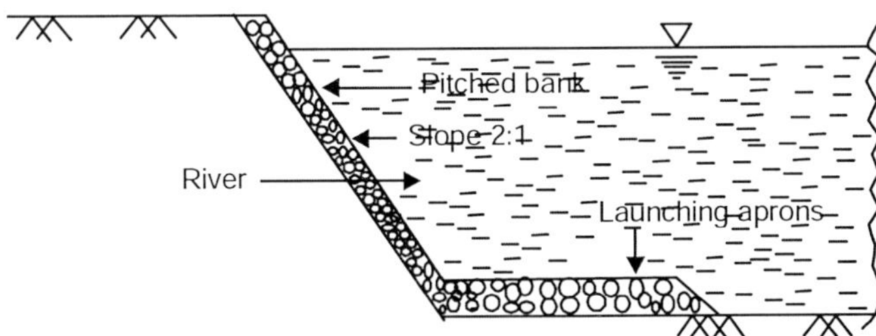

Figure 11.16 Pitched bank and launching apron on bed.

Dredging is possible in small rivers to increase the depth in the middle of the river.

Submerged sills of dikes are essential when river create deep channels near some structures. In such situations, submerged sills are placed across the scoured portion of the bed for safety of the important structure.

Closing dikes are used to close some particular flow so that river may be directed towards a desired direction.

Nanal is plant like cane. Nanal plants is planted along the bank. Their roots spread out quickly. Siltation in the area takes place. Bank may be saved from erosion.

11.11 CONCLUSION

Rivers are gifts of nature to the mankind, wild life, crops and plantation. The characteristics of river flow, its control against flood, erosion and sediment deposition are essential for engineers and experts in other fields. To harness this gift of nature for the overall development of the society, these rivers are required to control. This chapter deals with origin of rivers, their types and different control works to obtain the optimum benefits of this gift of nature.

EXERCISES

11.1 What are the different stages of rivers? Explain them with sketches where applicable.

11.2 Classify rivers and explain each of them.

11.3 Explain in details on meandering of rivers.

11.4 Explain sediment transport in rivers and some of the existing theories.

11.5 Explain in detail the Shield's sediment transport theory. Draw Shield's curve. How do you predict from it the movement of mobile bed?

11.6 Explain objectives and the classification of river training. Describe different river training works with sketches.

SUGGESTED FURTHER READINGS

Ahmed, M., Spacing and Projection of Spurs for Bank Protection, Civil Engg. and Pub. Works Review, March-April 1951.

C.B.I.P. (India) Manual of River Behaviour Control and Training, Publication 60, New Delhi, 1971.

Chitale, S.V., River Channel Patterns, *Jour. Hy. Div.*, *Proc.* ASCE, vol. 96, January 1970.

Einstein, H.A., Formulas for Transportation of Bed Load, Trans. AECE, 1942.

Engelund, F. and Hansen, E., Monograph on Sediment Transport in Alluvial Streams, Teknisk, Forlag, Denmark, 1967.

Garg, S.P., Asthana, B.N. and Jain, S.K., Design of Guide Bunds for Alluvial Rivers, *J. of Inst. of Engrs.* (India), vol. 52, September 1971.

Inglis, C.C., The Behaviour and Control of Rivers and Canals (with the aid of models), Parts. I and II, CWINRS, Pub. No. 13, 1949.

Joglekar, D.V., Manual of River Behaviour, Control and Training, CBIP, Pub 60, 1971.

Manual of River Behaviour, Control and Training Publication No. 60, Central Board of Irrigation and Power (CBIP), New Delhi, 1971.

Peterson, M.S., *River Engineering*, Prentice-Hall, Upper Saddle River, N.J., 1986.

Punmia, B.C. and Pande, B.B., *Irrigation and Water Power Engineering*, 12th ed., Laxmi Publications, 1999.

Ranga Raju, K.G., Resistance Relation in Alluvial Streams, La Houille Blenche, No. 1, 1970.

Spring, F.I.E., River Training and Control of Guide Bank System, Railway Board, Govt. of India, Tech. Paper No. 153, 1903.

Viessinan, W. and Lewis, Q.L., *Introduction to Hydrology*, Harper Collins, New York, 1996.

Vittal, N. and Mittal, M.K., Design of Spurs (groynes), Tech Report No. 36, Res. Sch. Appld. to River Valley Projects, CBIB, July 1987.

Chapter 12
Hydrologic Routing

12.1 INTRODUCTION

The flood in river can be treated simply as a uniformly progressive flow. However, if the river is irregular and resistance to flow is high, the configuration of this wave is modified as it moves with time and space. The determination of this modification is called **routing**. Thus, routing is a process used to determine the variation of flow rate for a flood wave as it moves through water course. In engineering hydrology, routing is an important technique necessary for the complete solution of flood control problems and for the satisfactory operation of a flood prediction service. For such purposes, routing is recognized as a procedure required determining the hydrograph at any point of the river at a certain time from known hydrograph of upstream point as a boundary condition. Routing technique is applied to both river and reservoir. Routing is classified into two methods:

1. Hydrologic routing
2. Hydraulic routing

In this chapter, only hydrologic routing will be discussed.

12.2 HYDROLOGIC ROUTING

Hydrologic routing is based on storage continuity equation. If I is the inflow rate, O is outflow rate, S is the storage, t is the time then the continuity equation is:

$$I - O = \frac{\Delta S}{\Delta t} \tag{12.1}$$

Equation (12.1) may be written from Figure 12.1(a) as:

$$\frac{I_1 + I_2}{2} - \frac{O_1 + O_2}{2} = \frac{S_2 - S}{\Delta t}$$

or

$$\left(\frac{I_1 + I_2}{2}\right)\Delta t - \left(\frac{O_1 + O_2}{2}\right)\Delta t = S_2 - S_1 \tag{12.2}$$

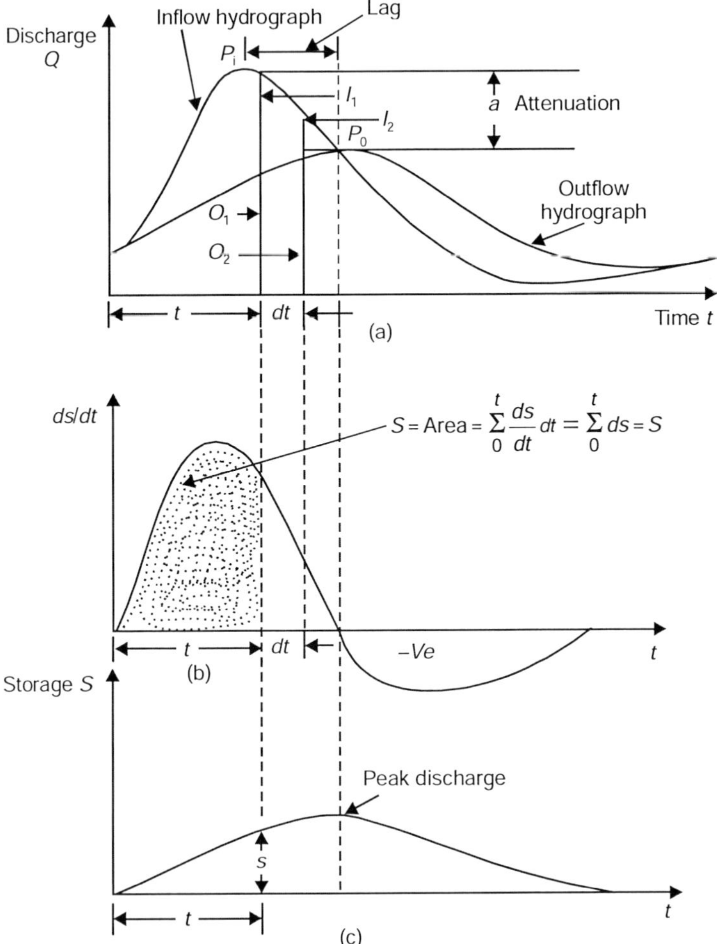

Figure 12.1 Relationship of inflow *I*, outflow *O*, storage *S* in a channel reach when a flood passes the reach.

Almost all storage or hydrologic routing methods are based on Eq. (12.2).

The peak of inflow hydrographs are attenuated and delayed as shown by attenuation and lag in Figure 12.1(a).

Assuming negligible loss or gain of water in the reach, total area under the two hydrographs is equal. The difference between ordinates of inflow and outflow hydrographs shown by dotted area is the storage up to time *t* [Figure 12.1(b)]. Figure 12.1(c) shows storage curve. Peak storage occurs when $I = O$.

12.3 RELATIONSHIP OF OUTFLOW AND STORAGE

Figure 12.2 shows the plot of outflow *O* against storage *S*. When outflow *Q* increases initially, storage also increases. But after sometime, it starts decreasing, thus the resulting curve represents a loop with rising and falling limbs as shown in Figure 12.2. The dotted line between the two limbs represents storage-outflow relationship of steady flow.

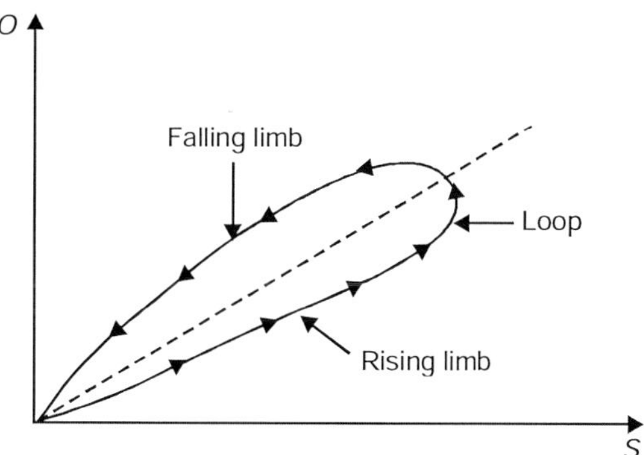

Figure 12.2 Outflow-storage relationship.

12.4 THE GENERAL STORAGE EQUATIONS

Flow in river is unsteady. In unsteady flow, storage in a channel reach is dependent on inflow I, outflow O, channel geometry, hydraulic characteristics and its control features. If y is the depth, following relationships of I, O, S_i (Storage at upstream inflow section), S_o (Storage at downstream outflow section) may be assumed:

$$I = ay^n \tag{12.3}$$

$$O = ay^n \tag{12.4}$$

$$S_i = by^m \tag{12.5}$$

$$S_o = by^m \tag{12.6}$$

Substituting the value of y from Eq. (12.3) in Eq. (12.5),

$$S_i = b\left(\frac{I}{a}\right)^{m/n} \tag{12.7a}$$

Similarly

$$S_o = b\left(\frac{O}{a}\right)^{m/n} \tag{12.7b}$$

Let X is the non-dimensional factor that defines the relative weights to inflow and outflow in determination of storage volume within the reach. Then storage S at any given time may be expressed as:

$$S = XS_i + (1 - X)S_o \tag{12.8}$$

The weighing effect X may be defined by the fact that if storage is a sole function of O only, then $X = 0$. When storage has the effect of back water at upstream, X is greater than 0. In uniform flow, since equal effect of both inflow and outflow is there and therefore, $X = 0.5$.

Substituting S_i and S_o from Eq. (12.7a) and (12.7b) in (12.8), it may be written as:

$$S = K > (XI^x + (1 - X)\, O^x) \tag{12.9}$$

where
$$K = \frac{b}{a^{m/n}} \quad \text{and} \quad x = \frac{m}{n}$$

Equation (12.9) is at the root of development of many hydrologic routings. Quite a number of methods of hydrologic flood routing have been developed. Graphical routings have been presented by Wilson[1] (1941), Cheng[2] (1946) and Kohler[3] (1958). Chow[4] (1951) presented his own solution of practical procedure. For routing through rivers storage-discharge relationship in the methods has been developed by Meyer[5] (1941), Puls[6] (1928), Wilson (1941) and Kohler[7] (1944).

12.5 MUSKINGUM ROUTING EQUATION

McCarthy[8] (1938) and others while working at Muskingum Conservancy District Flood Control Project of the US Army Corps of Engineers for Natural Channel, proposed the value of exponents in Eq. (12.9) to be unity. Equation (12.9) becomes

$$S = K[XI + (1 - X)O] \tag{12.10}$$

Hence Eq. (12.10) becomes Muskingum routing equation. The values of K and X are determined by the following method:

12.5.1 Steps Required for Determining *K* and *X*

1. Collect flood data of the channel reach from previous year floods.
2. Assume value of X (i.e., $0 < X < 0.3$)
3. Calculate values of the term $[XI + (1 - X)O]$ with chosen value of X (say $X = 0.3$).
4. Estimate storage S at different times from known values of inflows I and outflows O.
5. Plot S vs $[XI + (1 - X)O]$ as shown in Figure 12.3 for different assumed X values (say, $X = 0.3, 0.25, 0.20, 0.1$, etc.).

[1] Wilson, W.T., A Graphical Routing Method, *Trans Am. Geophys. Union*, vol. 23. Pt III, pp. 893–897, 1941.
[2] Cheng, H.M., A Graphical Solution of Flood Routing Problems, Civil Engg., vol. 6, March 1946.
[3] Kohler, M.A., Mechanical Analogs Aid Graphical Flood Routing, *J.Hy. Div.* ASCE, vol. 84, April 1958.
[4] Chow, V.T., Practical Procedure of Flood Routing, Civil Engg. and Public Works Review, London, vol. 46 No. 542, August 1951.
[5] Meyer, O.H., Simplified Flood Routing, Civil Engg., vol. II, No. 5, May 1941.
[6] Puls, L.G., Flood Regulation of Tennessee River, House Document, No. 185, 70th Congress, 1st Session, US Govt. Printing Office, Washington, D.C., 1928.
[7] Kohler, M.A., A Forecasting Technique for Routing and Combining Flow in terms of Stage *Trans. Am. Geophys. Union*, vol. 25, Pt 6, 1944.
[8] McCarthy, G.T., "Unit Hydrograph and Flood Routing" unpublished Manuscript, presented at North Atlantic Div., U.S. Army Corps of Engrs, and June, 1938.

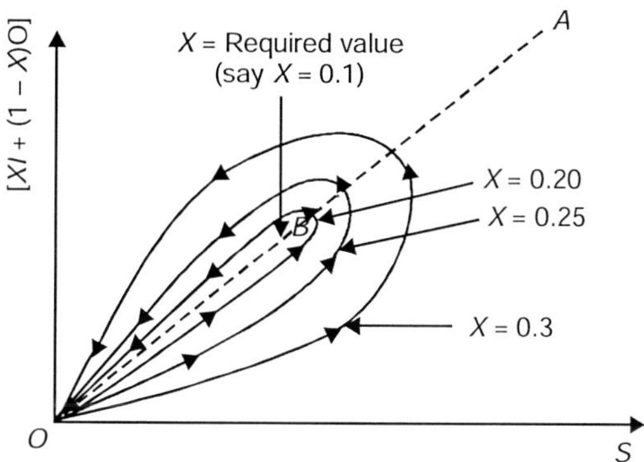

Figure 12.3 Determination of K and X.

6. It is seen that for values of X equal to 0.3, 0.25, 0.20, ... plot forms a loop. At some particular value of X, the curve rises and traces back almost the same path (say OB). The value of X at which this situation arises, is the value of X to be determined. For example, let in our case it happens when $X = 0.1$, i.e., X value is equal to 0.1. Extending OB to A, the slope of OA is the value of K. It is seen that the unit of K is time (hrs or days), and it is approximately equal to travel time of flood wave through the reach or basin lag.

12.6 MUSKINGUM METHOD OF ROUTING

Writing Eq. (12.10) as:

$$S_1 = K[XI_1 + (1 - X)O_1] \tag{12.11}$$

$$S_2 = K[XI_2 + (1 - X)O_2] \tag{12.12}$$

Now

$$S_2 - S_1 = K[X(I_2 - I_1)(1 - X)(O_2 - O_1)] \tag{12.13}$$

Equation (12.2) is:

$$S_2 - S_1 = \left[\left(\frac{I_1 + I_2}{2}\right)\Delta t - \left(\frac{O_1 + O_2}{2}\right)\Delta t\right]$$

Replacing $S_2 - S_1$ in (12.13) by (12.2),

$$\left(\frac{I_1 + I_2}{2}\right)\Delta t - \left(\frac{O_1 + O_2}{2}\right)\Delta t = K\left[X(I_2 - I_1) + (1 - X)(O_2 - O_1)\right]$$

Simplifying further, Muskingum routing equation is obtained as:

$$O_2 = C_1 I_2 + C_2 I_1 + C_3 O_1 \tag{12.14}$$

where

$$C_1 = \frac{0.5\Delta t - KX}{K - KX + 0.5\Delta t} \tag{12.15}$$

$$C_2 = \frac{0.5\Delta t + KX}{K - KX + 0.5\Delta t} \tag{12.16}$$

$$C_3 = \frac{K - KX - 0.5\Delta t}{K - KX + 0.5\Delta t} \tag{12.17}$$

and
$$C_1 + C_2 + C_3 = 1 \tag{12.18}$$

In Eq. (12.14), initial outflow O_1, I_1, I_2, C_1, C_2 and C_3 are known (as K, X, Δt are known), so O_2 can be computed. Next to compute $O_3 = C_1I_3 + C_2I_2 + C_3O_2$, RHS is known so O_3 is computed. Similarly with known O_3, O_4 is computed. Thus, other values of O_5, Q_6,..., O_n may be computed. Hence outflow hydrograph (O vs t) is known, and thus the flood is routed.

Figure 12.4 shows the known inflow hydrograph and the outflow or routed hydrograph by Muskingum method.

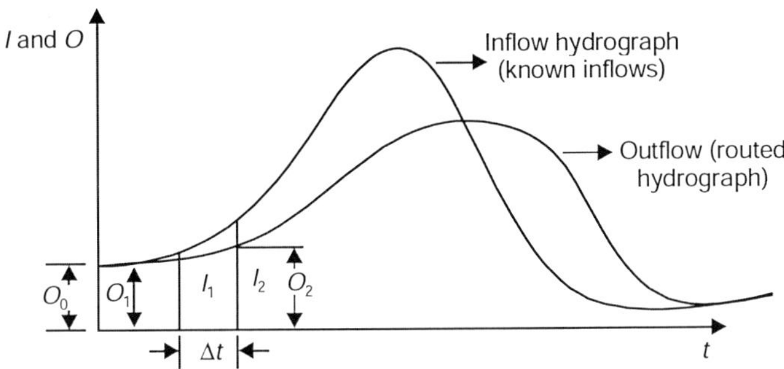

Figure 12.4 Inflow and outflow (routed) hydrographs.

12.6.1 Other Methods of Computing *K*

Some empirical equation and graphical methods have been given to determine the value of K of Muskingum equation. Empirical equations have their inherent weakness or limitations of the constant which varies in a wide range of values from catchment to catchment. Therefore, they cannot be used safely unless the catchments are identical almost in all respects. The graphical method already described in section 12.5.1 is quite reasonable and dependable. Yet the other methods are as follows:

Clark's method (1945): An empirical equation given by Clark[9] (1945) is:

$$K = \frac{CL}{\sqrt{s}} \tag{12.19}$$

Here

L = Length main stream in miles
s = mean slope of the channel
C = constant, varies from 0.8 to 2.2

[9] Clark, C.O., Storage and Unit Hydrograph, Trans. 110, pp. 1418–1488, 1945.

Linsley's method (1945): He suggested that

$$K = \frac{bL\sqrt{A}}{\sqrt{s}} \tag{12.20}$$

where

A = Drainage area in km^2

b varies from 0.01 to 0.03 when L is in km.

Agarwal and Das[10] (1987), analyzing the Himalayan catchment of Ramaganga river proposed that $C = 0.00427$ in Clark's equation and $b = 0.000123$ in Linsley's equation when L is in metre and A is in km^2.

Wilson's method (1969): Referring $X = O$ in Muskingum equation, i.e., pure reservoir,

$$I = O,$$

$$\therefore \quad S = KO \tag{12.21}$$

i.e., storage is a function of outflow only. Wilson[11] suggested that K can be determined as a ratio of volume under narrow band of recession limb of hydrograph as (Figure 12.5a):

$$K = \frac{\Delta V}{\Delta O} \tag{12.22}$$

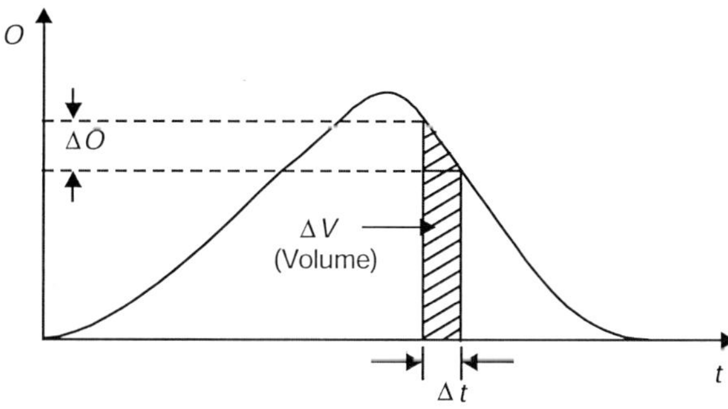

(a) Determination of K by Wilson method

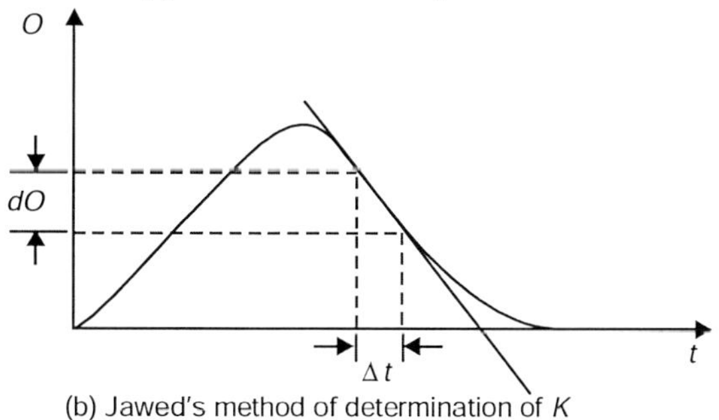

(b) Jawed's method of determination of K

Figure 12.5 Determination of K by Wilson and Jawed method.

[10] Agrwal, A. and Das, Ghansyam, Determination of Instantaneous Unit Hydrograph for Himalayan Watershed, *Jour. Inst of Engrs,* 67, CI 4, 1987.

[11] Wilson, E.M., *Engineering Hydrology*, Macmillan, London, 1969.

Jawed's method (1973): Differentiating Eq. (12.21),

$$\frac{dS}{dt} = K\frac{dO}{dt} \quad \text{(i)}$$

Again continuity gives

$$I - O = \frac{dS}{dt}$$

At the point of inflection on the recession limb, $I = O$

$$\frac{dS}{dt} = -O \quad \text{(ii)}$$

Comparing Eqs. (i) and (ii),

$$K\frac{dO}{dt} - O$$

$$K = \frac{-O}{\frac{dO}{dt}} \quad (12.23)$$

($\frac{dO}{dt}$ at the point is –ve, hence K is positive.)

Equation (12.22) can be solved graphically locating the point.

Example 12.1

The inflow and outflow hydrograph of a natural stream are as follows. Estimate the value of K and X in Muskingum equation.

Time (hrs)	0	6	12	18	24	30	36	42	48	54	60
Inflow (m^3/sec)	5	18	48	48	30	20	12	8	5	3	3
Outflow (m^3/sec)	5	4	10	27	36	33	27	21	15	11	7

Solution:

Here Δt = 6 hrs

(i) Find storage S against time t

Table 12.1 Computation of storage S

t	0	6	12	18	24	30	36	42	54	60
I	5	18	48	48	30	20	12	8	3	3
O	5	4	10	27	36	33	27	21	11	7
$I - O$	0	14	38	21	–6	–13	–15	–13	–8	–4
$(I - O)\ \Delta t = S$	0	84	228	126	–36	–78	–90	–78	–48	–24
$\Sigma\ \Delta S = S$	0	84	312	438	402	324	234	156	108	84

(ii) Calculate [(*XI* + (1 – *X*)*O*] for different *X* values (i.e., *X* = 0.3, *X* = 0.25)

Table 12.2 Computation of [*XI* + (1 – *X*)0]

***X* = 0.3**			***X* = 0.25**		
XI	(1 – *X*)*O*	[*XI* + (1 – *X*)*O*]	*XI*	(1 – *X*)*O*	[*XI* + (1 – *X*)*O*]
1.5	3.5	5	1.25	3.75	5
5.4	2.8	8.2	4.5	3.0	7.5
14.4	7	21.4	12	7.5	19.5
14.4	18.9	33.3	12	20.25	32.25
9.0	25.2	34.2	7.5	27	32.5
6.0	23.1	29.1	7.5	27	32.5
3.6	18.9	22.5	5.0	24.75	29.75
2.4	14.7	17.1	3.0	20.25	23.25
0.9	7.7	8.6	2	15.75	17.75
0.9	4.9	5.8	0.75	8.25	9.00
		0.75	5.25	6.00	

(iii) Now, plot *S* i.e., (Σ Δ*S*) vs [*XI* + (1 – *X*)*O*] when *X* = 0.3 and *S,* i.e., (Σ Δ*S*) vs [*XI* + (1 – *X*)*O*] when *X* = 0.25

From Tables 12.1 and 12.2

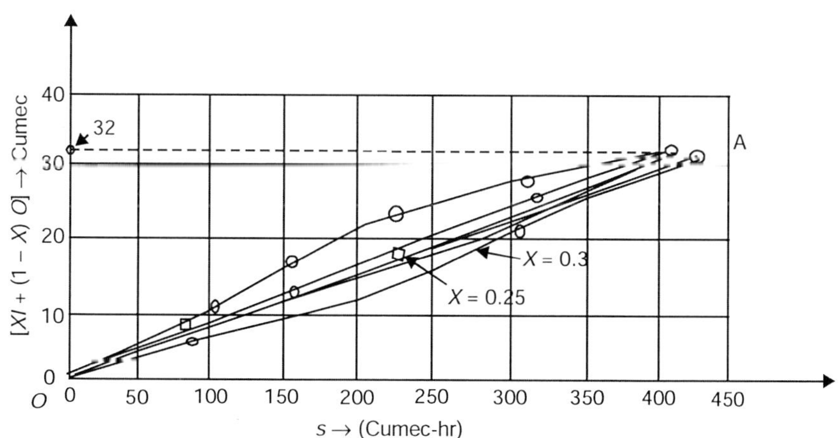

$$\therefore \qquad X = 0.25, \text{Slope of } OA, K = \frac{S}{[XI + (1-X)O]}$$

$$= \frac{450}{32} = 14.0625 \cong 14 \text{ hrs}$$

X = 0.25 and *K* = 14 hrs

EXAMPLE 12.2

In a river, the following inflow hydrograph was recorded. Route the hydrograph in the reach when K = 10 hrs, X = 0.25, initial outflow = 10 m^3/sec. Also find the attenuation of peak flow and lag of peak time.

Time (hrs)	0	6	12	18	24	30	36	42	48	54	60	66
Inflow	10	25	50	75	80	74	65	50	40	30	20	10

Solution: Muskingum routing equation is given as:

$$O_2 = C_1 I_2 + C_2 I_1 + C_3 O_1 \quad \text{(i)}$$

All I_1, I_2, I_3, …, I_n and O_1 are known.

Now, $C_1 = \dfrac{0.5\Delta t - KX}{K - KX + 0.5\Delta t}$ Here K = 10 hrs, X = 0.25 and Δt = 6 hrs

$$\therefore \quad C_1 = \frac{0.5 \times 6 - 10 \times 0.25}{10 - 10 \times 0.25 + 0.5 \times 6} = \frac{0.5}{10.5} = \left(\frac{1}{21}\right)$$

Then

$$C_2 = \frac{0.5\Delta t + KX}{K - KX + 0.5\Delta t} = \frac{5.5}{10.5} = \left(\frac{1.1}{2.1}\right)$$

and

$$C_3 = \frac{K - KX + 0.5\Delta t}{K - KX + 0.5\Delta t} = \frac{4.5}{10.5} = \left(\frac{0.9}{2.1}\right)$$

Using Eq. (i),

$$O_2 = \frac{1}{21} \times 25 + \frac{1.1}{2.1} \times 10 + \frac{0.9}{2.1} \times 10 = 10.74 \text{ m}^3/\text{sec}$$

and

$$O_3 = \frac{1}{21} \times 25 + \frac{1.1}{2.1} + \frac{0.9}{2.1} \times 10.74 = 20.068 \text{ m}^3/\text{sec}$$

Similarly

$$O_4 = 38.36 \text{ m}^3/\text{sec}, O_5 = 59.54 \text{ m}^3/\text{sec}, O_6 = 70.94 \text{ m}^3/\text{sec}$$

$$O_7 = 71.85 \text{ m}^3/\text{sec}, O_8 = 63.30 \text{ m}^3/\text{sec}, O_9 = 56.50 \text{ m}^3/\text{sec}$$

$$O_{10} = 46.58 \text{ m}^3/\text{sec}, O_{11} = 36.61 \text{ m}^3/\text{sec and } O_{12} = 26.64 \text{ m}^3/\text{sec}$$

Plot of inflow hydrograph and outflow hydrograph to find attenuation and lag time of peak flow is shown in Figure 12.6.

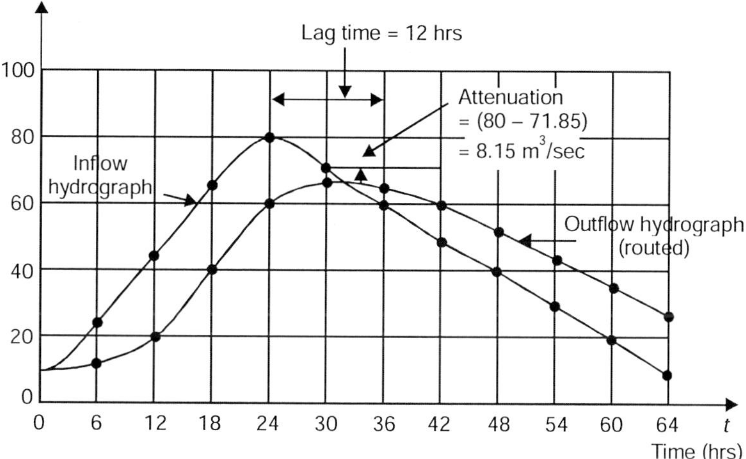

Figure 12.6 Answer of Example 12.2, routed floods, attenuation and lag time.

12.7 RESERVOIR ROUTING

The routing technique when applied to reservoir, is called **reservoir** or **storage routing**. In reservoir, inflow $I(t)$ comes from river which is known. Outflow $O(t)$ from reservoir is controlled by spillway gates. Thus, both storage $S(t)$ and outflow $O(t)$ vary with time which causes the variation of stage or elevation $H(t)$. When $I(t)$ is known, determination of $S(t)$, $O(t)$ and $H(t)$ is called **reservoir routing.**

Essential data required for reservoir routing are:

1. Field data to establish relationship among outflow $O(t)$, Storage $S(t)$ and stage or reservoir elevation $H(t)$.
2. Inflows $I(t)$ at any time step Δt, i.e., inflow hydrographs.
3. Known initial storage S_1, outflow O_1 at the start, i.e., at $t = 0$ as boundary condition.

12.8 METHODS OF RESERVOIR ROUTING

The most common and simple methods that are frequently used by field engineers are:

1. Goodrich method (1931)
2. Modified Pul's method (1928)

12.8.1 Goodrich Method (1931)

In Goodrich[12] method, continuity equation [Eq. (12.1)] is written as:

$$\left(\frac{I_1 + I_2}{2}\right) - \left(\frac{O_1 + O_2}{2}\right) = \frac{S_2 - S_1}{\Delta t}$$

or

$$\left(\frac{2S_2}{\Delta t} + O_2\right) = I_1 + I_2 + \left(\frac{2S_1}{\Delta t} - O_1\right) \tag{12.24}$$

Equation (12.24) is called **Goodrich equation** of reservoir routing.

[12] Goodrich, R.D., Rapid Calculation of Reservoir Discharge Civil Engg., vol., 1931.

Stepwise solution procedure:

1. Select time increment Δt.
2. From previous field data, plot stage or elevation H vs $\left(\frac{2S}{\Delta t}+O\right)$ curve and elevation H vs outflow O curve (Figure 12.7).

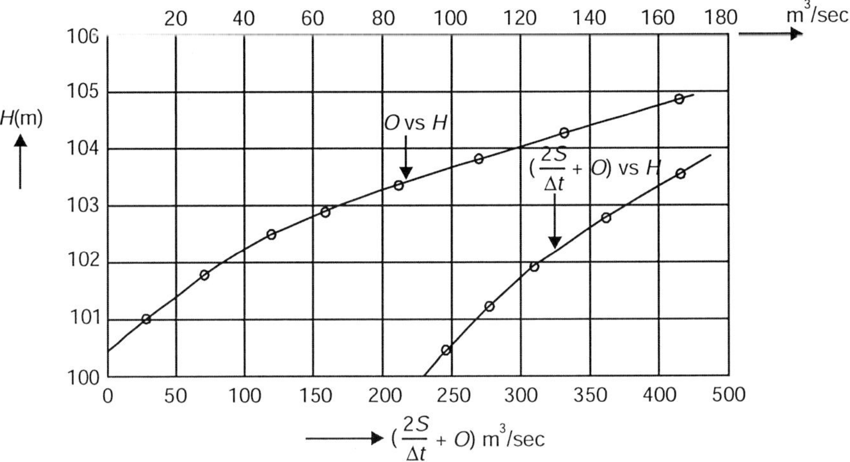

Figure 12.7 Plot of $\left(\frac{2S}{\Delta t}+O\right)$ vs H and O vs H curves of reservoir from field data.

3. In the beginning of the routing, the data of storage S_1, outflow O_1 and elevation are known so we can calculate $\left(\frac{2S_1}{\Delta t}+O_1\right)$.
4. In Eq. (12.24), R.H.S. is known, so $\left(\frac{2S_2}{\Delta t}+O_2\right)$ term is known.
5. From known $\left(\frac{2S_2}{\Delta t}+O_2\right)$, find out corresponding H(m) (is known) and from H, find O_2 from the curve O vs H.
6. Calculate $\left(\frac{2S_2}{\Delta t}-O_2\right)$ term subtracting $2O_2$ from $\left(\frac{2S_2}{\Delta t}+O_2\right)$ and again R.H.S. is known for calculating $\left(\frac{2S_3}{\Delta t}-O_3\right)$.
7. Keep repeating the procedures until all outflows $O_2, O_3, O_4, \ldots, O_n$ are obtained, i.e., all floods are routed.

12.8.2 Modified Pul's Method

From equation of continuity,

$$\left(\frac{I_1 + I_2}{2}\right) - \left(\frac{O_1 + O_2}{2}\right) = \frac{S_2 - S_1}{\Delta t}$$

or
$$\frac{1}{2}(I_1 + I_2)\Delta t + \left(S_1 - \frac{O_1}{2}\Delta t\right) = \left(S_2 + \frac{O_2}{2}\Delta t\right) \tag{12.25}$$

Modified Pul's method is based on Eq. (12.25). In the Pul's method, data of storage S, elevation H and outflow O are required to prepare the routing curves O vs E and $\left(S + \frac{O}{2}\Delta t\right)$ vs E as shown in Figure (12.8)

Steps required in routing:

1. Time interval Δt is selected within 20% to 40% of time of rise of inflow hydrograph.
2. From the available past data, routing curves O vs H and $\left(S_2 + \frac{O}{2}\Delta t\right)$ vs H are plotted as shown in Figure 12.8.

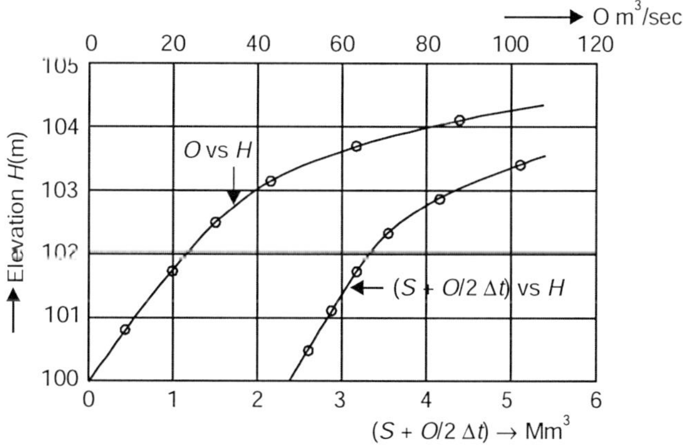

Figure 12.8 Routing curves: H vs O and $(S + O/2\ \Delta t)$ vs H in Pul's method.

3. Initial outflow O_1, S_1, Δt and all inflows are known. Therefore L.H.S. of Eq. (12.25) is known. Hence $\left(S_2 + \frac{O_2}{2}\Delta t\right)$ is known.
4. Determine corresponding H from known $\left(S_2 + \frac{O_2}{2}\Delta t\right)$ from the routing curve of $\left(S + \frac{O}{2}\Delta t\right)$ vs H.

5. Determine O_2 from O vs H curve.

6. Subtract $O_2\,\Delta t$ from known $\left(S_2 + \frac{O_2}{2}\Delta t\right)$ to $\left(S_2 - \frac{O_2}{2}\Delta t\right)$.

7. Again L.H.S. is known to compute $\left(S_3 + \frac{O_3}{2}\Delta t\right)$.

8. Repeat the process until all the outflows O_4, O_5,..., O_n are computed.

12.9 RUNGE-KUTTA METHOD

This numerical scheme was initially devised by **Carl Runge** (1894), a German mathematician and finally extended by W. Kutta, a few years later. Therefore, this scheme is called **Runge-Kutta method**. This method can be applied to many problems of science and engineering where iteration for a solution is required. In open channel flow, Runge-Kutta method is successfully applied in computation or solution of gradually varied flow.

Similarly, this method is applied to reservoir routing also. Carnahan et al[13] (1969), adopted third order Runge-Kutta Scheme. This is another alternative scheme of reservoir routing to solve the continuity equation of storage. But this method is more complicated than the method applied or used earlier. Only advantage in this method is storage vs elevation curve is not required. Stage H or increase or decrease of stage after time Δt is computed by Runge-Kutta method and after determining H, outflow O has to be interpolated. Computer program may be developed to determine H at every time step by interpolation and then outflow is to be interpolated from O vs H curve. Details of this method are referred to Chow et al[14] (1988).

12.10 LINEAR RESERVOIR MODELLING METHOD

This method is also known as **Nash modelling**. The routed outflow hydrograph is in the form of instantaneous unit hydrograph. Details of the method have already been discussed in Chapter 7 in Section 7.25.

12.11 COMPOSITE MODEL METHOD

In hydrologic model method, linear reservoirs linking in parallel can also be used for flood routing. Nash model[15] (1957) described in Chapter 7 (Section 7.22) was on linear reservoir placed in series. Kraijenhoff van der leur[16] (1958) showed that linear reservoirs may be used to

[13] Carnahan, B., Luther, H.A. and Wikes, J.O., *Applied Numerical Methods*, Wiley, New York, 1969.

[14] Chow, V.T., Maidant, D.R. and Mays, L.W., *Applied Hydrology*, McGraw-Hill, New York, 1988.

[15] Nash, J.E., The Form of the Instantaneous Hydrograph, Intl. Assoc. of Scientific Hydrology, Publ. 45 (3), pp. 114–21, 1957.

[16] Kraijenhoff van der Leur, D.A., A Study of non-steady Groundwater Flow with Special References to Reservoir Coefficient, De Ingenieur, vol. 70, No. 19, 1958.

model sub-surface water in saturated phase as well as surface water problems. Diskin et al[17] (1978) presented urban watershed parallel cascade linear reservoir model to route the flood. His input to the model was total rainfall hyetograph which feeds two parallel cascades of linear reservoir for impervious and pervious areas of the watershed, respectively.

Linear reservoir in series and parallel may be combined to model a hydrologic system. Doodge[18] (1959) developed a composite model of linear channels and linear reservoirs.

Randomized linear reservoir models have also been developed in which storage constant K is related to Horton[19] (1945) stream order. Boyd et al[20] (1979), Rodriguez-Iturbe and Valdes[21] (1979), Gupta et al[22] (1980), Gupta et al[23] (1968) developed geomorphic instantaneous unit hydrograph, the shape of which is related to the stream pattern of watershed. They considered the networks of streams draining a watershed as being a random combination of linear reservoirs, with the mechanism of the combination of linear reservoirs.

12.12 CONCLUSION

Different methods of hydrologic flood routing for both channel and reservoir routing are discussed in detail. The routing techniques help the hydrologists and meteorologists to predict or forecast the flood in river. Therefore, techniques are a presented step by step.

EXERCISES

12.1 The inflow and outflow hydrographs of in a certain river reach are as follows:

Time in days	1	2	3	4	5	6	7	8	9	10	11	12	13
I in m³/sec	25	65	130	200	175	130	105	82	63	58	52	46	38
O in m³/sec	24	32	51	90	135	148	108	130	110	92	78	65	52

Find the routing coefficient X and K. *[Ans: $X = 0.1$, $K = 2.35$ days)*

[17] Diskin, M.H., Ince, S. and Oben-Nyarke, K., Parallel Cascade Model for Urban Watersheds, *Jour. Div.*, ASCE, vol. 104, No. Hydro. 2, pp. 261–276, February 1978.

[18] Doodge, J.C.I., A General Theory of Unit Hydrograph, *Jour. Geophyhs. Res.*, vol. 64, No. 1, pp. 241–256, 1959.

[19] Horton, R.E., Erosional Development of Streams and their Drainage Basins: Hydro Physical Approach of Quantitative Morphology, Bull. Geol. Soc. Am., vol. 56, pp. 275–370, 1945.

[20] Boyd, M.J. Pilgrim, D.H. and Cordery, I., A Storage Routing Model based on Catchment Geomorphology, *Jour. Hydrology*, vol. 42, pp. 209–230, 1979.

[21] Rodriguer-Iturbe, I. and Valdes, J.B., The Geomorphological Structure of Hydrologic Response, Water Resour. Res., vol. 15, 1979.

[22] Gupta, V.K., Waymire, E. and Wang, C.T., Representation of Instantaneous Unit Hydrograph from Geomorphology, Water Resour. Res., vol. 16, No. 5, pp. 855–862, 1980.

[23] Gupta, V.K., Rodrigniz-Iturbe, I. and Wood, E.F., Scale Problem in Hydrology, D. Reidel Dordrecht, Holland, 1986.

12.2 Develop the outflow hydrograph of the same river (with $X = 0.1$ and $K = 2.35$ days) when inflow hydrograph is given as:

Time in days	1	2	3	4	5	6	7	8	9	10	11	12	13
I in m^3/sec	40	101	210	340	280	205	170	130	100	95	83	76	65

The outflow hydrograph is:

Time in days	1	2	3	4	5	6	7	8	9	10	11	12	13
O in m^3/sec	40.0	46.0	77.9	141.0	211.2	230.2	216.6	195.0	165.2	141.2	122.5	106.7	94.0

12.3 What is hydrologic routing? Show the relationship of outflow and storage. Deduce the general storage equation.

12.4 What is Muskingum routing equation? How will you route flood by this method?

12.5 Discuss Goodrich and Pul's methods of reservoir routing. Also give the stepwise solution of these methods.

SUGGESTED FURTHER READINGS

Chow, V.T. (Ed.), *Open Channel Hydraulics*, McGraw-Hill, New York, 1959.

_______, *A Handbook of Applied Hydrology*, McGraw-Hill, New York, 1964.

Jowed, K., Comparison of Methods of Deriving Unit Hydrographs, M.Sc. thesis of Civil Engg., Colorado State Unit, Fort Collins, 1973.

Linsley, R.K., Kohler, M.A. and Paulhus, *Applied Hydrology*, McGraw-Hill, New York, 1949.

Nemec, J., *Engineering Hydrology*, McGraw-Hill, London, 1972.

Prasad, R., Nonlinear Hydrologic System Response Model, *Jour. Hyd. Div.*, ASCE, vol. 93, pp. 201–222, July 1967.

World Meteorological Organisation (WHO), *Guide of Hydrological Practices*, WHO No. 168, 3rd. ed., Geneva, 1974.

Chapter 13

Hydraulic Routing

13.1 INTRODUCTION

The hydrologic routing discussed in the Chapter 12, can provide majority of routing problem in hydrology. But in some situations those procedures are not adequate. In the situation of stream where there may be reversal of flow, back water effect, flood routing in unusual events such as flood waves caused by failure of dam, probable maximum flood, hydrologic routing cannot serve the purpose. In those situations, only hydraulic method can give satisfactory result. Hydraulic method is based on solution of gradually varied unsteady continuity and momentum equations, i.e., Saint-Venant equations. Solutions of these two equations are complex and require computer to solve them by different numerical methods of solution.

13.2 SAINT-VENANT EQUATIONS

Saint-Venant equations are due to A.J.C. Barré de Saint-Venant[1] (1871). The following assumptions are made in the derivation of the equations:

1. Flow is one-dimensional.
2. Flow is assumed to be gradually varied along the channel so that hydrostatic pressure prevails and vertical acceleration may be neglected (Chow, 1959).
3. Channel bed slope is assumed to be small. Effect of scour and deposition are negligible.
4. Steady uniform turbulent flow resistance coefficient, i.e., Manning's n is applicable.
5. Fluid is incompressible, i.e., density ρ is constant.

[1] Saint-Venant, A.J.C Barré de, Theory of non permanent movement of water with application of floods of rivers and to the introduction of tides within their beds, Competes rendus des séances de L′Academic, vol. 73, pp. 147–154 and 237–240, 1871.

13.3 CONTINUITY EQUATION: IT'S DERIVATION

The continuity equation of an unsteady incompressible fluid may be written by Reynold's transport theorem, i.e.,

$$\frac{d}{dt}\iiint_{cv}\rho \; d\forall + \iint_{cs} \rho \; \mathbf{V}\cdot d\mathbf{A} = 0 \tag{13.1}$$

Here first term indicates rate of change of extensive property stored in the control volume (cv), and second term indicates inflow minus outflow, i.e., net outflow through the control surface (cs). When this theorem is used, inflows are always negative and outflows are positive.

Consider a small control volume of length dx in a channel (Figure 13.1). If Q is the inflow to this control volume,

$$\iint_{\text{inlet}} \rho \mathbf{V}\cdot d\mathbf{A} = -\rho Q \tag{13.2}$$

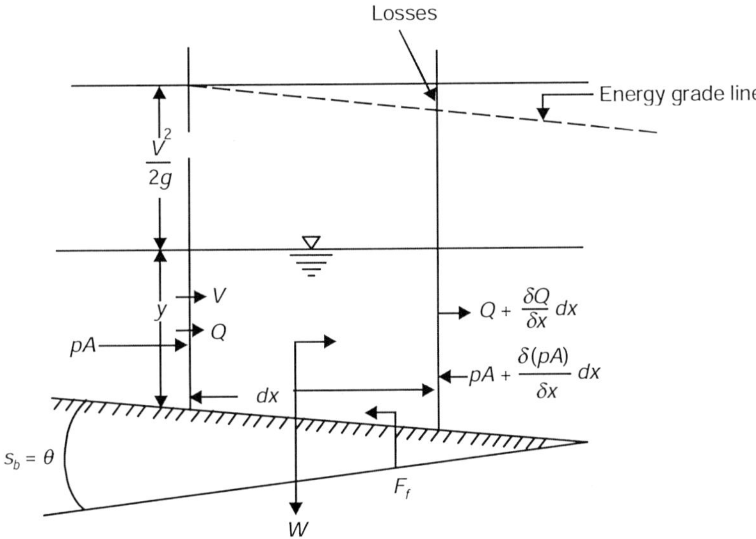

Figure 13.1 An elemental reach for derivation of Saint-Venant equation.

The right hand side is negative as inflow in negative in Reynold's transport theorem.

The mass outflow from the control volume is:

$$\iint_{\text{outlet}} \rho \; \mathbf{V}\cdot d\mathbf{A} = \rho\left(Q + \frac{\delta Q}{\delta x}dx\right) \tag{13.3}$$

If A is area (average) in the element dx,

$$\text{Volume} = Adx$$

$\therefore$ Rate of change of mass stored within the control volume is:

$$\frac{d}{dt}\iiint_{cv} \rho\forall = \frac{\delta(\rho A dx)}{dt} \tag{13.4}$$

Substituting the values from Eqs. (13.2), (13.3) and (13.4) in Eq. (13.1),

$$\frac{\delta(\rho A dx)}{\delta t} - \rho Q + \rho\left(Q + \frac{\delta Q}{\delta x}dx\right) = 0$$

Dividing by ρdx,

$$\frac{\delta A}{\delta t} + \frac{\delta Q}{\delta x} = 0 \tag{13.5}$$

Here Eq. (13.5) is the equation of continuity in the conservative form without any inflow to the reach considered. For solution of Saint-Venant equation, the non-conservative form of equation is mostly used in which the average velocity V is a dependent variable instead of Q. This form can be derived for a unit width of flow within the channel, neglecting the lateral inflow as follows:

For unit width of flow,

$$A = y \times 1 = y \text{ and } Q = VA = VY$$

Substituting $Q = VY$ in (13.5),

$$\frac{\delta y}{\delta t} + \frac{\delta(Vy)}{\delta x} = 0$$

or

$$\frac{\delta y}{\delta t} + y\frac{\delta V}{\delta x} + v\frac{\delta y}{\delta x} = 0 \tag{13.5a}$$

Equation (13.5a) is the continuity equation in non-conservation form. If q_l is the lateral inflow or outflow the reach, (13.5) may be written as:

$$\frac{\delta A}{\delta t} + \frac{\delta Q}{\delta x} = \pm q_l \tag{13.5b}$$

In a channel, top width T, average bottom width B is continuous function of depth y, the average inflow area δA for small change in flow depth δy may be approximated as $B\delta y$. Hence Eq. (13.5b) may be written as:

$$\frac{\delta Q}{\delta x} + B\frac{\delta y}{\delta t} = q_l \tag{13.5c}$$

Similarly, we may write $Q = AV$ and $\dfrac{\delta A}{\delta x} = B\dfrac{\delta y}{\delta x}$ then simplifying Eq. (13.5c) without q_l,

$$\frac{\delta(AV)}{\delta x} = B\frac{\delta y}{\delta t} = 0$$

$$A\frac{\delta V}{\delta x} + V\frac{\delta A}{\delta x} + B\frac{\delta y}{\delta t} = 0$$

$$A\frac{\delta V}{\delta x} + VB\frac{\delta y}{\delta x} + B\frac{\delta y}{\delta t} = 0$$

Dividing by B,

$$\left(\frac{A}{B}\right)\frac{\delta V}{\delta x}+\frac{VB}{B}\frac{\delta y}{\delta x}+\frac{B}{B}\frac{\delta y}{\delta t}=0$$

$$\therefore \quad \frac{\delta y}{\delta t}+D\frac{\delta V}{\delta x}+V\frac{\delta y}{\delta x}=0 \tag{13.5d}$$

where Hydraulic depth $D=\left(\frac{A}{B}\right)$

Another form of Eq. (13.5) is:

$$V\frac{\delta V}{\delta x}+V\frac{\delta A}{\delta x}+T\frac{\delta y}{\delta t}-q_l=0 \tag{13.5e}$$

Here

$$\delta A = T\delta y$$

13.4 DERIVATION OF MOMENTUM EQUATION

Newton's second law is written in the form of Reynolds transport theorem, i.e.,

$$\Sigma F=\frac{d}{dt}\iiint_{cv} V\rho d\forall+\iint_{cs} V\rho V\cdot dA \tag{13.6}$$

i.e., sum of the forces applied is equal to the rate of change momentum stored within the control volume plus the momentum across the control surface. In Saint-Venant equations, gravity force F_g, force of resistance or friction F_f and pressure force (hydrostatic) P are mainly considered.

Force of gravity W acting vertically downward for the control volume in dx (Figure 13.1) and its component along flow direction $= W \sin\theta = F_g = \rho gAdx \sin\theta = \rho gAdxs_b$ (13.7)

$$\text{Channel bed slope } s_b=-\frac{dz}{dx}$$

if z is datum, z decreases as x increases, hence s_b is negative.

Force of friction on the control volume is $-\lambda_0\,(Pdx)$, where p is the wetted perimeter $= Adx$ and

$$\lambda_0 = wRs_f = \rho g\,(A/P)\,s_f$$

$$\therefore \quad F_f=-\rho g\left(\frac{A}{P}\right)s_f Pdx=-\rho gAs_f dx \tag{13.8}$$

Pressure Force

$$P=pA-\left(pA+\frac{\partial(pA)}{\partial x}dx\right)=-\frac{\partial(pA)}{\partial x}dx$$

$$\because \quad p=\rho gy,\ P=-\frac{\partial(\rho gy)A}{\partial x}dx=-\rho gA\frac{\partial y}{\partial x}dx \tag{13.9}$$

Assuming momentum coefficient β to be unity,

$$\iint_{\text{inlet}} V\rho\mathbf{V}\cdot d\mathbf{A} = -\rho QV$$

and at outlet,

$$\iint_{\text{outlet}} V\rho\mathbf{V}\cdot d\mathbf{A} = \rho QV + \frac{\partial(\rho QV)}{\partial x}dx$$

Thus net moment outflow across control surface (cs) is:

$$\iint_{cs} V\rho\mathbf{V}\cdot d\mathbf{A} = -\rho QV + \rho QV + \frac{\partial(\rho QV)}{\partial x}dx = \frac{\partial(\rho QV)}{\partial x}dx \quad (13.10)$$

The time rate of change of momentum stored in the control volume is obtained as:

Volume of the elemental area dx (Figure 13.1) of average area $A = Adx$

∴ Momentum is:

$$\rho(Adx)V = \rho(AV)dx = \rho Qdx$$

Then
$$\frac{d}{dt}\iiint_{cv} V\rho d\forall = \frac{\partial}{\partial t}(\rho Qdx) = \rho\frac{\partial Q}{\partial t}dx \quad (13.11)$$

Equating the sum of Eqs. (13.7), (13.8) and (13.9) with sum of Eqs. (13.10) and (13.11), we get

$$\rho gAdxs_b - \rho gAdxs_f - \frac{\rho gA\partial y}{\partial x}dx = -\frac{\partial(\rho QV)}{\partial x}dx + \rho\frac{\delta Q}{\delta t}dx$$

Dividing by ($\rho\, dx$),

$$gAs_b - gAs_f - \frac{gA\partial y}{\partial x} = \frac{\partial(QV)}{\partial x} + \frac{\partial Q}{\partial t}$$

Replacing V by $\dfrac{Q}{A}$,

$$gAs_b - gAs_f - gA\frac{\partial y}{\partial x} = \frac{\partial(Q^2/A)}{\partial x} + \frac{\partial Q}{\partial t}$$

or
$$\frac{\partial Q}{\partial t} + \frac{\partial(Q^2/A)}{\partial x} + gA\left(\frac{\partial y}{\partial x} + s_f - s_b\right) = 0 \quad (13.12)$$

Equation (13.11) is the Saint-Venant moment equation in conservation form.

To obtain the non-conservative form of moment of equation like the continuity equation, it can be shown that

Non-conservative form of Saint-Venant momentum equation per unit to be

$$\frac{\partial V}{\partial t} + \frac{V\partial V}{\partial x} + g\left(\frac{\partial y}{\partial x} + s_f - s_b\right) = 0 \quad (13.13)$$

Strelkoff[2] (1969) has shown that Saint-Venant momentum equation can be derived from energy approach also. The following are two easy methods of deriving momentum equation in non-conservative form.

Derivation by energy approach: Consider Figure 13.2 and writing energy equation at (1) and (2),

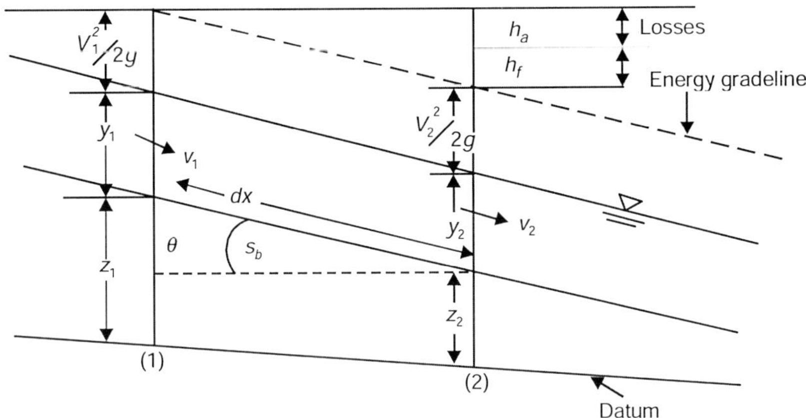

Figure 13.2 Simplified energy representation in unsteady flow.

$$z_1 + y_1 + \frac{V_1^2}{2g} = z_2 + y_2 + \frac{V_2^2}{2g} + \text{Losses} \tag{13.14}$$

Loss in unsteady flow due to friction and acceleration, i.e.

$$\text{Losses} = h_f + h_a$$

But

$$\frac{h_f}{\partial x} = s_f$$

$\therefore$

$$h_f = s_f \partial x \tag{13.15}$$

In unsteady flow, velocity varies with t, i.e., $\dfrac{\partial v}{\partial t}$ exists. This $\dfrac{\partial v}{\partial t}$ is an acceleration and due to creation of this acceleration, a force is required to produce causing loss of energy which is h_a, i.e., in head form per unit weight.

[2] Strelkoff, T., One Dimensional Flow of Open Channels, *Jour. Hydro. Div.*, ASCE, Civil Engineers, 95, Hy3 pp. 861–66, 1969.

Force produced due to this acceleration/unit weight

$$= \left(\frac{\text{Mass}}{\text{Unit weight}}\right) \times \text{Acceleration}$$

$$= \left(\frac{1}{g}\right)\frac{\partial v}{\partial t}$$

Work done by this force = Force × Distance

i.e.,
$$h_a = \frac{1}{g}\frac{\partial v}{\partial t}\partial x$$

Substituting these two losses h_f and h_a in Eq. (13.14),

$$z_1 + y_1 + \frac{V_1^2}{2g} = z_2 + y_2 + \frac{V_2^2}{2g} + \frac{1}{g}\frac{\partial v}{\partial t}\partial x + s_f \partial x$$

or
$$(z_2 - z_1) + (y_2 - y_1) + \left(\frac{V_2^2}{2g} - \frac{V_1^2}{2g}\right) + \frac{1}{g}\frac{\partial v}{\partial t}\partial x + s_f \partial x = 0$$

or
$$\partial z + \partial y + \partial\left(\frac{V^2}{2g}\right) + \frac{1}{g}\frac{\partial v}{\partial t}\partial x + s_f \partial x = 0$$

Dividing by ∂x,

$$\frac{\partial z}{\partial x} + \frac{\partial y}{\partial x} + \frac{\partial}{\partial x}\left(\frac{V^2}{2g}\right) + \frac{1}{g}\frac{\partial v}{\partial t} + s_f = 0$$

Now,
$$\frac{\partial z}{\partial x} = -s_b \quad \text{and} \quad \frac{\partial}{\partial x}\left(\frac{V^2}{2g}\right) = \frac{2V\partial V}{2g\partial x} = \frac{V}{g}\frac{\partial V}{\partial x}$$

or
$$-s_b + \frac{\partial y}{\partial x} + \frac{V}{g}\frac{\partial V}{\partial x} + \frac{1}{g}\frac{\partial V}{\partial t} + s_f = 0$$

Multiplying by g,

$$-gs_b + g\frac{\partial y}{\partial x} + V\frac{\partial V}{\partial x} + \frac{\partial V}{\partial t} + gs_f = 0$$

or
$$\frac{\partial V}{\partial t} + V\frac{\partial V}{\partial x} + g\left(\frac{\partial y}{\partial x} + s_f - s_b\right) = 0$$

It is the same equation (non-conservative form) as obtained in Eq. (13.13).

Derivation by momentum approach: On the control volume between (1) and (2) of the reach length, Δx of Figure 13.3, it may be written as:

$$pA - \left(p + \frac{\partial p}{\partial x}\Delta x\right)A + W\sin\theta - F_f = \text{Mass} \times \text{Acceleration} \tag{13.16}$$

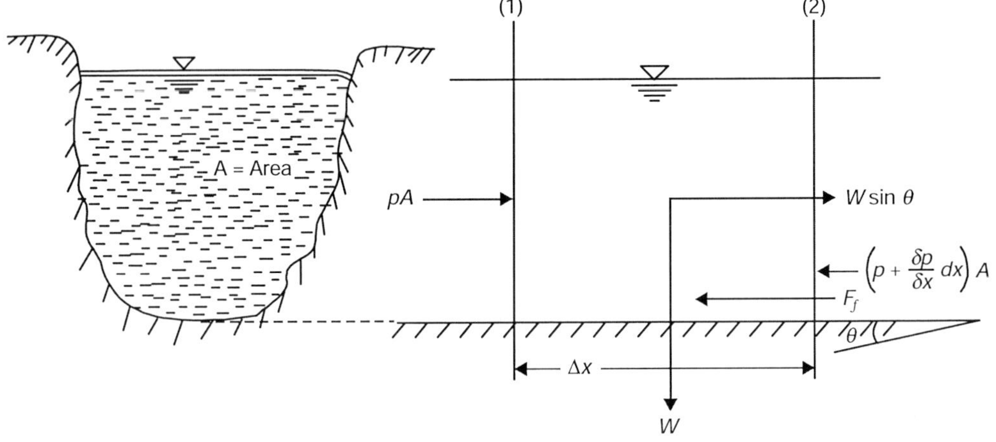

Figure 13.3 Simplified momentum representation in unsteady flow.

In unsteady flow,

$$\text{Acceleration} = \left(\frac{\partial V}{\partial t} + V\frac{\partial V}{\partial x}\right)$$

$$F_f = (wRs_f)(P\Delta x) = \rho gRs_f P\Delta x$$

$$W\sin\theta = \rho g(A\Delta x)s_b \qquad \because \sin\theta = s_b \text{ (small)}$$

$$\text{Mass} = \rho(A\Delta x)$$

Equation (13.16) becomes

$$-\frac{\partial p}{\partial x}\Delta xA + \rho gA\Delta xs_b - \rho gRs_f P\Delta x = \rho A\Delta x\left(\frac{\partial V}{\partial t} + V\frac{\partial V}{\partial x}\right)$$

Dividing by (ΔxA),

$$-\frac{\partial p}{\partial x} + \rho gs_b - \rho g\frac{A}{P}s_f\frac{P}{A} = \rho\left(\frac{\partial V}{\partial t} + V\frac{\partial V}{\partial x}\right) \qquad \because p = \rho gy$$

$$-\rho g\frac{\partial y}{\partial x} + \rho gs_b - \rho gs_b = \upsilon\left(\frac{\partial V}{\partial t} + V\frac{\partial V}{\partial x}\right)$$

Dividing by ρg,

$$\frac{-\partial y}{\partial x} + s_b - s_f = \frac{1}{g}\left(\frac{\partial V}{\partial t} + V\frac{\partial V}{\partial x}\right)$$

$$\therefore \qquad \frac{\partial V}{\partial x} + V\frac{\partial V}{\partial x} + g\left(\frac{\partial y}{\partial x} - s_b + s_f\right) = 0$$

Solution of Saint-Venant continuity equation and momentum equation in a river, i.e., to solve for flood hydrograph downstream from known inflow hydrograph at upstream is the hydraulic flood routing. But exact mathematical solution of these equations is not possible; therefore, different numerical solutions are used to solve these equations.

13.5 SOLUTION BY PREVIOUS INVESTIGATORS

As routing involves the solution of nonlinear hyperbolic partial differential equations of Saint-Venant, graphical procedures had been developed by Massu[3] (1899), Craya[4] (1946), Bergeron[5] (1937, 1953), Levin[6] (1942) before the advent of computer. Also method of approximate solution and trial and error were reported by Thomas[7] (1934). Choudhury[8] (1994) reported that in 1789, Mongue developed a graphical procedure for integration of Saint-Venant equations. He called this method or procedure the **method of characterizes** (MoC) which was used by Massu (1899) and Craya (1946) with the modification for analyzing flood waves in rivers. Craya's development had been used by Isaacson[9] et al (1954, 1956 and 1958) in flood prediction and river regulation problems. Further works on these equations for flood routing and unsteady flow conditions have been done by Stoker[10] (1953, 1957) on flood regulation of water waves in river. Brutsaert[11] (1971) verified Saint-Venant equations experimentally. A lot of works have been done by Huang and Song[12] (1985) on dynamic flood routing schemes, Amien and Fang[13] (1970)

[3] Massu, J., L′ integration graphique and appendice au memrie sur I′ integration graphique de grand, Annales 13, pp. 185–244, Belgium, 1899.

[4] Craya, A., Calcul Graphique des regimes variables des canaux, La Houille Blanche, No. 1, Nov, 1945, January; 1946, pp. 79–138, No. 2, March 1946, pp. 117–130.

[5] Bergeron, L. (General graphical method of commutation of the propagation plane waves), Memoires Société, des Ingeniures Civils de la, pp. 407–419, July–August, France, 1937.

———, (Graphical method for the computation of the Translatory waves) Société Francaise des Mecaniciena, Bulletin No. 7, Paris, 1953.

[6] Levin, Léon (Graphical method of computation of unsteady flow in open channel), Le Génié civil, vol. 119, No. 11–12, pp. 109–113, March 1942.

[7] Thomas, A.H., Hydraulics of Flood Measurement in River, Carnegie Institute of Engineering Bulletin 1934 (reprinted in 1936 and 1937).

[8] Choudhury, M.H., *Open Channel Flow,* Prentice-Hall of India, New Delhi, 1994.

[9] Isaacson, E., Stoker, J.J. and Troesch, B.A., Numerical Solution of Flood Prediction and River Regulation Problem (Ohio-Mississippi floods) Report II, Inst. Math. Science report, IMM-NYV-205, New York University, New York, 1954, 1950, 1958.

[10] Stoker, J.J., Numerical Solution of Flood Prediction and River Regulation Problems I, Derivation of Basic Theory and Formulation of Numerical Methods of Attack, New York University, Inst. of Math. Sci., Report no IMM – 200, 1953.

———, *Water waves,* Wiley Interscience, New York, 1957.

[11] Brutsaert, W.H., Saint-Venant equations experimentally verified, J. HY. Div. ASCE, vol. 97, No. Hy9, pp. 1387–1401, 1971.

[12] Huang, J. and Song, C.C.S., Stability of Dynamic Flood Routing Schemes, Jour. Hydro. Engg. ASCE, III, No. 12, pp. 1497–1505, 1985.

[13] Amien, M., and Fang, C.S., Implicit routing in natural channels, Jour. Hydro. Div. ASCE, pp. 2481–2500, December 1970.

on implicit method of flood routing, Tergidis and Strelkoff[14] (1970) on open channel surges, Fread[15] (1973, 1974, 1975, 1976, 1988, 1993) on flood routing, Conge[16] et al (1980) on river hydraulics, Choudhury[17] (1987) on applied hydraulic transients and Price[18] (1974) on four numerical schemes for flood routing. Dam-break flood routing works have been reported by Fennema and Choudhury[19] (1987), Das[20] (1978), Barr and Das[21] (1980, 1981), Sarma and Das[22] (2001), Su and Barness[23] (1970), Sakkas and Strelkoff[24] (1973, 1976), Balloffet et al[25] (1974), Sudobicher[26] (1972), Strelkoff[27] (1969, 1970), U.S. Army Corps of Engineers[28] (1960, 1961), Ligget and Woolhiser[29] (1967), Escande et al[30] (1961), Ponce et al[31] (2003), Katapodes and Strelkoff[32] (1978), Hromodka[33](1985), Bellos and Sakkas[34] (1987), Aknabi and Katapodes[35]

---

[14] Tergidis, G. and Strelkoff, T., Computation of Open Channels Surges and Shocks, *Jour. Hydro. Div.*, ASCE, 96, No. 2, pp. 2581–2610, 1970.

[15] Fread, D.L., Flood Routing in *Handbook of Hydrology*. (D.R. Maidment, Ed.), McGraw-Hill, New York, 1993.

[16] Conge, J.A., Holly, F.M. and Verway, A., *Practical Aspects of Computed River Hydraulics*, Pitman, London, 1980.

[17] Choudhury, M.H., *Applied Hydraulic Transients*, 2nd ed., Von Nostrant Reinhold, New York, 1987.

[18] Price, R.K., Computation of Four Numerical Methods for Flood Routing, *Jour Hydro. Div. Proc.*, ASCE, pp. 879–899, July 1974.

[19] Fennema, R.J. and Choudhury, M.H., Simulation of 1-D Dam Break Flows, *Jour. Hydro. Research Inter. Assoc. for Hydro. Research*, 25, No. 1, pp. 41–51, 1987.

[20] Das, M.M., Resistance to Flow in Steady and Unsteady Free Surface Profiles, PhD Thesis, University of Strathclyde, Glasgow, 1978.

[21] Barr, D.I.H. and Das, M.M., Numerical Simulation of Dam-burst and Reflections against Laboratory Data, Inst. of Civ. Engrs., Part 2, 69, London, pp. 379–373, June 1980.
———, Simulation of Surges after Removal of Separating Barrier Between Shallower and Deeper Water Bodies, Inst. of Civ. Engrs., Part 2, 71, pp. 911–919, London, 1981.

[22] Sarma, A.K. and Das, M.M., Mathematical Model for Simulating Flood Propagation on Downstream of River Dike, Dept. of Mathematics, Univ. of Roorkee, pp. 569–573, January 2001.

[23] Su, S.T. and Barness, A.H., Geometric and Frictional Effects on Sudden Release, *Proc. ASCE, Hydro. Div.*, November 1970.

[24] Sakkas, G.J. and Strelkoff, T., Dam-break Flood in a Prismatic Dry Channel, *Proc. ASCE, Hy. Div.*, May 1973.

[25] Balloffet, J., Nahas, N. and Balloffet, A.F., Dam Collapse Waves in River, *Proc. ASCE, Hy. Div.*, May 1974.

[26] Sudobicher, V.G., Motion of Breach Wave along Dry Channel, Translated from Gidrotech. Stroit, No. 11, pp. 44–46, No. 11, pp. 44–46, Hydro Technic Construction, VDC, 532, 59, 629, 81, November 1972.

[27] Strelkoff, T., Numerical solution of Saint-Venant equations, *Jour. Hy. Div.*, ASCE, 96, No. 1, pp. 229–52, 1970.
———, One dimensional Equation of Open Channel Flow, *J. Hy. Div.*, ASCE, CIV. Engg., vol. 95, No. Hy. 3, pp. 861–876, 1969.

[28] U.S. Army Corps of Engineers, Flood Routing from Sudden Breached Dam, Mice. Paper No. 2–374, Report 1, Smooth Channel, 1960, Report 2, Rough Channel, 1961.

[29] Ligget, J.A. and Woolhiser, Difference Solution of Shallow Water Equation, *Jour. of Engg. Mechanics Div.*, Proc., ASCE, pp. 39–41, April 1967.

[30] Escande, L., Nongora, F., Castex, L. and Barthil, H., Influence of Certain Parameters on Sudden Flood Wave Downstream from A dam, La Houillie Blanche, No. 5, pp. 565–575, 1961.

[31] Ponce, V.M., Taher-Shamsi, A and Shetty, A.V., Dam Breach Flood Wave Propagation Using Dimensionless Parameters, *Jour. of Hy. Engg*, vol. 129, No. 10, pp. 777–782, 2003.

[32] Katapodcs, N. and Strclkoff, T., Computing 2D Dam-brcak Flood Wavcs, *Jour. Hy. Div.*, ASCE 104 (9), pp. 1269–1288, 1978.

[33] Hromodka, A two Dimensional Dam-break Flood Plain Model, Adv. Water Resources, vol. 8, 7–14, 1985.

[34] Bellos, C.S. and Sakkao, J.G., 1D Dam-break Flood Propagation on Dry Bed, *Jour. Hy. Eng.*, ASCE, vol 113, No. 2, p. 1510, 1987.

[35] Aknabi, A.A. and Katapodes, N.D., Model for Flood Propagation on Initially Dry Bed, *J. Hy. Eng.*, ASCE, 114, No. 7, 1988.

(1988), Savic and Holly[36] (1993), Wang et al[37] (2000), Tseng and Chu[38] (2000), Zoppou and Roberts[39] (2003), Sarma and Saikia-Das[40] (2006), Saikia-Das and Sarma[41] (June 2006, August 2006, October 2006, November 2006, December 2007).

Numerical solution methods of Saint-Venant equations are reported by Abbott and Basco[42] (1990), Abbott[43] (1980), Choudhury (1994), Katapodes[44] (1984), Gabutti[45] (1983), Lax[46] (1954), McCormack[47] (1969), Fennema and Choudhury[48] (1986), Morretti[49] (1979), Preissmann and Cunge[50] (1961), Vasiliev[51] et al (1965), Stoker (1953), Sudobicher and Shugrin[52] (1968) and many more investigators who have worked on this problem.

---

[36] Savic, L.J. and Holly, F.M., Dam-break Flood Wave Computed by Modified Gudunov Method, *Jour. Hy. Research,* Delft, Netherland, 311 (2), pp. 187–204, 1993.

[37] Wang, J.S., Ni, H.G. and He, H.S., Finite Difference TVD Scheme for Computation of Dam-break Problems, *J. Hy. Eng.,* 126(4), pp. 253–262, 2000.

[38] Tseng, M.H. and Chu, C.R., The Simulation of Dam-break Flows by an Improved Predictor-corrector TVD Scheme, Adv. Water Resource, 23, pp. 637–643, 2000.

[39] Zappou, C. and Roberts, S. Explicit Schemes for Dam-break Simulations, *Jour. Hy. Eng.*, 129 (1), pp. 11–34, 2003.

[40] Sarma, A.K. and Saikia-Das, M., Dam-break Hydraulics in Natural Channel, World Environ. and Water Resource Congress, 21–25 May, Omaha, Nebraska, USA, EWRI, ASCE, 2006.

[41] Saikia-Das, M. and Sarma, A.K., Numerical Simulation Model for Computation of Dambreak Flood in Natural Flood Plain Topography, *Jour., Dam Eng.,* vol. 17, No. 7, pp. 31–50, London, June 2006.

———, Dam Flood Forecasting with Simple Numerical Model, 3rd APHW, Held in Bangkok, 16–18 October 2006.

———, Dam-break Flood Simulation for Analysis of Risk Management and Damage Estimation, 1st Indian Disaster Management Congress, 28–30 November, Vigan Bhavan, New Delhi, 2006.

———, Numerical Model for Simulating Flow and River Bed Profiles in Natural River under Dam-failure Conditions, Proc. 15th IASTED conf., Rhodes, vol. 522, Greece, 26–28 June 2006.

———, Simulation Modeling for Planning and Management of River Basin's Under Flood Disaster Proc. Inter. Perspective on Environ. and Water Research, December 18–20, New Delhi, EWRI, ASCE.

———, Analysis for Adopting Logical Channel Section for 1D Dam-break Analysis in Natural Channels, *ARPW Jour. of Engg. and Appl. Sci.,* vol. 1, No. 2, August 2006.

[42] Abbott, M.B. and Basco, D.R., *Computational Fluid Dynamics,* John Wiley, New York, 1990.

[43] Abbott, M.B., *Computational Hydraulics*, Pitman, London, 1980.

[44] Katapodes, N., A Dissipative Galerkin Scheme for Open Channel Flow, *Jour. Hy. Eng*., ASCE, 110, 4, pp. 450–460, 1984.

[45] Gabutti, B., On Two Upwind Finite Difference Schemes for Hyperbolic Equations in Non-conservative Form, Computers and Fluids, 11, No. 3, pp. 207–230, 1983.

[46] Lax, P.D., Weak Solutions of Nonlinear Hyperbolic Partial Differential Equations and their Numerical Computations, Comm. on Pure and Appl. Math, 7, 1954.

[47] MacCormack, R.W., Effect of Viscosity in Hyper Velocity Impact Clattering, Am. Inst. Aero. Astros., Paper 69–354, Cincinnati, OH, 1969.

[48] Fennema, R.J. and Choudhury, M.H., Explicit Numerical Scheme for Unsteady Free Surface Flows with Shocks, Water Res., 22, No. 13, pp. 1923–30, 1986.

[49] Morretti, G., The Lambda Scheme, Computers and Fluids, 7, pp. 191–205, 1979.

[50] Preissmann, A. and Cunge, J., Calcul du mascaret sur mechanics electroniques, La Houillie Blanche, No. 5, pp. 588–96, 1961.

[51] Vasiliev, O.F., Gladyshev, M.T., Pritvis, N.A. and Sudobicher, V.G., Numerical Method of Calculation of Shock Wave Propagation in Open Channels, Proc. 11th congress, Inter. Assoc. for Hy. Research, vol. 12, Paper 3, 1965.

[52] Sudobicher, V.G. and Shugrin, S.M., Flow Along Dry Channel, Izvestiya, So. AN SSSR, No. 3, 1968.

13.6 DIFFERENT NUMERICAL SCHEMES

Different numerical schemes used to solve Saint-Venant equation both in non-conservative and conservative forms are:

1. Method of characteristics
2. Finite difference explicit schemes
   - (i) Central difference scheme
   - (ii) Lax diffusive scheme
   - (iii) Lambda scheme
   - (iv) Gabutti scheme
   - (v) MacCormack scheme
3. Finite difference implicit schemes
   - (i) Preissmann scheme
   - (ii) Beam and warming scheme
   - (iii) Vasiliev scheme or method

Some of the above methods in conservative and non-conservative forms of Saint-Venant equations are discussed as follows:

13.6.1 Method of Characteristics (MoC)

According to Choudhury (1994), it was Mongue in 1789 that developed a graphical method to integrate the Saint-Venant equations in non-conservative form. He called this procedure as **method of characteristics (MoC)**. Massu, Craya, Isaacson et al used this method for analyzing surges in river. Stoker and Abbott further modified this MoC. To describe this method, following equations are used:

$$\frac{\partial y}{\partial t} + D\frac{\partial V}{\partial x} + V\frac{\partial y}{\partial x} = 0 \tag{13.16a}$$

$$\frac{\partial y}{\partial t} + V\frac{\partial V}{\partial x} + g\frac{\partial y}{\partial x} = g(s_b - s_f) \tag{13.17}$$

Multiply Eq. (13.16a) by an unknown multiplier λ to get

$$\lambda\frac{\partial y}{\partial t} + \lambda D\frac{\partial V}{\partial x} + \lambda V\frac{\partial y}{\partial x} = 0 \tag{13.18}$$

Adding Eqs. (13.18) and (13.17),

$$\left[\frac{\partial V}{\partial t} + (V + \lambda D)\frac{\partial V}{\partial x}\right] + \lambda\left[\frac{\partial y}{\partial t} + \left(V + \frac{g}{\lambda}\right)\frac{dy}{dx}\right] = g(s_b - s_f) \tag{13.19}$$

Since $V(x, t)$ and $y(x, t)$ are functions of x and t, total derivatives of V and y are:

$$\left.\begin{aligned} \frac{dV}{dt} &= \frac{\partial V}{\partial t} + \frac{\partial V}{\partial x}\frac{dx}{dt} \\ \frac{dy}{dt} &= \frac{\delta V}{\delta t} + \frac{\delta V}{\delta x}\frac{dx}{dd} \end{aligned}\right\} \tag{13.20}$$

Equation (13.20) shows that the two terms inside the big brackets of Eq. (13.19) are total derivatives of unknown multiplier λ which is defined as:

$$\left.\begin{aligned} (V+\lambda D) &= \left(V+\frac{g}{\lambda}\right)=\frac{dx}{dt} \\ \text{or} \quad V+\lambda D &= V+\frac{g}{\lambda} \end{aligned}\right\} \tag{13.21}$$

$\therefore$
$$\lambda^2 = \frac{g}{D}$$

or
$$\lambda = \pm\sqrt{\frac{g}{D}}$$

i.e.
$$\lambda_1 = \sqrt{\frac{g}{D}} = \sqrt{g\frac{B}{A}} \text{ and } \lambda_2 = -\sqrt{\frac{g}{D}} = -\sqrt{g\frac{B}{A}}$$

But celerity
$$C = \sqrt{gD} = \sqrt{g\frac{A}{B}}$$

$\therefore$
$$\lambda_1 = \frac{g}{C} \quad \text{and} \quad \lambda_2 = -\frac{g}{C} \tag{13.22}$$

Equation (13.19) can be written as:

$$\frac{dV}{dt} + \frac{g}{C}\frac{dy}{dt} = g(s_b - s_f) \tag{13.23}$$

If
$$\frac{dx}{dt} = V + C \tag{13.24}$$

Similarly, for $\lambda_2 = -g/c$, Eq. (13.19) becomes

$$\frac{dV}{dt} - \frac{g}{C}\frac{dy}{dt} = g(s_b - s_f) \tag{13.25}$$

If
$$\frac{dx}{dt} = V - C \tag{13.26}$$

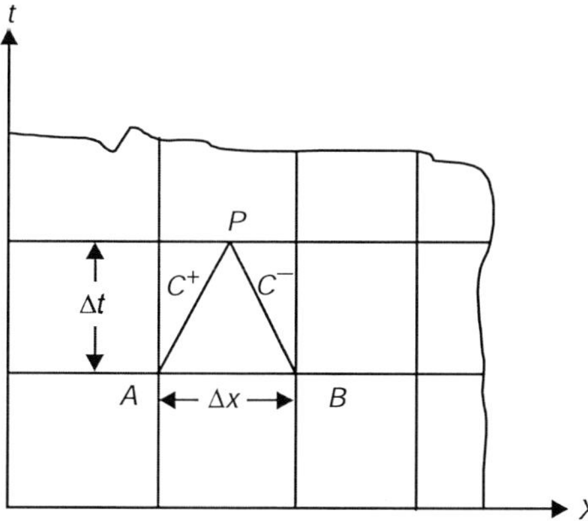

Figure 13.4 Positive and negative characteristics along *AP* and *BP*.

Equation (13.24) plots a curve in x–t plane, and this is called **positive characteristic** C^+. Equation (13.26) also plots a curve in x–t plane, and is called **negative characteristic** C^-. Equations (13.23) and (13.25) are valid if Eqs. (13.24) and (13.26) are satisfied respectively. That is equation (13.23) is valid along positive characteristics AP, and Eq. (13.25) is valid along negative characteristics BP. Equations (13.23) and (13.25) are called **compatibility equations**. With these algebraic manipulations, space variable x is eliminated from the governing partial differential equations (13.16) and (13.17), and they are converted to ordinary differential equations. Equations (13.16) and (13.17) are valid for any value of x and t; however, transformed equations are valid along the two characteristics. Equations (13.23) and (13.25) are multiplied by dt and integrated along AP and BP which gives

$$\int_A^P dV + \int_A^P \frac{g}{C} dy = \int_A^P g(s_b - s_f) dt \tag{13.27}$$

and

$$\int_B^P dV - \int_B^P \frac{g}{C} dy = \int_B^P g(s_b - s_f) dt \tag{13.28}$$

Integrating Eqs. (13.27) and (13.28), we obtain

$$V_P - V_A + \left(\frac{g}{C}\right)_A (y_P - y_A) = g(s_b - s_{fA})(t_P - t_A) \tag{13.29}$$

$$V_P - V_B + \left(\frac{g}{C}\right)_B (y_P - y_A) = g(s_b - s_{fB})(t_P - t_B) \tag{13.30}$$

Again

$$\text{Celerity } C = \sqrt{gD} = \sqrt{g\frac{A}{B}} = \left(g\frac{A}{B}\right)^{1/2}$$

Differentiating C with respect to t (assuming B to be constant),

$$\frac{dC}{dt} = \frac{1}{2} \cdot \frac{g}{C} \frac{dy}{dt}$$

$$\therefore \qquad \frac{d}{C} \cdot \frac{dy}{dt} = \frac{d(2C)}{dt} \tag{13.31}$$

Substituting the value of $\frac{g}{C} \cdot \frac{dy}{dt}$ in Eqs. (13.23) and (13.25), we have

$$\frac{d}{dt}(V + 2C) = g(s_b - s_f) \tag{13.32}$$

and

$$\frac{d}{dt}(V - 2C) = g(s_b - s_f) \tag{13.33}$$

The right hand side of the above two equations is called **source term**. If channel is horizontal and frictionless source term is zero.

Equation (13.32) is valid if $\dfrac{dx}{dt} = v + C$

Similarly, Eq. (13.33) is valid if $\frac{dx}{dt} = v - C$

We assume that flow depth and velocity are known at A and B as initial condition at $t = 0$ (Figure 13.4). Multiplying Eqs. (13.32), (13.33), (13.26) by dt and integrating within the limit,

$$(V + 2C)_P = (V + 2C)_A + g\int_A^P (s_b - s_f)dt \tag{13.34}$$

$$(V - 2C)_P = (V - 2C)_B + g\int_B^P (s_b - s_f)dt \tag{13.35}$$

$$x_P = x_A + \int_A^P (V + C)\,dt \tag{13.36}$$

$$x_P = x_B + \int_B^P (V - C)\,dt \tag{13.37}$$

Using trapezoidal rule to evaluate Eqs. (13.34) to (13.37), they may be written as:

$$V_P = V_A + 2(C_A - C_P) + \frac{1}{2}g(t_P - t_A)\,[(s_b - s_f)_P + (s_b - s_f)_A] \tag{13.38}$$

$$V_P = V_B + 2(C_A - C_P) + \frac{1}{2}g(t_P - t_B)\,[(s_b - s_f)_P + (s_b - s_f)_B] \tag{13.39}$$

$$x_P = x_A + (t_P - t_A)\left[\frac{1}{2}(V_P + C_P) + \frac{1}{2}(V_A + C_A)\right] \tag{13.40}$$

$$x_P = x_B + (t_P - t_B)\left[\frac{1}{2}(V_P + C_P) + \frac{1}{2}(V_B + C_B)\right] \tag{13.41}$$

Equations (13.38) to (13.41) are four equations with four unknowns, i.e., $v_{p,}$- y_p, x_p and t_p whose corresponding values at points A and B are known as **initial conditions**. Substituting the boundary conditions at both ends and taking these calculated values of v_p, y_p, x_p and t_p as initial values for the next time step Δt, values of v, y, x and t in the next time can be computed. The procedure is repeated until time $t_p = t_{\text{final}}$. The whole computations are normally done by writing a computer program.

13.6.2 Finite Difference Explicit Scheme

Central Difference Scheme: Saint-Venant equations have been solved by many of the investigators in finite difference (FD) methods in non-conservative form. Cunge et al (1980) are of the opinion that conservative form of the equations gives the better result. Conservative form will be discussed later in this chapter.

In the central difference scheme, variables V and y with respect to time and space may be written as follows with respect to Figure 13.5.

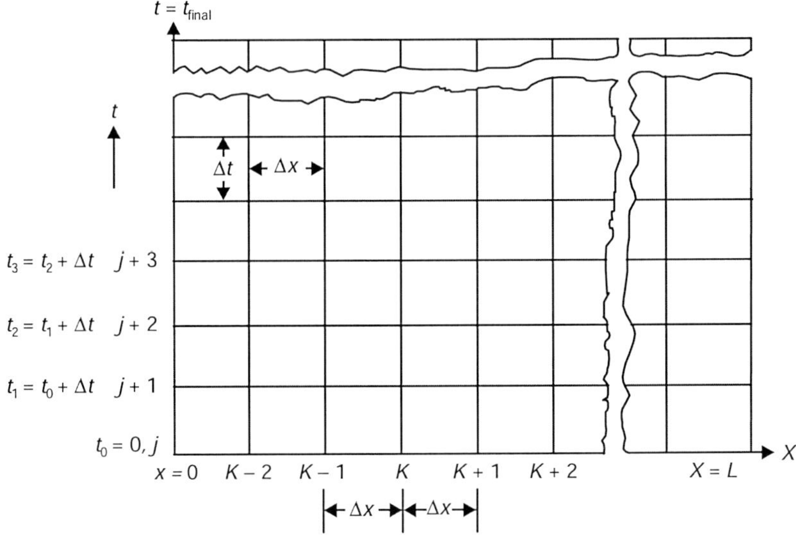

Figure 13.5 Flow domain in $x - t$ plane with finite difference grid.

$$\left.\begin{aligned}\frac{\partial V}{\partial x} &= \frac{V_{k+1}^{j} - V_{k-1}^{j}}{2\Delta x} \\ \frac{\partial y}{\partial x} &= \frac{y_{k+1}^{j} - y_{k-1}^{j}}{2\Delta x} \\ \frac{\partial V}{\partial t} &= \frac{V_{k}^{j+1} - V_{k}^{j}}{\Delta t} \\ \frac{\partial y}{\partial t} &= \frac{y_{k}^{j+1} - V_{k}^{j}}{\Delta t}\end{aligned}\right\} \tag{13.42}$$

Substitution of expressions of variables V and y from Eq. (13.42) in Saint-Venant equations, we have the following two equations of unknown y_k^{j+1} and V_k^{j+1}after simplification at next time step (j + 1), i.e.,

$$t_1 = t_0 + \Delta t$$

$$y_k^{j+1} = \frac{1}{2}\left(y_{k-1}^{j} + y_{k+1}^{j}\right) - \frac{1}{2}\frac{\Delta t}{\Delta x} D_k^{j}\left(V_{k+1}^{j} - V_{k-1}^{j}\right) - \frac{1}{2}\frac{\Delta t}{\Delta x} V_k^{j}\left(y_{k+1}^{j} + y_{k-1}^{j}\right) \tag{13.43}$$

and $$y_k^{j+1} = \frac{1}{2}\left(V_{k-1}^{j} - V_{k+1}^{j}\right) - \frac{1}{2}\frac{\Delta t}{\Delta x} g\left(y_{k+1}^{j} - y_{k-1}^{j}\right) - \frac{1}{2}\frac{\Delta t}{\Delta x} V_k^{j}\left(V_{k+1}^{j} - V_{k-1}^{j}\right) + g\Delta t\left(s_b - s_f\right) \tag{13.44}$$

The values of the variables on right hand side of Eqs. (13.43) and (13.44) are known from initial conditions. So values of y and V in the next time step $t_1 = (0 + \Delta t)$ in all interior grid points may be calculated. Boundary conditions or values of y and V at $x = 0$, $x = L$ and $t = t_1$ are given.

Now, all the values of y and V at all grid points at $t = t_1$ are known, and these values now become the initial values for calculating y and V in all interior points at time $t_2 = t_1 + \Delta t$ with the help of Eqs. (13.43) and (13.44). Again boundary conditions of y and V values are given at $x = 0$ and at $x = L$. All the values at t_2 become the initial values to calculate y and V at time $t_3 = t_2 + \Delta t$. This process of calculation and imposing boundary conditions are repeated until time $t = t_{final}$. The process of progressing the solution with time is called **marching procedure**. All these calculations at grid points are done by the computer program.

Stability conditions: In every numerical scheme, stability analysis is essential as the error in the solution may grow without limit depending on the sizes of Δx and Δt. For stability of solution, any one of following courant conditions must be maintained:

$$\Delta t \le \frac{\Delta x}{V_o + c}$$

i.e.

$$\Delta t \le \frac{\Delta x}{V_o + \sqrt{gD_o}} \tag{13.45}$$

and

$$\Delta t \le \frac{\sqrt{1 + 2\left|\frac{V_o}{C_o}\right|} - 1}{\left|\frac{V_o}{C_o}\right| \frac{gs_b}{V_o}} \tag{13.46}$$

Koren[53] (1967) while working on Lax diffusive scheme (will be discussed later) concluded that courant criteria given in Eq. (13.46) gives additional stability. V_o and C_o are the values at initial conditions.

French[54] (1986) worked in this scheme and concluded that the scheme is weak and inherently unstable. Care must be taken to choose very small Δx given by courant criteria for stability if the scheme is used. Making Δt much smaller stability may be achieved, but computer runtime will be more. But since it is the explicit solution, comparatively smaller runtime will be required.

Lax diffusive scheme: Lax (1954) presented a slight change of the central difference. This scheme is called **Lax diffusive scheme**. Choudhury (1987) claimed that this difference explicit scheme is one of the simplest of all to program and yields satisfactory results.

Variables V and Y in this scheme with respect to x and t are written as:

$$\frac{\partial V}{\partial x} = \frac{V_{k+1}^{j} - V_{k-1}^{j}}{2\Delta x} \tag{13.47a}$$

$$\frac{\partial y}{\partial x} = \frac{y_{k+1}^{j} - y_{k-1}^{j}}{2\Delta x} \tag{13.47b}$$

[53] Koren, V.I., The Analysis of Stability of Some Explicit Schemes for the Integration of Saint-Venant Equations, Meteorlogiyai Gidrologia, No. 1, 1967.

[54] French, R.H., *Open Channel Hydraulics*, McGraw-Hill, New York, 1986.

$$\frac{\partial V}{\partial t} = \frac{V_k^{j-1} - V^*}{\Delta t} \tag{13.47c}$$

when
$$V^* = \frac{1}{2}(V_{k-1}^j + V_{k+1}^j) \tag{13.47d}$$

$$\frac{\partial y}{\partial t} = \frac{y_k^{j+1} - y^*}{\Delta t} \tag{13.47e}$$

when
$$y^* = \frac{1}{2}(y_{k-1}^j + y_{k+1}^j) \tag{13.47f}$$

$$D^* = \frac{1}{2}(D_{k-1}^j + D_{k+1}^j) \tag{13.47g}$$

$$s_f^* = \frac{1}{2}(s_{f\,k-1}^{\;j} + s_{f\,k+1}^{\;j}) \tag{13.47h}$$

Differential formulations given by Eq. (13.47) when substituted in Saint-Venant equations (in non-conservative form), equations of depth y and velocity V in the next time step are obtained after simplification as:

$$y_k^{j+1} = \frac{1}{2}(y_{k-1}^j + y_{k+1}^j) - \frac{1}{2}\frac{\Delta t}{\Delta x}D_k^*(V_{R-1}^j - V_{k+1}^j) - \frac{1}{2}\cdot\frac{\Delta t}{\Delta x}V_k^*(y_{k+1}^j - y_{k-1}^j) \tag{13.48}$$

$$V_k^{j+1} = \frac{1}{2}(V_{k-1}^j + V_{k+1}^j) - \frac{1}{2}\cdot\frac{\Delta t}{\Delta x}g(y_{k+1}^j - y_{k-1}^j) - \frac{1}{2}\cdot\frac{\Delta t}{\Delta x}V_R^*(V_{R+1}^j - V_{k-1}^j) + g\Delta t(s_b - s_{f\,k}^*) \tag{13.49}$$

Equations (13.48) and (13.49) can be used to calculate y and V in the next time $t_1 = 0 + \Delta t$ in all the interior grid points and boundary conditions at $x = 0$ and at $x = L$ are imposed to calculate again y and V in the next time step $t_2 = t_1 + \Delta t$. The process of computation in the next time step and insertions of boundary values are being continued until time $t = t_{\text{final}}$. Selection of values for Δt and Δx must be made as per courant criteria in Eq. (13.46). According to Koren (1967), a courent criterion given by Eq. (13.46) gives a stable solution. This simplest method gives better result making Δt and Δx even smaller than given by courant criteria.

Lambda scheme: This scheme was proposed by Moretti (1979). This scheme has been used by Fennema and Choudhury (1986) and the scheme can take care of state of flow from subcritical to supercritical.

In this scheme, the Saint-Venant equations are converted to λ-form and then discretised them according to the sign of the characteristic directions thereby enforcing the correct signal direction, and eventually both predictor and corrector formulae have been applied to get the final form of y_k^{j+1} and v_k^{j+1} in marching procedure.

To formulate the scheme, Eq. (13.13a) is multiplied by $\dfrac{C}{g}$ to have,

$$\frac{C}{g}\frac{\partial V}{\partial t} + \frac{C}{g}V\frac{\partial V}{\partial x} + C\frac{\partial y}{\partial x} = C(s_b - s_f) \tag{13.50}$$

Equation (13.5d) is:

$$\frac{\partial y}{\partial t} + D\frac{\partial V}{\partial x} + V\frac{\partial y}{\partial x} = 0$$

We know

$$C = \sqrt{gD}$$

∴

$$\frac{C^2}{g} = D$$

$$\frac{\partial y}{\partial t} + \frac{C^2}{g}\frac{\partial V}{\partial x} + V\frac{\partial y}{\partial x} = 0 \tag{13.51}$$

Adding Eqs. (13.50) and (13.51), we have

$$\left[\frac{\partial y}{\partial t} + (V + C)\frac{\partial y}{\partial x}\right] + \frac{C}{g}\left[\frac{\partial V}{\partial t} + (V + C)\frac{\partial V}{\partial x}\right] = C(s_b - s_f) \tag{13.52}$$

Put $(V + C) = \lambda^+$ Eq. (13.52) becomes

$$\frac{\partial y}{\partial t} + \lambda^+\frac{\partial y}{\partial x} + \frac{C}{g}\left(\frac{\partial V}{\partial t} + \lambda^+\frac{\partial V}{\partial x}\right) = C(s_b - s_f) \tag{13.53}$$

Similarly, Eq. (13.50) from Eq. (13.51), and putting $V - C = \lambda^-$, we get

$$\frac{\partial y}{\partial t} + \lambda^-\frac{\partial y}{\partial x} - \frac{C}{g}\left(\frac{\partial V}{\partial x} + \lambda^+\frac{\partial V}{\partial x}\right) = -C(s_b - s_f) \tag{13.54}$$

If Eqs. (13.53) and (13.54) are added, it yields

$$\frac{\partial y}{\partial t} + \frac{1}{2}\left[\lambda^+\left(\frac{\partial y}{\partial x}\right)^+ + \lambda^-\left(\frac{\partial y}{\partial x}\right)^-\right] + \frac{1}{2}\frac{C}{g}\left[\lambda^+\left(\frac{\partial V}{\partial x}\right)^+ - \lambda^-\left(\frac{\partial v}{\partial x}\right)^-\right] = 0 \tag{13.55}$$

Subscripts + and – indicate evaluation of partial derivatives along +ve and –ve characteristics respectively. Again subtracting Eq. (13.54) from (13.53) and multiplying (g/C) yields

$$\frac{\partial V}{\partial t} + \frac{1}{2}\frac{g}{C}\left[\lambda^+\left(\frac{\partial y}{\partial x}\right)^+ - \lambda^-\left(\frac{\partial y}{\partial x}\right)^-\right] + \frac{1}{2}\left[\lambda^+\left(\frac{\partial V}{\partial x}\right)^+ + \lambda^-\left(\frac{\partial V}{\partial x}\right)^-\right] = g(s_b - s_f) \tag{13.56}$$

Predictor scheme. The following spatial finite differences are written for three and two grid points to use in the predictor scheme.

$$\left(\frac{\partial V}{\partial x}\right)^+ = \frac{2V_k^j - 3V_{k-1}^j + V_{k-2}^j}{\Delta x}$$

$$\left(\frac{\partial y}{\partial x}\right)^{+} = \frac{2y_k^j - 3y_{k-1}^j + y_{k-2}^j}{\Delta x}$$

$$\left(\frac{\partial V}{\partial x}\right)^{-} = \frac{V_{k+1}^j - V_k^j}{\Delta x} \tag{13.57}$$

$$\left(\frac{\partial y}{\partial x}\right)^{-} = \frac{y_{k+1}^j - y_k^j}{\Delta x}$$

And let the quantities computed during the predictor are denoted by *. Then,

$$\frac{\partial y}{\partial t} = \frac{y^* - y_k^j}{\Delta t}$$

$$\frac{\partial V}{\partial t} = \frac{V^* - V_k^j}{\Delta t} \tag{13.58}$$

Applying Eqs. (13.57) and (13.58) in Eq. (13.55) and (13.56), Eqs. (13.55) and (13.56) become

$$y_k^* = y_k^j - \frac{1}{2}\frac{\Delta t}{\Delta x}\Big[\lambda^+(2y_k^j - 3y_{k-1}^j + y_{k-2}^j) + \lambda^-(y_{k+1}^j - y_k^j)$$

$$-\frac{1}{2}C_k^j \frac{\Delta t}{g\Delta x}\Big[\lambda^+(2V_k^j - 3V_{k-1}^j + V_{k-2}^j) - \lambda^-(V_{k+1}^j - V_k^j)\Big]\Big] \tag{13.59}$$

$$V_k^* = V_k^j - \frac{1}{2}\frac{\Delta t}{\Delta x}\cdot\frac{g}{C_k^j}\Big[\lambda^+(2y_k^j - 3y_{k-1}^j + y_{k-2}^j) - \lambda^-(y_{k+1}^j - y_k^j)\Big]$$

$$-\frac{1}{2}\cdot\frac{\Delta t}{\Delta x}\Big[\lambda^+(2V_k^j - 3V_{k-1}^j + V_{k-2}^j) + \lambda^-(V_{k+1}^j - V_k^j)\Big] + g\Delta t(s_b - s_{f_k}^{\,j}) \tag{13.60}$$

Corrector scheme. Finite difference approximations used are:

$$\left(\frac{\partial V}{\partial x}\right)^{+} = \frac{V_k^* - V_{k-1}^*}{\Delta x}$$

$$\left(\frac{\partial y}{\partial x}\right)^{+} = \frac{y_k^* - y_{k-1}^*}{\Delta x}$$

$$\left(\frac{\partial V}{\partial x}\right)^{-} = \frac{-2V_k^* + 3V_{k-1}^* - V_{k+2}^*}{\Delta x} \tag{13.61}$$

$$\left(\frac{\partial y}{\partial x}\right)^{-} = \frac{-2y_k^* + 3y_{k-1}^* - y_{k+2}^*}{\Delta x}$$

$$\frac{\partial v}{\partial t} = \frac{v_k^{**} - v_k^j}{\Delta t}$$

$$\frac{\partial y}{\partial t} = \frac{y_k^{**} - y_k^j}{\Delta t}$$

And

$$\frac{\partial V}{\partial t} = \frac{V_k^{**} - V_k^j}{\Delta x}$$

$$\frac{\partial y}{\partial t} = \frac{y_k^{**} - y_k^j}{\Delta t} \tag{13.62}$$

Substituting Eqs. (13.61) and (13.62) in Eqs. (13.55) and (13.56), we get

$$y_k^{**} = y_k^j - \frac{1}{2} \cdot \frac{\Delta t}{\Delta x}\left[\lambda^+(y_k^* - y_{k-1}^j) + \lambda^-(-2y_k^* + 3y_{k+1}^* - y_{k+1}^j)\right]$$

$$-\frac{1}{2} \cdot \frac{\Delta t}{\Delta x}\frac{C_k^j}{g}\left[\lambda^+(V_k^* - V_{k-1}^*) - \lambda^-(2V_k^* + 3V_{k+1}^* - V_{k+1}^*)\right] \tag{13.63}$$

$$V_k^{**} = V_k^j - \frac{1}{2}\frac{g}{c_k^j}\frac{\Delta t}{\Delta x}\left[\lambda^+(y_k^* - y_{k-1}^j) + \lambda^-(-2y_k^* + 3y_{k+1}^* - y_{k+2}^*)\right] - \frac{1}{2} \cdot \frac{\Delta t}{\Delta x}$$

$$\left[\lambda^+(V_k^* - V_{k-1}^*) + \lambda^-(-2V_k^* + 3V_{k+1}^* - V_{k+2}^*)\right] + g\Delta t(s_b - s_f^*) \tag{13.64}$$

Now, values of y and v at $j + 1$ time step may be obtained from the following two equations:

$$y_k^{j+1} = \frac{1}{2}(y_k^* + y_k^{**}) \tag{13.65}$$

$$V_k^{j+1} = \frac{1}{2}(V_k^* + V_k^{**}) \tag{13.66}$$

Equations (13.65) and (13.66) now give the values of y and v respectively in all the interior grid point in the next time $t_1 = 0 + \Delta t$. Boundary values at $x = 0$ and $x = L$ are given and marching procedure is carried out till $t = t_{\text{final}}$. Courant criteria are maintained for stability.

Gabutti scheme: Gabutti (1983) has extended the Lambda scheme and, therefore, it is called **Gabutti scheme**. Gabutti scheme also takes care of all flow states, i.e., subcritical to supercritical. In the dam-break flood analysis with Saint-Venant equations, Fennma and Choudhury (1987) have used this scheme. Here also both predictor corrector schemes are applied.

Predictor part I. Spatial derivatives are approximated as:

$$\left.\begin{aligned}\left(\frac{\partial V}{\partial x}\right)^{+} &= \frac{V_k^j - V_{k-1}^j}{\Delta x}\\ \left(\frac{\partial V}{\partial x}\right)^{-} &= \frac{V_{k+1}^j - V_k^j}{\Delta x}\\ \left(\frac{\partial y}{\partial x}\right)^{+} &= \frac{y_k^j - y_{k\ 1}^j}{\Delta x}\\ \left(\frac{\partial y}{\partial x}\right)^{-} &= \frac{y_{k+1}^j - v_k^j}{\Delta x}\end{aligned}\right\} \tag{13.67}$$

and time derivatives are:

$$\left.\begin{aligned}\left(\frac{\partial V}{\partial t}\right) &= \frac{V_k^* - V_k^j}{\Delta t}\\ \left(\frac{\partial y}{\partial t}\right) &= \frac{y_k^* - y_k^j}{\Delta t}\end{aligned}\right\} \tag{13.68}$$

Substituting spatial derivatives of Eq. (13.67) and time derivatives of Eq. (13.68) in Eq. (13.55) and (13.56) on simplifying the resulting equation, the following two equation for y_k^* and V_k^* are obtained:

$$y_k^* = y_k^j - \frac{1}{2}\frac{\Delta t}{\Delta x}\left[\lambda^{+}(y_k^j - y_{k-1}^j) + \lambda^{-}(y_{k+1}^j - y_k^j)\right] + \frac{1}{2}\frac{\Delta t}{\Delta x}\cdot\frac{c_k^j}{g}$$

$$\left[\lambda^{+}(V_k^j - V_{k-1}^j) - \lambda^{-}(V_{k-1}^j - V_k^j)\right] \tag{13.69}$$

$$V_k^* = V_k^j - \frac{1}{2}\frac{\Delta t}{\Delta x}\cdot\frac{g}{c_k^j}\left[\lambda^{+}(y_k^j - y_{k-1}^j) - \lambda^{-1}(y_{k+1}^j - y_k^j)\right] + \frac{1}{2}\frac{\Delta t}{\Delta x}$$

$$\left[\lambda^{+}(V_k^j - V_{k-1}^j) + \lambda^{-}(V_{k+1}^j - V_k^j)\right] + g\Delta t(s_b - s_{f_k}^{\ j}) \tag{13.70}$$

In which subscript * refers to predicted values.

Predictor part II.

$$\left.\begin{aligned}\left(\frac{\partial V}{\partial x}\right)^{+} &= \frac{2V_k^j - 3V_{k-1}^j + V_{k-2}^j}{\Delta x}\\ \left(\frac{\partial y}{\partial x}\right)^{+} &= \frac{2y_k^j - 3y_{k-1}^j + y_{k-2}^j}{\Delta x}\\ \left(\frac{\partial V}{\partial x}\right)^{-} &= \frac{-2V_k^j + 3V_{k+1}^j - V_{k+2}^j}{\Delta x}\\ \left(\frac{\partial y}{\partial x}\right)^{-} &= \frac{-2y_k^j - 3y_{k+1}^j - y_{k+2}^j}{\Delta x}\end{aligned}\right\} \tag{13.71}$$

Substituting Eq. (13.71) in Eqs. (13.55) and (13.56), the following predicted values of y_t^* and v_t^* are determined. It is noted that values at known time level j are used. Thus, process will yield

$$y_t^* = -\frac{1}{2}\cdot\frac{1}{\Delta x}\left[\lambda^+(2y_k^j - 3y_{k-1}^j + y_{k-2}^j) + \lambda^-(-2y_k^j + 3_{k+1}^j - y_{k+2}^j)\right] - \frac{1}{2}\cdot\frac{1}{\Delta x}$$

$$\frac{c_k^j}{g}\left[\lambda^+(2V_k^j - 3V_{k-1}^j + V_{k-2}^j) - \lambda^-(-2V_k^j + 3V_{k+1}^j - V_{k+2}^j)\right] \tag{13.72}$$

$$V_t^* = -\frac{1}{2}\cdot\frac{1}{\Delta x}\cdot\frac{g}{C_K^J}\left[\lambda^+(2y_k^j - 3y_{k-1}^j + y_{k-2}^j) + \lambda^-(-2y_k^j + 3_{k+1}^j - y_{k+2}^j)\right] - \frac{1}{2}\cdot\frac{1}{\Delta x}$$

$$\left[\lambda^+(2V_k^j - 3V_{k-1}^j + V_{k-2}^j) + \lambda^-(-2V_k^j + 3V_{k+1}^j - V_{k+2}^j)\right] + g(s_b - s_{f_k}^{\ j}) \tag{13.73}$$

Corrector part. The first predicted values y^*_k and v^*_k in Eqs. (13.69) and (13.70) and corresponding values of coefficients and spatial derivatives by finite differences are used.

$$\left.\begin{aligned} \left(\frac{\partial V}{\partial x}\right)^+ &= \frac{V_k^* - V_{k-1}^*}{\Delta x}, \quad \left(\frac{\partial V}{\partial x}\right)^- = \frac{V_{k+1}^* - V_k^*}{\Delta x} \\ \left(\frac{\partial y}{\partial x}\right)^+ &= \frac{y_k^* - y_{k-1}^*}{\Delta x}, \quad \left(\frac{\partial y}{\partial x}\right)^- = \frac{y_{k+1}^* - y_k^*}{\Delta x} \end{aligned}\right\} \tag{13.74}$$

In Eqs. (13.55) and (13.56), derivatives of (10.58) are substituted to obtain y_t^{**} and v_t^{**} as:

$$y_t^{**} = -\frac{1}{2}\cdot\frac{1}{\Delta x}\left[\lambda^+(y_k^* - y_{k-1}^*) + \lambda^-(y_{k-1}^* - y_k^j)\right] - \frac{1}{2}\cdot\frac{1}{\Delta x}\cdot\frac{c_k^j}{g}$$

$$\left[\lambda^+(V_k^* - V_{k-1}^*) - \lambda^-(V_{k+1}^* - V_k^*)\right] \tag{13.75}$$

$$v_t^{**} = -\frac{1}{2}\cdot\frac{1}{\Delta x}\cdot\frac{g}{c_k^j} - \left[\lambda^+(y_k^* - y_{k-1}^*) - \lambda^-(y_{k+1}^* - y_k^*)\right] - \frac{1}{2}\cdot\frac{1}{\Delta x}$$

$$\left[\lambda^+(V_k^* - V_{k-1}^*) + \lambda^-(V_{k+1}^* - V_k^*)\right] + g(s_b - s_{f_k}^{\ j}) \tag{13.76}$$

The final values of y_k^{j+1} and v_k^{j+1} in the next time step now may be computed from the following equations:

$$y_k^{j+1} = y_k^j + \frac{1}{2}\Delta t(y_t^* + y_t^{**}) \tag{13.77}$$

$$V_k^{j+1} = V_k^j + \frac{1}{2}\Delta t(V_t^* + V_t^{**}) \tag{13.78}$$

The boundary values at $x = 0$ and $x = L$ are inserted after calculating values of y and V in the next time step by Eqs. (13.77) and (13.78) at all interior points. Then again values of y and V in the interior points are calculated for the next time step, and values at two boundaries are inserted. Thus, marching procedure is continued until $t = t_{final}$. For stability, courant criteria are to be maintained.

MacCormack scheme: Before using MacCormack method, Saint-Venant equations are written in the conservative form.

Continuity equation (derived in Section 13.3) is:

$$\frac{\partial A}{\partial t}+\frac{\partial Q}{\partial x}=0 \tag{13.79}$$

In conservative form, it is written as:

$$\frac{\partial A}{\partial t}+\frac{\partial (VA)}{\partial x}=0 \tag{13.80}$$

And Dynamic equation derived already in Section (13.4) is:

$$\frac{\partial Q}{\partial t}+\frac{\partial (Q^2/A)}{\partial x}+gA\left(\frac{\partial y}{\partial x}+s_f-s_b\right)=0 \tag{13.81}$$

or

$$\frac{\partial (VA)}{\partial t}+\frac{\partial (Q^2/A)}{\partial x}+gA\frac{\partial y}{\partial x}+gA(s_f-s_b)=0$$

or

$$\frac{\partial (VA)}{\partial t}+\frac{\partial}{\partial x}\left(\frac{Q^2}{A}+gAy\right)-gA(s_b-s_f)=0$$

Writing y to be average values,

$$y=\bar{y} \text{ and } \frac{Q^2}{A}=\frac{A^2V^2}{A}=V^2A$$

$$\therefore \quad \frac{\partial (VA)}{\partial t}+\frac{\partial}{\partial x}(V^2A+gA\bar{y})-gA(s_b-s_f)=0 \tag{13.82}$$

When equation of continuity [Eq. (13.80)] and dynamic or momentum equation [Eq. (13.82)] are expressed in conservative form, they are written as:

$$\frac{\partial \mathbf{U}}{\partial t}+\frac{\partial \mathbf{F}}{\partial x}+\mathbf{S}=0 \tag{13.83}$$

where

$$\mathbf{U}=\begin{bmatrix} A \\ VA \end{bmatrix}$$

$$\mathbf{F}=\begin{bmatrix} VA \\ V^2A+gA\bar{y} \end{bmatrix} \tag{13.84}$$

$$\mathbf{S}=\begin{bmatrix} 0 \\ -gA\left(s_b-s_f\right) \end{bmatrix}$$

Now,

$$A\bar{y} = \int_0^{y(x)} \left[y(x) - \eta\right]\sigma(b)d\eta \tag{13.85}$$

where $A\bar{y}$ is moment of flow area about water surface. This moment is computed by Eq. (13.85) and σ is the water surface width at depth η.

MacCormack method. It was initially started by MacCormack (1969) and then by Anderson et al[55] (1984). The method is second order accurate both in space and time. Fennema and Choudhury (1986, 1987) used this method in dam-break analysis. It is also used by Dammuller et al[56] (1989), Sarma and Das (2001), Saikia-Das and Sarma [(2006 October), 2006 (June), 2006 (June), 2006 (August), 2006 (December), 2006 (May)] and others.

In the two-step procedure, backward difference is utilized for special derivative in the predictor step, and forward difference is utilized for the special derivative in the corrector part. Saikia-Das and Sarma's method is the combination of first order diffusive scheme; second order modified 2 step predictor corrector scheme and Total Variation Diminishing (TVD) MacCormack scheme using Veun Leer Flux Limiter.

In the predictor part, the partials are defined as:

$$\left.\begin{aligned} \frac{\partial \mathbf{U}}{\partial t} &= \frac{\mathbf{U}_k^* - \mathbf{U}_k^j}{\Delta t} \\ \frac{\partial F}{\partial x} &= \frac{\mathbf{F}_k^j - \mathbf{F}_{k-1}^j}{\Delta x} \end{aligned}\right\} \tag{13.86}$$

where * refers to variables computed during the predictor part. Substituting Eq. (13.86) in Eq. (13.83), we get

$$\frac{\mathbf{U}_k^* - \mathbf{U}_k^j}{\Delta t} + \frac{\mathbf{F}_k^j - \mathbf{F}_{k-1}^j}{\Delta x} + \mathbf{S}_k^j = 0$$

$$\mathbf{U}_k^* = \mathbf{U}_k^j - \frac{\Delta t}{\Delta x}\left(\mathbf{F}_k^j - \mathbf{F}_{k-1}^j\right) - \mathbf{S}_k^j \Delta t \tag{13.87}$$

This computed value of U^* gives A^* (and y^*) and V^* (and Q), i.e., values of V^* and y^* are determined in all computational grid point. These values are used in the corrector part to compute $\mathbf{F}^*$ and $\mathbf{S}^*$.

In the corrector part, partials are defined as:

$$\frac{\partial \mathbf{U}}{\partial t} = \frac{\mathbf{U}_k^{**} - \mathbf{U}_k^j}{\Delta t}$$

---

[55] Anderson, D.A., Tonnehill, J.C. and Pletcher, R.H., *Computational Fluid Mechanics and Heat Transfer*, McGraw-Hill, New York, 1984.

[56] Dammuller, D., Bhallmudi, S.M. and Choudhury, M.H., Modeling of Unsteady Flow in Curved Channel, *Jour. Hy. Eng.*, ASCE, 115, No. 11, pp. 1479–95, 1989.

$$\frac{\partial \mathbf{F}}{\partial x} = \frac{\mathbf{F}^*_{k+1} - \mathbf{F}^*_k}{\Delta x} \tag{13.88}$$

Substituting the partials from Eq. (13.88) in Eq. (13.83), we get

$$\frac{\mathbf{U}^{**} - \mathbf{U}^j_k}{\Delta t} + \frac{\mathbf{F}^*_{k+1} - \mathbf{F}^*_k}{\Delta x} + \mathbf{S}^*_k = 0$$

Solving for $\mathbf{U}^{**}$,

$$\mathbf{U}^{**} = \mathbf{U}^j_k - \frac{\Delta t}{\Delta x}(\mathbf{F}^*_{k+1} - \mathbf{F}^*_k) - \mathbf{S}^*_k \Delta t \tag{13.89}$$

Subscripts ** refers the values of variables after the corrector step. The value of U_k^{j+1} at unknown next time step j+1 is given by

$$U_k^{j+1} = ½ \, (U_k^{**} + U_k^{*}) \tag{13.90}$$

Equation (13.90) gives the values of V and y in the interior grid points. Values at boundary $x = 0$ and $x = L$ are given just like the previous schemes. Then the values computed at time $t_1 = 0 + \Delta t$ are ready as initial values for calculating the values in the next time step $t_2 = t_1 + \Delta t$. Thus marching procedure is continued until $t = t_{\text{final}}$.

Lax diffusive scheme in conservative form: Partial differential in central difference are:

$$\frac{\partial \mathbf{F}}{\partial x} = \frac{\mathbf{F}^j_{k+1} - \mathbf{F}^j_{k-1}}{2\Delta x}, \quad \frac{\partial \mathbf{U}}{\partial t} = \frac{\mathbf{U}^j_{k+1} - \mathbf{U}^j_{k-1}}{\Delta x} \tag{13.91}$$

where $\mathbf{U}^* = \frac{1}{2}(\mathbf{U}^t_{k-1} + \mathbf{U}^j_{k+1})$ and $\mathbf{S}^* = (\mathbf{S}^j_{k-1} + \mathbf{S}^j_{k+1})$

Substituting the partials of Eq. (13.91) in Eq. (13.83)

$$\mathbf{U}^{j+1}_k = \frac{1}{2}(\mathbf{U}^j_{k-1} + \mathbf{U}^j_{k+1}) - \frac{1}{2} \cdot \frac{\Delta t}{\Delta x}(\mathbf{F}^j_{k+1} - \mathbf{F}^j_{k-1}) - \frac{\Delta t}{2}(\mathbf{S}^j_{k+1} + \mathbf{S}^j_{k-1}) \tag{13.92}$$

Equation (13.92) gives the values of A and VA at j+1 time level, and then we can determine the values of our interest, i.e., y and V in all the interior grids solution. It can be advanced with time with the insertion of boundary values. It is already said that this method is simplest to programs, and the method gives good results.

13.6.3 Finite Difference Implicit Schemes

Preissmann scheme: In implicit method, the spatial derivatives and co-efficient are replaced in terms of the values at unknown time level. These unknown variables, therefore, appear implicitly in the algebraic equations and hence the method is called **implicit scheme** or **method**. All the algebraic equations in this method are required to be solved simultaneously.

This scheme has been used since 1961 by Preissmann and Cunge (1961), Abbot and Ionesq[57] (1967), Strelkoff (1970), Amein and Fang (1970), Terzidis and Strelkoff (1979), Price (1974), Ligget and Yevjevich (1975), Mahmood and Yevjevich[58] (1975), Choudhury (1987), Fennema and Choudhury (1987), Hou Zang et al (1993) and others.

Price (1974) studied four implicit numerical schemes and concluded that four-point methods by Amien (1970) are most efficient from the point of stability.

An extensive bibliography on solution of unsteady open channel flow equations by implicit method is available in the literature of Miller and Yevjevich[59] (1975) and Fread[60] (1973a, 1974a, 1974b, 1975, 1976a, 1985).

Also comprehensive literature survey of finite difference solutions both by explicit and implicit till 2007 have presented by Saikia Das (2007) in dam-break routing solution.

In this section, discussion of implicit scheme developed by Preissmann and Cunge (1961), commonly known as **Preissmann method**, will be presented for implicit solution. This scheme has a lot of advantages like variable spatial grid, assumption of weight coefficient α, simulation of the steep wave front of flood by varying α. Oscillation behind the steep wave front is taken care of when α = 0.6 to 0.7. α makes the scheme stable when $0.52 < \alpha < 1.0$.

The partial derivatives and other coefficients are approximated as:

$$\frac{\partial \mathbf{U}}{\partial t} = \frac{(\mathbf{U}_k^{j+1} + \mathbf{U}_{k+1}^{j}) - (\mathbf{U}_k^{j} + \mathbf{U}_{k+1}^{j})}{2\Delta t}$$

$$\frac{\partial \mathbf{F}}{\partial x} = \frac{\alpha(\mathbf{F}_{k+1}^{j+1} - \mathbf{F}_k^{j+1})}{\Delta x} + \frac{(1-\alpha)(\mathbf{F}_{k+1}^{j} - \mathbf{F}_k^{j})}{\Delta x} \tag{13.93}$$

$$\mathbf{S} = \frac{1}{2}\alpha(\mathbf{S}_{k+1}^{j+1} + \mathbf{S}_k^{j+1}) + (1-\alpha)(\mathbf{S}_{k+1}^{j} + \mathbf{S}_k^{j})$$

[57] Abbott, M.B. and Ionesq, F., On the Numerically Computation Nearly Horizontal Flows, *Jour. Hy. Res. IAHR*, vol. 5, No. 2, pp. 96–117, 1967.

[58] Mahmood, K. and Yevjevich, V., Unsteady Flow in Open Channels, vol. I and II, Water Resources Publications, Fort Collins, Colorado, 1975.

[59] Miller, W.A. and Yevjevich, V., Unsteady Flow in Open Channels, vol. III, Bibliography, Water Resources Publication, Fort Collins, Colorado, 1975.

[60] Fread, D.L., Effect of Time Step Size in Implicit Dynamic Flood Routing, *Water Resource Bull.*, vol. 9, No. 2, pp. 358–351, 1973a.

———, Channel Routing, Chapter 14, in *Hydrological Forecasting*, M.G. Anderson and T.P. Burt., Wiley, New York, pp. 437–503, 1985.

———, Discussion of Comparison of Four Numerical Methods for Flood Routing by R.K. Price., *J. Hy. Div.*, ASCE, vol. 101, No. Hy3, pp. 505–507, 1975.

———, Flood Routing in Meandering River with Flood Plains, Rivers 176, Symposium. Inland Waterways for Niagara, Flood Control, and Water Diversions, August 10–12, 1976, Colorado State Univ. Fort Collins, Colo., vol. 1, pp. 16–35, 1976a.

———, Implicit Dynamic Routing of Floods and Surges in the Lower Mississippi, presented at Am. Geophys, Union Nat. Meeting, April, 1974, Washington D.C., 1974b.

———, Numerical Properties of Implicit Four-point Finite Difference Equations of Unsteady Flow, NOAA Tech. Memorand. NWS, HYDRO-18, Nat. Weather. Ser., NOAA, US Dept. of Commerce. Silver Spring, Md., 1974a.

Substituting the derivatives of Eq. (13.93) in Eq. (13.83) and rearranging the terms, we get

$$\mathbf{U}_k^j + \mathbf{U}_{k+1}^j = \mathbf{U}_k^{j+1} + \mathbf{U}_{k+1}^{j+1} + 2\frac{\Delta t}{\Delta x}\left[\alpha(\mathbf{F}_{k+1}^{j+1} - \mathbf{F}_k^{j+1}) + (1-\alpha)(\mathbf{F}_{k+1}^{j} - \mathbf{F}_k^{j})\right] + \Delta t$$

$$\left[\alpha(\mathbf{S}_{k+1}^{j+1} + \mathbf{S}_k^{j+1}) + (1-\alpha)(\mathbf{S}_{k+1}^{j} + \mathbf{S}_k^{j})\right] \tag{13.94}$$

In Eq. (13.94), there are four unknowns, namely $V_k^{j+1}, V_{k+1}^{j+1}, Y_k^{j+1}$ and Y_{k+1}^{j+1}. If these two equations of each grid point are written, there will be $2N$ equations, N being the number of small reaches on the channel total reach under consideration. These equations cannot be written for the grid points at the downstream end. However, there are $2(N + 1)$ unknowns, i.e., two unknowns for each grid point. Thus, for a unique solution, two more equations are needed. These are provided by the end or boundary conditions which may be reservoir at upstream or downstream or sluice gate or a rating curve or series junction depending upon physical conditions of the problem.

Stability. Regarding stability, the solution is unconditionally stable as long as the weighing coefficient $\alpha > 0.5$, i.e., variables are weighted towards $j + 1$ time level. If there is no restriction on size of Δx and Δt, it is called **unconditionally stable**. But yet, it is better to maintain Courant number C_N close to unity. Samuels and Skeels[61] (1990), Evans[62] (1977) and Yen and Lin[63] (1986), investigated Vendernikov[64] (1945) number for stability in their numerical experimentations. Vendernikov number may be expressed as:

$$V_n = \frac{x\gamma V}{V_w - V} \tag{13.95}$$

where x is the exponent of hydraulic radius R in uniform flow formula, being $x = 1/2$ for Chezy's formula, 2/3 for Manning's formula and 2 for laminar flow. V is the average velocity, V_w is the absolute velocity of disturbance of waves in river, γ is a shape factor of river section defined by

$$\gamma = 1 - R\left(\frac{dP}{dA}\right) \tag{13.96}$$

P is the wetted perimeter and A is the water area. Thus $\gamma = 1$ for a very wide channel, $\gamma = 0$ for narrow channel. It is seen that $(V_w - V)$ is equal to celerity C of the waves or the critical velocity V_c. Since Froude number $F_r = V/V_c$, Vendernikov number V_n, i.e., Eq. (13.95) is reduced to

$$V_n = x\gamma F_r \tag{13.97}$$

When $V_n < 1$, any wave in the channel is stable and wave will be depressed and when $V_n > 1$, it becomes unstable.

---

[61] Samuels, P.G. and Skeels, C.P., Stability Limits for Preissmann's Scheme, *Jour. Hy. Engg.* ASCE, 116, No. 8, pp. 997–1012, 1990.

[62] Evans, E.P., The Behavior of Mathematical Model of Open Channel Flow, Paper A97, 17th Congress, Inter. Assoc. for Hy. Research, Baden, Germany, 1977.

[63] Yen, C.L. and Lin, C.H., Numerical Stability in Unsteady Superficial Flow Simulation, 5th Congress, Asian Specific Regional Div., Inter. Assoc. for Hy. Research, August 1986.

[64] Vedernikov, V.V., The Conditions at the Front of a Translation Wave Distributing Motion of Real Fluid, Competes rendus (Doklady) de I' Academia des Sciences de P.U.R.S.S., vol. 48, No. 4, pp. 234–242, 1945.

Solution procedure. Expanding the solution of Eq. (13.94) in A, VA, α, s_b and s_f terms in different grid points, it yields

$$A_k^j + A_{k+1}^j = A_k^{j+1} + A_{k+1}^{j+1} + 2\frac{\Delta t}{\Delta x}[\alpha\{(VA)_{k+1}^{j+1} - (VA)_k^{j+1}\} + (1-\alpha)\{(VA)_{k+1}^j\ (VA)_k^j\}]$$

$$(VA)_k^{j+1} + (VA)_{k+1}^{j+1} + 2\frac{\Delta t}{\Delta x}\left[\alpha\left\{\left(V^2A + gA\bar{y}\right)_{k+1}^{j+1} - \left(V^2A + gA\bar{y}\right)_k^{j+1}\right\}\right] \tag{13.98}$$

$$-gA\Delta t\left[\alpha\left\{(s_b - s_f)_{k+1}^{j+1} + (s_b - s_f)_k^{j+1}\right\}\right] = gA\Delta t\left[(1+\alpha)\left\{(s_b - s_f)_{k+1}^j + (s_b - s_f)_k^j\right\}\right] + (VA)_k^j$$

$$+ VA_{k+1}^j - (1-\alpha)\frac{2\Delta t}{\Delta x}\left[\left(V^2A + gA\bar{y}\right)_{k+1}^j - \left(V^2A + gA\bar{y}\right)_k^j\right] \tag{13.99}$$

Equations (13.98) and (13.99) for each grid points and boundary conditions give a set of nonlinear algebraic equation which can be solved by iterative technique. Solution of iterative technique may be done by

1. Newton-Raphson method
2. Double sweep method

Double sweep method is used by Cunge et al (1980), Newton-Raphson method was used by Choudhury (1987). Newton-Raphson method is discussed here in brief.

In this method, values of unknown variables V and y at each node are estimated and then iterated to refine the solution. To determine the correction for the iteration, the partial derivatives of Eqs. (13.98) and (13.99), and of the boundary conditions with respect to y_k^{j+1}, Y_{k+1}^{j+1}, V_k^{j+1}, $V_{k+!}^{j+1}$ are needed. If these partial derivatives are derived and arranged, the governing equation becomes

$$\mathrm{A}_x = \mathbf{b} \tag{13.100}$$

Here A is matrix comprising the partial derivatives. **X** is column unknown vector comprising the correction Δy_k and $\Delta V_{k,}(k = 1, 2, 3, \ldots, N+1)$, **b** is right hand column vector. Thus, there are $2(N + 1)$ equations and $2(N + 1)$ unknowns. These unknowns are the corrections. Next, it is required to check for $\sum_{k=1}^{n+1} |\Delta y_k| + |\Delta Y y_k| \le$ a specified tolerance. If this sum of the correction is less than a specified tolerance, then applying the correction, solution is proceeded to the next step. Otherwise, iteration of the given procedure is again repeated. It is noted that matrix A is banded with a bandwidth of 4. The fact is utilized while solving Eq. (13.100) since a banded matrix solution routine requires less storage space and gives more accurate results. According to Choudhury (1987) in branching and parallel systems, the nodes can be numbered such that the resulting matrix is banded.

Beam and warming scheme: This is another better implicit method of solution which takes care of supercritical flow. For further details of this scheme, refer to Anderson et al (1984), Fennema and Choudhury (1986) and Choudhury (1994).

13.7 KINEMATIC ROUTING

It is the routing of kinematic wave. For kinematic wave or flow in river, the acceleration pressure terms in the momentum equation are negligible, so the wave motion is mainly described by the equation of continuity. The name kinematic refers to the study when force, acceleration and mass terms are neglected.

Thus, the kinematic wave routing depends on the following equation:

$$\frac{\partial Q}{\partial x}+\frac{\partial A}{\partial t}=q \tag{13.101}$$

Here q is positive
And momentum equation is:

$$s_b = s_f$$

$$\left(\text{i.e. negecting } \frac{\partial V}{\partial t}, V\frac{\partial V}{\partial t}, g\frac{\partial y}{\partial x} \text{ terms}\right)$$

Momentum equation can also be expressed in terms of A and Q (without any force term) as:

$$A = \alpha Q^{\beta} \tag{13.102}$$

When $s_b = s_f$, $R=\dfrac{A}{P}$, Manning's equation is written as:

$$Q=\frac{1}{n}\frac{A^{5/3}}{P^{2/3}}s_b^{1/2}$$

The solution for A is:

$$A = \left(\frac{nP^{2/3}}{\sqrt{s_b}}\right)Q^{3/5} \tag{13.103}$$

Comparing Eqs. (13.102) and (13.103),

$$\alpha=\left(\frac{nP^{2/3}}{\sqrt{s_b}}\right) \text{and } \beta=3/5$$

Equations (13.102) and (13.103) contain two dependable variables, A and Q. A can be eliminated by differentiating (13.102),

$$\frac{\partial A}{\partial t}=\alpha\beta Q^{\beta-1}\left(\frac{\partial Q}{\partial t}\right)$$

Substituting $\dfrac{\partial A}{\partial t}$ in Eq. (13.101),

$$\frac{\partial Q}{\partial x}+\alpha\beta Q^{\beta-1}\left(\frac{\partial Q}{\partial t}\right)=q \tag{13.104}$$

Thus, Eq. (13.104) is obtained by combining continuity and momentum to route the flood in kinematic wave modelling.

13.7.1 Numerical Explicit Solution of Kinematic Routing

Derivatives of Eq. (13.104) with space and time are:

$$\frac{\partial Q}{\partial x} = \frac{Q_{k+1}^{j} - Q_{k}^{j}}{\Delta x}, \quad Q = \frac{Q_{k}^{j} + Q_{k+1}^{j}}{2}$$

$$\frac{\partial Q}{\partial t} = \frac{Q_{k+1}^{j+1} - Q_{k}^{j}}{\Delta t}, \quad q = \frac{q_{k}^{j} + q_{k+1}^{j}}{2} \tag{13.105}$$

Substituting the derivatives from Eq. (13.105) in Eq. (13.104),

$$\frac{Q_{k+1}^{j} - Q_{k}^{j}}{\Delta x} + \alpha\beta\left(\frac{Q_{k}^{j} + Q_{k+1}^{j}}{2}\right)^{\beta-1}\left(\frac{Q_{k+1}^{j+1} - Q_{k}^{j}}{\Delta t}\right) = \left(\frac{q_{k}^{j} + q_{k+1}^{j}}{\Delta t}\right)$$

Solving for values of unknown Q_{k+1}^{j+1},

$$\frac{Q_{k+1}^{j+1} - Q_{k}^{j}}{\Delta t} = \frac{Q_{k}^{j} - Q_{k+1}^{j}}{\Delta x} - \alpha\beta\left(\frac{Q_{k}^{j} + Q_{k+1}^{j}}{2}\right) + \left(\frac{q_{k}^{j} + q_{k+1}^{j}}{2}\right)$$

$$\therefore \quad Q_{k+1}^{j+1} = Q_{k}^{j} + \frac{\Delta t}{\Delta x}\left(Q_{k}^{j} - Q_{k+1}^{j}\right) - \frac{\alpha\beta\Delta t}{2}\left(Q_{k}^{j} + Q_{k+1}^{j}\right) + \Delta t\left(\frac{q_{k}^{j} + q_{k+1}^{j}}{2}\right) \tag{13.106}$$

Equation (13.106) is the routing equation. Values of Q, α, β on RHS at time t (j line) and along x (k line) are known as **initial conditions**. From these values, Q values at next time step t_1, i.e., at (j + 1) are computed in all the interior grids. Boundary conditions at $x = 0$ and $x = L$ are inserted. Now the values at (j + 1) time are ready to calculate Q by Eq. (13.106) for the next time $t_2 = t_1 + \Delta t$ are ready. Thus, marching procedure is repeated like the previous case until $t = t_{\text{final}}$. Henderson (1966) is of the opinion that relative error in computation is less when Q is taken as dependent variable. Here Q is taken as a dependable variable to minimize the error. This can be shown from Eq. (13.102), i.e., $A = \alpha Q^{\beta}$. Taking $\log_e$ of this equation,

$$\log_e A = \log_e \alpha + \beta \log_e Q$$

Differentiating,

$$\frac{dA}{A} = O + \beta\frac{dQ}{Q}$$

$$\therefore \quad \frac{dQ}{Q} = \frac{1}{\beta}\frac{dA}{A} \tag{13.107}$$

Equation (13.107) shows if Manning's equation is used,

$$\beta = 3/5 = 0.6$$

$$\therefore \quad 1/\beta > 1$$

It shows that discharge estimation will be magnified if A becomes the dependent variable instead of Q.

13.8 CONCLUSION

This chapter mainly dealt in the derivation of unsteady Saint-Venant equations of continuity and momentum at the beginning. Then different numerical solutions of these equations under natural flood and dam-break flood routing situations, kinematic wave routing situations are presented. Major portions of the chapter include the different numerical solutions of the equations by different existing methods for both on conservative and non-conservative form. Researchers are still working on these equations using different methods of solutions on different physical unsteady problems.

EXERCISES

13.1 Derive the unsteady continuity equation and the momentum equation in any one of the methods you like best.

13.2 Describe the method of characteristics (MoC) for the solution of Saint-Venant equations numerically.

13.3 What are the different explicit finite difference schemes for the solution of the Saint-Venant equations? Describe the Lax diffusive scheme.

13.4 While measuring the flow in a river with unsteady flow, the depth y was found to increase at a rate of 0.06 m/hour. The surface width of the river is 30 m and discharge at this section is 30 m^3/sec. Estimate the discharge at a section 1 km upstream.

(*Ans*: 35.5 m^3/sec)

13.5 In the measurement of discharge in a river, it was found that depth decreases at a rate of 0.05 m/hour. If the discharge in that section is 15 m^3/sec and surface width is 15 m, estimate the discharge at a section 1.2 km upstream.

(*Ans*: 17.5 m^3/sec)

SUGGESTED FURTHER READINGS

Abbott, M.B., *Computational Hydraulics*: *Elements of Theory of Surface Flow*, Pitman, London, 1979.

Amien, M. and Chu, H.L., Implicit Numerical Modeling of Unsteady Flow, *Jour. Hy. Div.*, ASCE, pp. 1717–1731, January 1975.

Ligget, J.A. and Cunge, J.A., Numerical Methods of Solutions of Unsteady Flow Equations in Unsteady Open Channel Flow, edited by K. Mahmood and V. Yevjevich, Water Resource Publications, Fort Collin's, Colo., 1975.

Chapter 14
Groundwater Hydrology

14.1 INTRODUCTION

The science of occurrence, distribution and movement of water below the surface of earth is defined as *groundwater hydrology*. Water contained in the saturated zones is important for engineers for engineering works, geologic studies, water supply and irrigation development. Water in unsaturated zone is major concern of agriculture, botany and soil science. Ground water is relatively free of pollution and therefore, very important for domestic, industrial small firm use. Besides its important use, groundwater is an important phase of hydrologic cycle. Most of the sources of perennial rivers are from groundwater or subsurface water. Two fluid-flow systems involving oil and water, and three fluid-flow systems involving gas, oil and water, occur frequently in the development of petroleum and hence groundwater flow hydrology is important in petroleum engineering like the geologists. It is estimated that total groundwater storage is about 42 million km^3 which is approximately 35 times more than the fresh water available in lakes, reservoirs and stream. In arid regions, groundwater is the only source of water for domestic and irrigation purposes. A large part of precipitation and water from streams and lakes joins the groundwater storage by way of infiltration and then percolation. Thus there exists a direct relationship between surface water hydrology and subsurface and groundwater hydrology.

14.2 HISTORICAL BACKGROUND

Biswas[1] (1971) had claimed that groundwater development dates from ancient times. In ancient times, other than dug wells, groundwater had been extracted from quant or karez, or foggara or falaj. These quants are still available in present day in the arid region of southern Asia. Beaumont[2] (1971) nicely presents how quants are made with the help of a sloping tunnel beneath the water table to cause the ground water to flow by gravity at the outlet. There are still ancient 22,000 quants still available in Iran supplying 75% of all water used in the country. Adams[3]

[1] Biswas, A.K., History of Hydrology, Amer Elsevier, New York, 1970.
[2] Beaumont, P., Quant Systems in Iran, Bull. Intl. Assocsci. Hydrology, vol. 16, pp. 39–50, 1971.
[3] Adams, F.D., Origins of Springs and Rivers – A Historical Review, *Fennia*, vol. 50, No. 1, 1928.

(1928), Baker and Horton[4] (1936) said that the writings of Greek and Roman philosophers explained the theory of origin of groundwater are nearly correct. Early Greek philosophers like Homer, Plato hypothesized that springs were formed by seawater conducted by channels below the mountains. Aristotle believed that air enters the cold dark caverns of mountains where it condenses to water and contributes to spring. Roman architect Vitruvius forwarded an important step. He believed that mountains collect rainwater that percolates through rock strata and emerges at the base to form streams. There was no advance till the middle ages until Pierre Perrault (1611–1680) measured rainfall and runoff in the Upper Seine River drainage basin and concluded that precipitation is six times more than river discharge. Perrault[5] (1957) is of the opinion that Pierre Perrault findings demonstrated the early false assumption of inadequate rainfall. Edme Mariotte (1620–1684) had confirmed Perrault findings. Meinzer[6] (1934) stated "Mariotte deserves to be regarded as founder of groundwater hydrology".

During eighteenth century, groundwater hydrology had been developed the understanding of occurrence and movement of groundwater. The French hydraulic engineer Henry Darcy[7] (1803–1858) developed the famous Darcy's law in 1856 in the nineteenth century. In the same century, hydraulics of groundwater was further developed due to contributions made by J. Boussinesq, J. Dupuit, P. Forchheimer and A. Thiem. In the twentieth century, activity in groundwater hydrology has been increased in all phases. Significant are due to R. Dachler, J. Kozeny, G. Thiem and others. O.E. Meinger, through his dynamic leadership in US Geological survey has advanced the field of groundwater hydrology quite a bit. Noteworthy contributions in recent decades are due to M.S. Hantush, C.E. Jacob[8], G.B. Maxey, R.W. Stallman[9], P.Y. Poluvarinova-Kochina[10] and others.

14.3 OCCURRENCE OF GROUNDWATER

Figure 14.1 shows a sectional view of earth's crust showing the occurrence of groundwater. Just below the ground level (GL) it is zone of aeration in which soil pores contain both air and water in varying proportion. Just above the water table (WT) a part of water is held by capillary action. This zone is called **capillary fringe**. Below the water table and in between an impervious layer groundwater exists which is called **unconfined aquifer** or **water table aquifer**. Between the impermeable layer of unconfined aquifer and an another impermeable layer below, water may be confined and this groundwater is called confined aquifer or artesian water. Confined aquifers are usually under pressure because of the weight of overburden and hydrostatic or piezometric head. If well penetrates the confined aquifer, water level in the well becomes equal to

---

[4] Baker, M.N. and Horton, R.E., Historical Development of Ideas regardings the origin of springs and groundwater, Trans. Am. Geophysical Union, vol. 17, pp. 395–400, 1936.

[5] Perrault, P., On the Origin of Springs, Trans. by A. LaRocque, Hafimer, New York, 1957.

[6] Meinzer, O.E., The History and Development of Groundwater hydrology, Jour. Washington Acad. Science, vol. 24, pp. 6–32, 1934.

[7] Darcy, H., Les Fontains Publiques de La Ville de Dijon, Dalmont, Paris, 1856.

[8] Jacob, C.E., Flow of Groundwater in *Engineering Hydraulics* (edited by Hunter Rouse), John Wiley & Sons, New York, 1950.

[9] Poluvarinova – Kochina, P.Y., Theory of Groundwater Movement, Translated from Russian at Princeton Univ. Press, New York, 1962.

[10] Stallman, R.W., The Flow in the Zone of Aeration, in Chow, V.T. (Ed.), *Advances in Hydroscience*, Academic Press, New York, 1967.

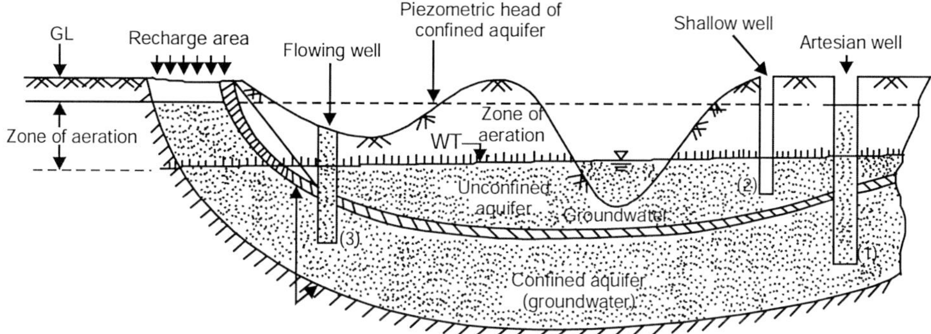

Figure 14.1 Groundwater in confined and unconfined aquifers.

piezometric level [shown in figure in well number (1)]. If a well penetrates only the unconfined aquifer, level of water is equal to WT [well number (2)] and if a well penetrates the confined aquifer in which ground level is below the piezometric head, well discharges as a flowing well [well number (3)]. Thus, groundwater occurs in both unconfined and confined aquifer depending on geologic formation below the ground surface. An unconfined aquifer is one in which water table changes in undulating form in slope, depending on areas of recharge and discharge, pumping from well and aquifer's permeability. Confined aquifers are known as **pressure aquifer** or **artesian type**. The word artesian is originated from French word *artésien*. Originally artesian well, meant freely flowing well, but in recent times, it is the well perpetrating the confined aquifer.

14.4 AQUIFER PARAMETERS

Before discussing further on groundwater, definition of some parameters relating to aquifer or groundwater is essential.

Porosity (*n*): It is defined as the ratio of the volume of voids V_v in a soil mass to the total volume V_t and is expressed in percentage as:

$$n = \frac{V_v}{V_t} \times 100$$

n varies from 0.25 to 0.75 depending on soil texture.

Coefficient of permeability (*K*): It is defined as the rate of flow of water through the aquifer per unit cross-sectional area per unit hydraulic gradient, i.e., it indicates the velocity through soil (cm/sec) and is denoted by K.

Coefficient of transmissibility (*T*): It is defined as the rate of flow (m^3/day) through soil strip of aquifer of unit width extending the full saturation height under unit hydraulic gradient. If the aquifer thickness is b, K is the coefficient of permeability, then

$$T = bK$$

Storage coefficient (S_c): It is defined as the volume of water that an aquifer releases from or takes into storage per unit area of aquifer per unit head change in the component of head normal to that surface. This coefficient is a dimensionless quantity involving a volume of water per volume of aquifer. In most confined aquifer, values fall in the range $0.00005 < S_c < 0.005$ and in unconfined aquifer $0.01 < S_c < 0.3$.

Soil moisture content (θ): It is the ratio of volume of water in soil to the total volume.

$$\theta = \frac{\text{Volume of water}}{\text{Total volume}} \qquad 0 \le \theta \le n$$

i.e., equal to porosity when soil is saturated.

14.5 DARCY'S LAW

Groundwater in its natural state moves with established hydraulic principles. This flow through natural porous media can be expressed by Darcy's law. Hydraulic conductivity K, which is a measure of the permeability of the media, is an important parameter in the flow equation. Henry Darcy (1856) published the results of his experiment (Figure 14.2) that he had performed in Paris. He proposed a law which is called as the Darcy's law. This law is the basic principle of flow through porous media.

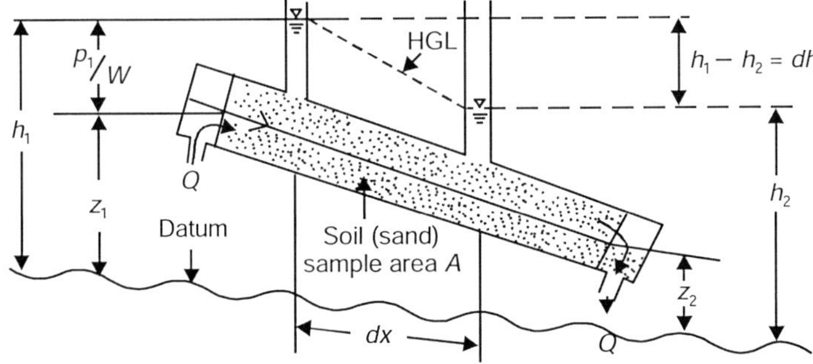

Figure 14.2 Darcy's experimental setup.

Darcy used an experimental setup almost similar to Figure 14.2 to study the flow through sand filter bed. He varied dx, Q and head difference $(h_1 - h_2)$, i.e., dh and discovered that flow rate Q varies directly with dh and inversely with dx having the same cross-sectional area A.

i.e.
$$Q \propto \frac{dh}{dx} A$$

or
$$\frac{Q}{A} \propto -\frac{dh}{dx}$$ (– ve sign indicates dh decreases when x increases)

In general form,
$$V = -K\frac{dh}{dx} \tag{14.1}$$

Here K is the hydraulic conductivity or Darcy's coefficient of permeability having the dimension of velocity L/T.

14.5.1 Range of Validity of Darcy's Law

1. Flow is laminar as velocity through the sample is very low.

2. Reynolds number $R_e = \dfrac{\rho VD}{\nu} < 1$

Here D is the average diameter of soil grain. Various investigators like Ahmed and Sunada (1969), Rumer and Drikler (1966), Smith and Sayre (1964), Ward (1964) invested on Reynolds number and concluded that law can be applied up to $R_e = 10$, although at high Reynolds number flow tends to be turbulent.

14.6 UNSTEADY GROUNDWATER FLOW EQUATION

Consider an unconfined aquifer in a flood plain of river as shown in Figure 14.3(a) and its sectional view in Figure 14.3(b). During low flow, aquifer discharge into the river. A control volume for 2D unconfined groundwater flow is shown in Figure 14.3(c). Assume homogenous isotropic aquifer. Water table in the aquifer changes with time. The expression for the conservation of mass for the control volume is:

$$\rho\left(u - \frac{\partial u}{\partial x}\frac{\Delta x}{2}\right)\Delta y(h-z) - \rho\left(u + \frac{\partial u}{\partial x}\frac{\Delta x}{2}\right)\Delta y(h-z)$$
$$+ \rho\left(v - \frac{\partial v}{\partial y}\frac{\Delta y}{2}\right)\Delta x(h-z) - \rho\left(v + \frac{\partial v}{\partial y}\frac{\Delta y}{2}\right)\Delta x(h-z)$$
$$= \rho S \Delta x \Delta y \frac{\Delta h}{\Delta t}$$

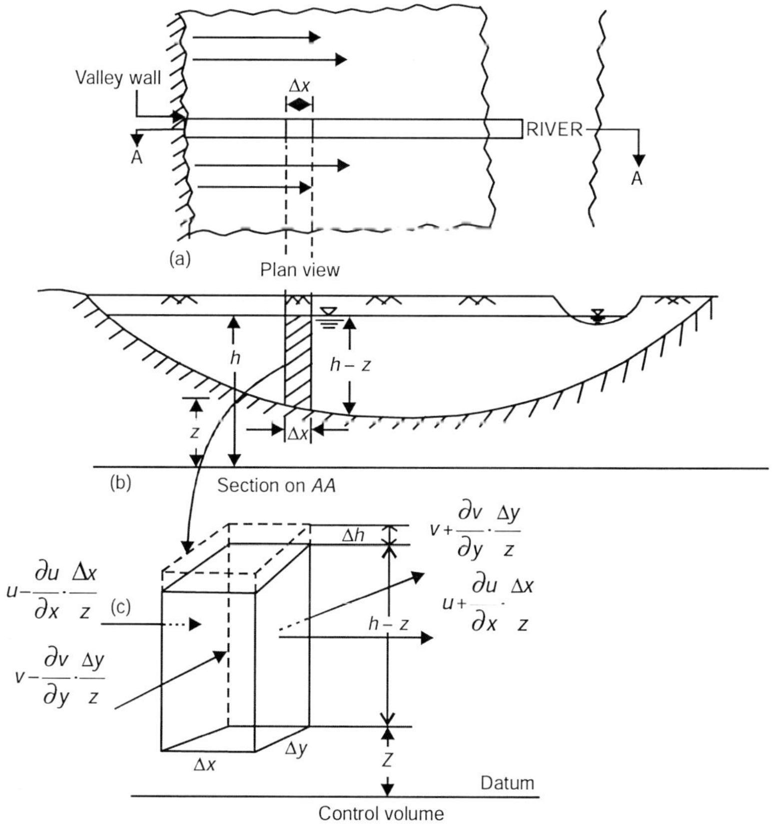

Figure 14.3 Control volume for 2D unconfined groundwater.

Simplifying

$$-(h-z)\frac{\partial u}{\partial x}-(h-z)\frac{\partial v}{\partial y}=S\frac{\partial h}{\partial t} \tag{14.2}$$

By Darcy's law, $u=-K\frac{\delta h}{\delta x}$, $v=-K\frac{\partial h}{\partial y}$ then Eq. (14.2) becomes

$$[(h-z)K]\frac{\partial^2 h}{\partial x^2}+[(h-z)K]\frac{\partial^2 h}{\partial y^2}-S\frac{\partial h}{\partial t} \tag{14.3}$$

$\therefore$ $$[(h-z)K]=T$$

Here T is coefficient of transmissibility (L^3/T). Now, from Eq. (14.3),

$$T\frac{\partial^2 h}{\partial x^2}+T\frac{\partial^2 h}{\partial y^2}=S\frac{\partial h}{\partial t}$$

$\therefore$ $$\frac{\partial^2 h}{\partial x^2}+\frac{\partial^2 h}{\partial y^2}=\frac{S}{T}\frac{\partial h}{\partial t} \tag{14.4}$$

Equation (14.4) is the 2D unsteady flow in unconfined aquifer. Equation (14.4) in three dimensions (3D) can be written as:

$$\frac{\partial^2 h}{\partial x^2}+\frac{\partial^2 h}{\partial y^2}+\frac{\partial^2 h}{\partial z^2}=\frac{S}{T}\frac{\partial h}{\partial t} \tag{14.4a}$$

or $$\nabla^2 h=\frac{S}{T}\frac{\partial h}{\partial t} \tag{14.4b}$$

Figure 14.4 shows the control volume for 2D confined aquifer below the ground level (G.L.)

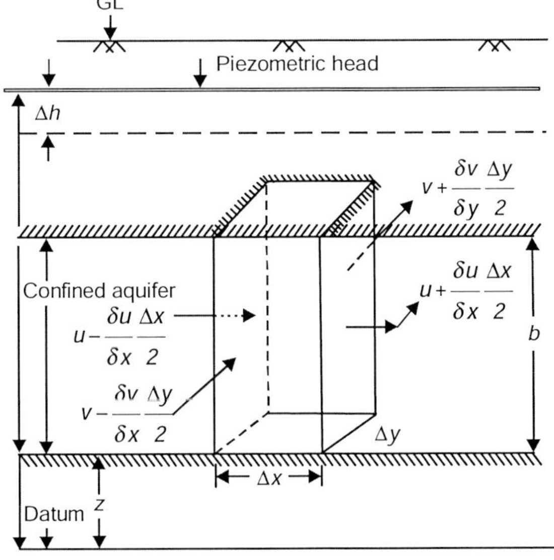

Figure 14.4 Control volume for 2D confined aquifer.

If the flow is steady, $\frac{\partial h}{\partial t} = 0$

Equation (14.4) can be written as:

In 3*D*, $$\frac{\partial^2 h}{\partial x^2} + \frac{\partial^2 h}{\partial y^2} + \frac{\partial^2 h}{\partial z^2} = 0 \tag{14.4c}$$

or $$\nabla^2 h = 0 \tag{14.4d}$$

Equation (14.4c) is known as **Laplace equation**. Equation (14.4), which is developed for unconfined aquifer, can also be applied to confined aquifer. Only here specific volume S_s is defined is:

$$S_s = \frac{S}{b}$$

Therefore, equation for confined aquifer in two-dimensional flow becomes

$$\frac{\partial^2 h}{\partial x^2} + \frac{\partial^2 h}{\partial h^2} = \frac{S_s}{K}\frac{\partial h}{\partial t} \tag{14.5}$$

or $$\nabla^2 h = \frac{S_s}{K}\frac{\partial h}{\partial t} \tag{14.5a}$$

When flow is steady in two-dimension,

$$\frac{\partial^2 h}{\partial x^2} + \frac{\partial^2 h}{\partial y^2} = 0 \tag{14.5b}$$

or $$\nabla^2 h = 0$$

Solution of Equation (14.4) or (14.5) can be solved with proper boundary condition by analytical, numerical or analog methods.

14.7 UNSTEADY FLOW EQUATION IN WELL

Consider a well in a confined equation with pumping of discharge Q as shown in Figure 14.5. During initial stage of pumping, cone of depression continues to g`row as shown in the figure, and flow to well at this stage is unsteady. To derive an equation of unsteady flow to the well at this stage, an incremental control volume for unsteady well flow shown in Figure 14.5, i.e., a cylinder with radius *r*, thickness *dr* and depth *b*, is considered.

During an incremental time step $(t_2 - t_1) = \partial t$, the change in the draw down in well is ∂h. Let S be the storage coefficient which is the volume of water per unit area of aquifer that the aquifer will yield per unit drop of piezometric head. Yield of water from the aquifer storage during the incremental time step ∂t of the control volume is:

$$\text{Yield} = -S\,(2\pi r dr)\,\partial h$$

$$q_o = q_l - S\,(2\pi r dr)\,\frac{\partial h}{\partial t}$$

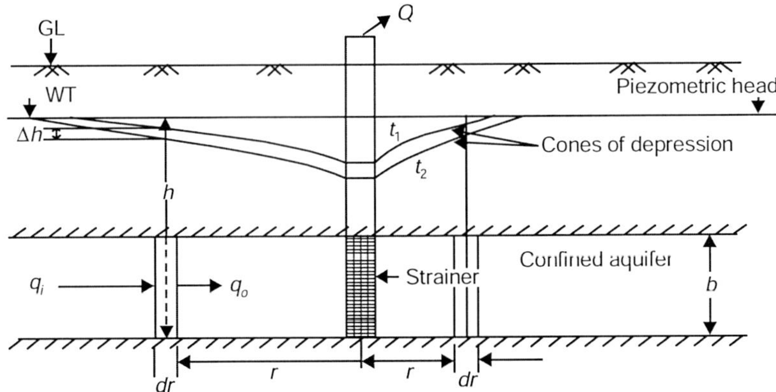

Figure 14.5 Unsteady flow condition in well of a confined aquifer or depth b.

$$\therefore \qquad q_l - q_o = S\,(2\pi r dr)\frac{\partial h}{\partial t} \tag{14.6}$$

Here q_i and q_o are shown in figure as inflow from aquifer and outflow to the well from the control volume. If q is the aquifer flow rate based on Darcy's law,

$$q = \text{Area} \times \text{Velocity} = (2\pi rb)K\frac{\partial h}{\partial r} \tag{14.7}$$

Change in aquifer flow rate in the elemental volume is:

$$q_l - q_o = \frac{\partial q}{\partial r}\partial r = 2\pi Kb\left(r\frac{\partial^2 h}{\partial r^2} + \frac{\partial h}{\partial r}\right)dr \tag{14.8}$$

Equating Eqs. (14.6) and (14.8), $S(2\pi r dr)\dfrac{\partial h}{\partial t} = 2\pi Kb\left(\dfrac{r\partial^2 h}{\partial r^2} + \dfrac{\partial h}{\partial r}\right)dr$

or

$$rS\frac{\partial h}{\partial t} = T\left(r\frac{\partial^2 h}{\partial r^2} + \frac{\partial h}{\partial r}\right)$$

$$\therefore \qquad Kb = T$$

Simplifying,

$$\frac{\partial^2 h}{\partial r^2} + \frac{1}{r}\frac{\partial h}{\partial r} = \frac{S}{T}\frac{\partial h}{\partial t} \tag{14.9}$$

Equation (14.9) is the unsteady flow equation to a well when pumping starts.

14.7.1 Assumption in Derivation of Eq. (14.9)

1. Aquifer is homogenous, isotropic and of uniform thickness and infinite aerial extent.
2. Before pumping, piezometric head or surface is horizontal.
3. Rate of pumping is assumed to be constant.
4. The well penetrates the entire aquifer and flow is everywhere horizontal within the aquifer to the well.
5. The well diameter is very small so that storage within the well is neglected.
6. Water removed from the storage discharges instantaneously with the decline of head.

14.7.2 Solution of Eq. (14.9)

1. Theis[11] (1935) was the first to solve this equation for well flow in an isotropic and homogenous confined aquifer. Theis solution is:

$$s = \frac{Q}{4\pi T}\int_u^{\alpha} \frac{e^{-u}}{u} du = \frac{Q}{4\pi T} W(u) \tag{14.10}$$

where s is the drawdown of piezometric head at a radial distance r, Q is the constant discharge, $W(u)$ is called well function of u or Theis function. u is non-dimensional term which is defined as:

$$u = \frac{r^2 S}{4Tt} \tag{14.11}$$

The integral of Eq. (14.10) is not directly integrable, but evaluated by the series as:

$$\int_u^{\alpha} \frac{e^{-u}}{u} du = W(u) = -0.5772 - \log_e(u) + u - \frac{u^2}{2.2!} + \frac{u^3}{3.3!} \tag{14.12}$$

Equation (14.10) is:

$$s = \left(\frac{Q}{4\pi T}\right) W(u)$$

Rearranging Eq. (14.11) as:

$$\frac{r^2}{t} = \left(\frac{4T}{S}\right) u$$

It is seen that relation between $W(u)$, u, s and $\left(\frac{r^2}{t}\right)$ in Eqs. (14.10) and (14.11) is similar as the two terms on right hand sides of the two equations under parenthesis are constants. Based on this observation, Theis suggested an approximate solution of s and T based on Graphic method of superposition. Details of method with examples and Table of well function or Theis function $W(u)$ and (u) are referred to Todd[12] (2001).

---

[11] Theis, C.V., The Relation between Lowering of the Piezometric Surface and Rate of Duration of Discharge of a Well using Groundwater Storage, Trans. *Am. Geophys. Union,* vol. 16, pp. 519–524, 1935.

[12] Todd, D.K., *Groundwater Hydrology*, 2nd., John Wiley & Sons, New York, 2001.

2. Cooper and Jocob[13] (1946) noted that for small values of r and large values of t, u is small, hence the series terms in Eq. (14.12) becomes negligible after the first two terms. As a result, the drawdown can be expressed as:

$$s = \frac{Q}{4\pi T}\left(-0.5772 - \log_e \frac{r^2 S}{4Tt}\right) \tag{14.13}$$

or
$$s = \frac{Q}{4\pi T}\left(-0.5772 - 2.3026 \log_{10} \frac{r^2 S}{4Tt}\right)$$

or
$$s = \frac{Q}{4\pi T}\left(-0.5772 + 2.3026 \log_{10} \frac{4Tt}{r^2 S}\right)$$

or
$$s = \frac{Q}{4\pi T}\left(2.3026 \log_{10} 0.56146 + 2.3026 \log_{10} \frac{4T^t}{r^2 S}\right)$$

or
$$s = \frac{2.3026Q}{4\pi T}\left[\log_{10} 0.56146 + \log_{10} 4Tt - \log_{10} r^2 S\right]$$

or
$$s = \frac{2.3026Q}{4\pi T}\left[\log_{10} \frac{(0.5616 \times 4Tt)}{r^2 S}\right]$$

$$s = \frac{2.3026Q}{4\pi T} \log_{10} \frac{2.2464Tt}{r^2 S} \tag{14.14}$$

Now, a plot of s versus the logarithm t can be drawn, which is a straight line as shown in Figure 14.6. Projecting this line to s = 0, when $t = t_0$,

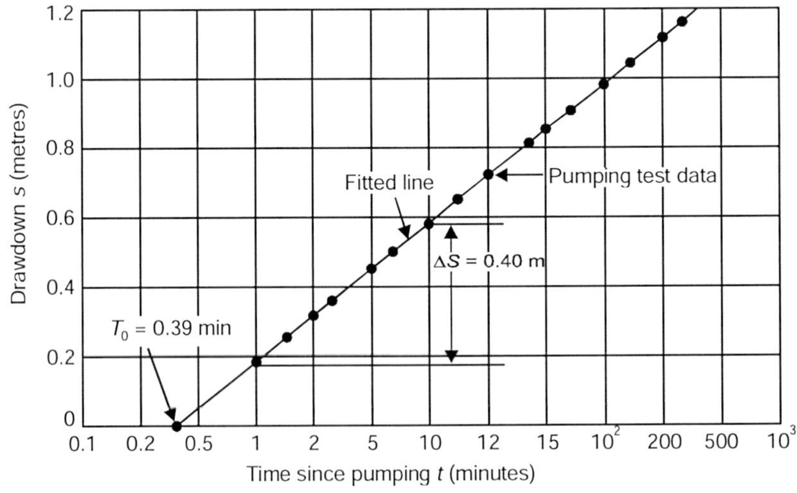

Figure 14.6 Cooper-Jacob method for solution of the equation.

[13] Cooper, H.H. Jr. and Jacob, C.E., A generalised graphical method for evaluating formation constants and summering well field history, *Am. Geophys Union*, vol. 27, pp. 526–534, 1946.

$$0 = \frac{2.3026Q}{4\pi T} \log_{10} \frac{2.2464 t_0 T}{r^2 S}$$

If
$$\frac{2.2464 t_0}{r^2 S} = 1$$

$$S = \frac{2.2464 T t_0}{r^2} \cong \frac{2.25 T t_0}{r^2} \tag{14.14a}$$

A value of T can be obtained by making $t/t_0 = 10$, then $\log_{10}(t/t_0) = \log_{10}10 = 1$. Replacing s by Δs, where Δs is the drawdown difference per log cycle of t, from Eq. (14.11), T becomes as per Kruseman and Ridder[14] (1970)

$$T = \frac{2.3026Q}{4\pi \Delta S} \tag{14.15}$$

Solve for T, with the help of Eq. (14.15), and then solve for S by Eq. (14.14). The straight line approximate values of $u(u < 0.01)$ is restricted to avoid errors.

3. In Chow's method[15] (1952), measurements of drawdown s against time are made in an observation well near the pumped well. Observational data are plotted in a semi log paper just like in Cooper and Jacob method. In the plotted curve, an arbitrary point is chosen and a tangent is drawn at that point.

Then find a new function

$$F(u) - s/\Delta S$$

A figure is drawn relating $W(u)$, $F(u)$ and u as shown in Figure 14.7. Find corresponding values of $W(u)$ and u from Figure 14.7. Finally compute T from Eq. (14.10) and s from Eq. (14.11).

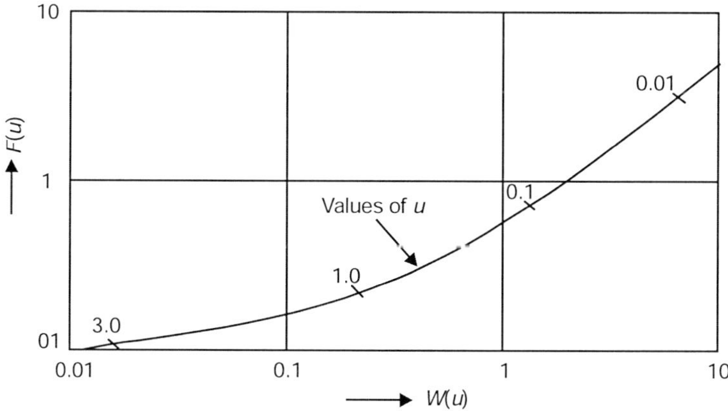

Figure 14.7 Relation among $F(u)$, $W(u)$ and u (After Chow, 1952).

[14] Kruseman, G.P. and Ridder N.A. de, Analysis and Evaluation of Pumping Test Data, Intl. Inst. for Land Reclamation and Improvement, Bulletin 11, Wageningen, 1970.

[15] Chow, V.T., On the Determination of Transmissibility and Storage Coefficients from Pumping Test Data, Trans *Am Geophys Union*, vol. 33, pp. 397–404, 1952.

14.8 STEADY RADIAL FLOW TO WELL IN UNCONFINED AQUIFER

When a well penetrates in an extensive homogenous unconfined aquifer in which water table was initially horizontal, circular cone of depression in the water table near the well takes place when water is pumped since no flow to the well takes place without a gradient towards the well. This depression is called **cone of depression** (Figure 14.8). The decrease of water table (WT) from h_e to h_w is called **drawdown** or **drop of piezometric head.** Water from the well in the unconfined aquifer (shown in Figure 14.8) is pumped out to form the cone of depression. As soon as the cone of depression is formed, water starts flowing towards the well due to gradient when amount of pumping will be equal to the amount of radial flow to the well, water level in the well will be constant. This flow to the well is steady radial discharge Q.

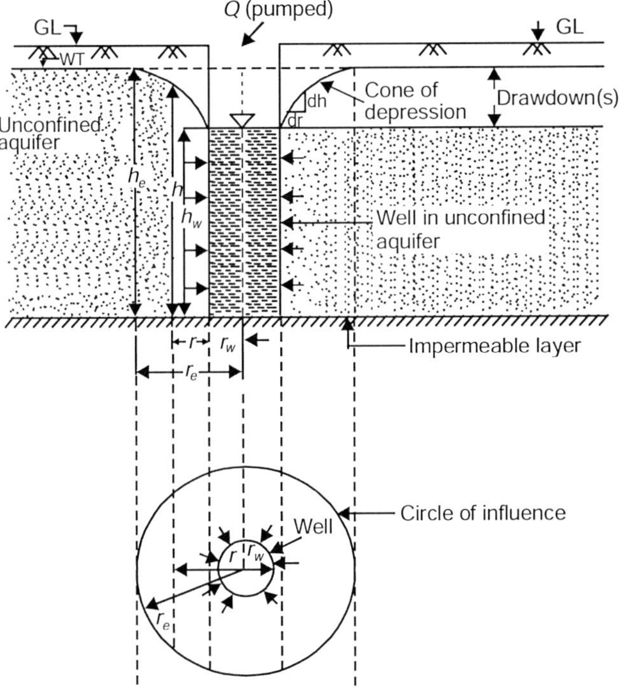

Figure 14.8 Steady radial discharge to a well in unconfined aquifer.

The analysis of such radial flow towards the well was originally proposed by A.J.E.J. Dupuit[16] (1863) and subsequently modified by G. Thiem[17] (1906).

Expression for steady radial discharge *Q*: Let h be the height of water table at a distance r from the centre of the well. By continuity,

[16] Dupuit, A.J.E.J., *Eludes theoriques sur la mouvement des eaux dans le canaux decouverts et a travers les terrains permeables*, 2nd ed., Dunod, Paris, 1863.

[17] Thiem, G., *Hydrologische Methoden*, Gebhardt, Lcipzig, 1906.

$$Q = \text{Area of flow at radius } r \times \text{Velocity at radius } r$$

$$Q = (2\pi rh)\left(K\frac{dh}{dr}\right)$$

or

$$\frac{Qdr}{r} = 2\pi Khdh$$

or

$$Q\int \frac{dr}{r} = 2\pi K\int hdh \tag{14.16}$$

When

$$r = r_w, \quad h = h_w$$
$$r = r_e, \quad h = h_e$$

$$Q\int_{r_w}^{r_e} \frac{dr}{r} = 2\pi K\int_{h_w}^{h_e} hdh$$

∴

$$Q\log_e\left(\frac{r_e}{r_w}\right) = \pi K\left(h_e^2 - h_w^2\right)$$

or

$$Q = \frac{\pi K\left(h_e^2 - h_w^2\right)}{\log_e\left(\frac{r_e}{r_w}\right)} \tag{14.17}$$

Equation (14.17) is known as **Dupuit's equation** for steady radial flow to unconfined aquifer. The same equation may be written as:

$$Q = \frac{\pi K(h_e + h_w)(h_e - h_w)}{\log_e\left(\frac{r_e}{r_w}\right)} \qquad 14.17(a)$$

But

$$(h_e - h_w) = s \text{ (drawdown)}$$

and

$$h_e + h_w = (h_e - h_w) + 2h_w = s + 2h_w$$

∴

$$Q = \frac{\pi Ks(s + 2h_w)}{\log_e\left(\frac{r_e}{r_w}\right)} \qquad 14.17(b)$$

14.8.1 Assumptions of Dupuit's Theory

1. Velocity of flow is proportional to the tangent of hydraulic gradient instead of sine.
2. Aquifer is assumed to be homogenous and infinite aerial extent.
3. Flow is laminar and Darcy's law is valid.
4. The flow is assumed to be horizontal and uniform everywhere in the vertical section.
5. Natural groundwater regime affecting the aquifer remains constant with time.
6. K is constant at all places and at all times.
7. Well receives water from the entire thickness of the aquifer.

14.8.2 Limitations of Dupuit's Theory

1. In Darcy's law, velocity varies with sin θ, but here it is assumed to be tan θ, so long the slope of drawdown is small; assumption is all right but near the well, slope is more, hence sin $\theta \neq$ tan θ.
2. Flow is assumed to be laminar, but near to the well flow no longer remains laminar.
3. The flow does not remain horizontal in all the sections of the aquifer.

14.8.3 Observation Wells

Consider two observation wells as shown in Figure 14.9.

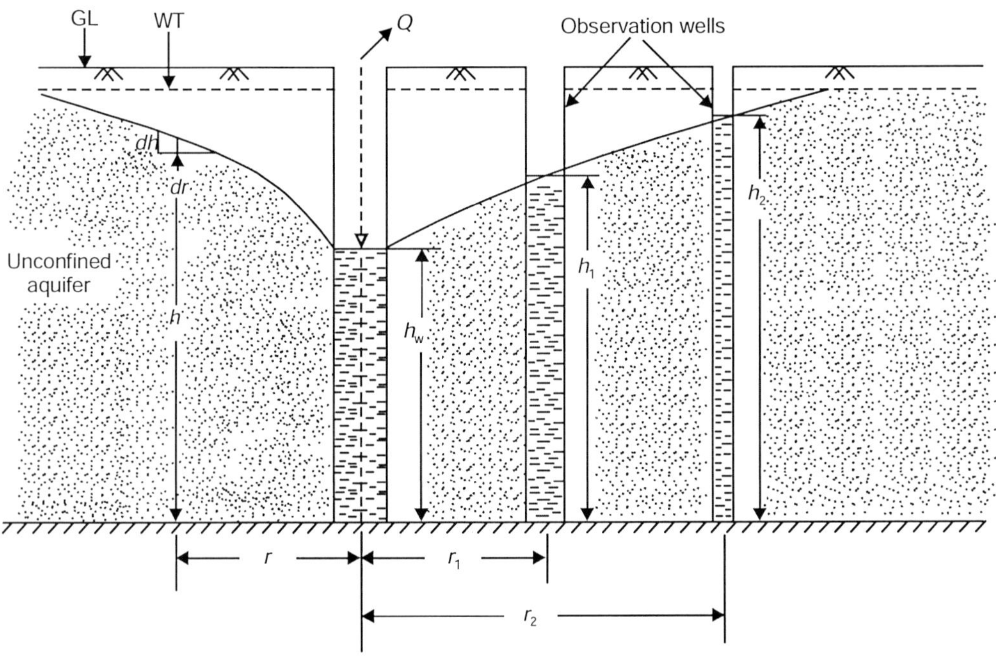

Figure 14.9 Observation wells.

Integrating Eq. (14.16) in limit taking from observation wells, i.e., $r = r_1$, $h = h_1$, $r = r_2$, $h = h_2$,

$$Q \int_{r_1}^{r_2} \frac{dr}{r} = 2\pi K \int_{h_1}^{h_2} h dh$$

$$\therefore \qquad Q = \frac{\pi K \left(h_1^2 - h_2^2\right)}{\log_e \left(\frac{r_2}{r_1}\right)} \qquad (14.17b)$$

i.e., steady radial discharge Q can be expressed in terms of observation well parameters.

14.9 STEADY RADIAL DISCHARGE TO WELL IN CONFINED AQUIFER

Figure 14.10 shows a well that perpetrates the confined aquifer of depth b. If Q is the steady flow to the well, then

$$Q = (2\pi rb)\left(K\frac{dh}{dr}\right)$$

$$Q\frac{dr}{r} = 2\pi Kb\frac{dh}{dr} \tag{14.18}$$

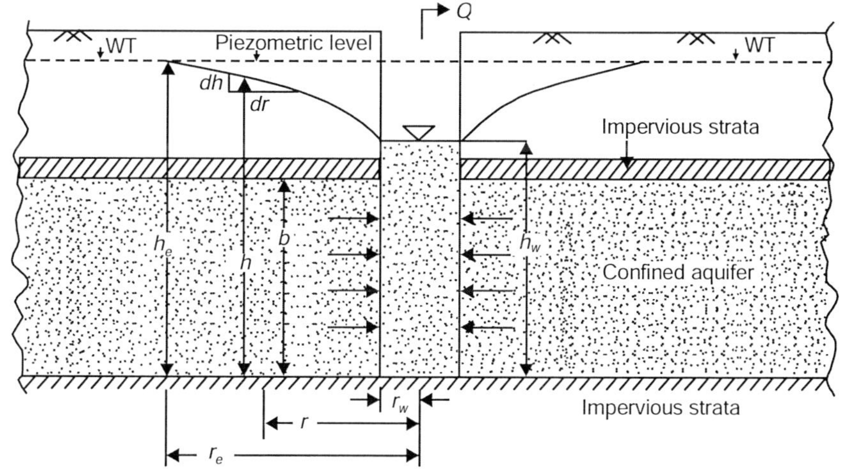

Figure 14.10 Steady radial discharge from confined aquifer.

When $r = r_w$, $h = h_w$

$r = r_e$, $h = h_e$

Then integrating Eq. (14.18),

$$Q\int_{r_w}^{r_e} \frac{1}{r}dr = 2\pi K\int_{h_w}^{h_e} hdh$$

$$\therefore \qquad Q = \frac{2\pi Kb(h_e - h_w)}{\log_e\left(\frac{r_e}{r_w}\right)} \tag{14.9}$$

But $(h_e - h_w)$ = drawdown s

and $T = kb$

$$\therefore \qquad Q = \frac{2\pi Ts}{\log_e\left(\frac{r_e}{r_w}\right)} \tag{14.19a}$$

Equation (14.19) is the steady radial discharge to well in confined aquifer.

14.10 INTERFERENCE OF WELL IN CONFINED AQUIFER

Figure 14.11 shows two wells constructed in confined aquifer near each other at a distance B apart, and Q_1 and Q_2 are the discharges of the two wells respectively as shown. Their drawdown curves intersect within their radius of zero drawdown. Both the wells have same diameter, drawdown and discharge (i.e., $Q_1 = Q_2$) over a same period of time. Let b be the depth of confined aquifer.

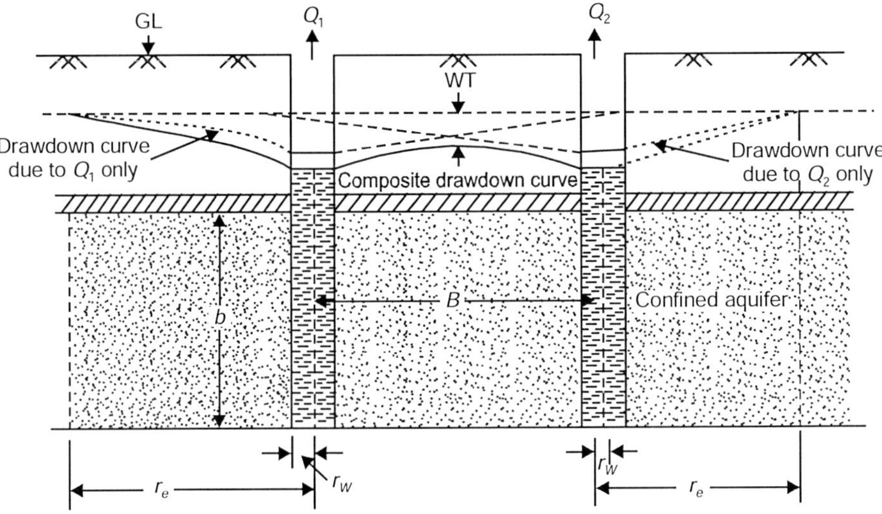

Figure 14.11 Interference between two wells in confined aquifer.

It can be shown by the method of complex variable that discharge in each well is given by

$$Q_1 = Q_2 = \frac{2\pi kb(r_e - r_w s)}{\log_e \dfrac{r_e^2}{r_w B}} \tag{14.20}$$

where $r_e >> B$

$$\therefore \qquad \frac{r^2}{r_w B} > \frac{r_e}{r_w}$$

Hence Q given by Eq. (14.19) is greater than Q_1 or Q_2, where $Q_1 = Q_2$, i.e., Eq. (14.19) $Q > Q_1$ or $Q > Q_2$. Thus, discharge in each well decreases, due to interference of wells.

14.11 YIELD OF WELL

The constant quantity of water that may be pumped per unit time is the yield of the well. Yield of well is determined by the following methods:

Recuperation test: In Figure 14.12, water level initially before pumping is at *aa,* i.e., in water table (WT) level. In this test, water level is depressed by pumping below normal level. Then pumping is stopped. Time required to recuperate to normal level is recorded. From this data, discharge or yield of the well is determined from the following equation:

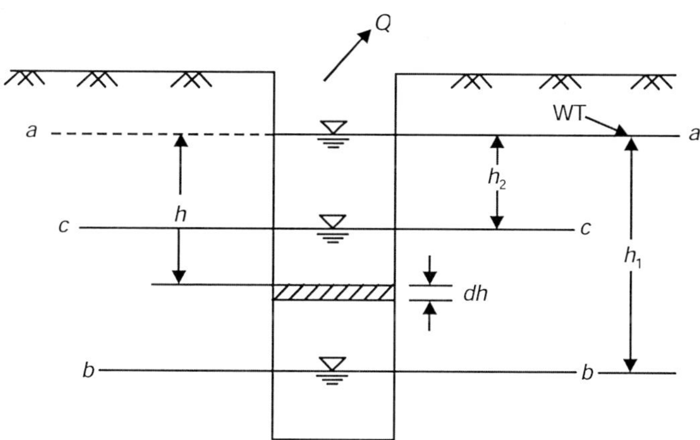

Figure 14.12 Recuperation test.

Let *aa* be the static level (WT) in the well, h_1 is the depression in the well when pumping is stopped, *cc* is the level reached in *T* hrs after pumping, h_2 is the depression of *cc* below normal after *T* hr of pumping.

At any instant, a small head *dh* between *cc* and *bb* is considered. This *dh* is the decrease in head in time *dt* hrs. The depression head of *dh* is *h*.

Now, volume of water recuperated or entered when head is recuperated in time *dt* is:

$$d\forall = Adh \tag{14.21}$$

where AS is the area of the well.

If *Q* is the rate of discharge when depression head is *h* after *t* hrs, the volume of water enters at instant in time *dt* is:

$$Qdt = d\forall$$

But $$Q \propto h$$

∴ $$Q = kh$$

where *k* is some constant dependent on soil characteristics.

∴ $$d\forall = khdt \tag{14.22}$$

Eqs. (14.21) and (14.22),

$$Adh = -khdt \tag{14.23}$$

Negative sign on RHS of Eq. (14.23) indicates when *h* decreases, *t* increases.
Integrating Eq. (14.23),

$$A\int_{h_1}^{h_2} \frac{dh}{kh} = -\int_{o}^{T} dt$$

As boundary conditions are:

When $t = 0, \quad h = h_1$

When $t = T, \quad h = h_2$

or
$$T = -\frac{A}{k}[\log_e h_2 - \log_e h_1]$$

or
$$T = \frac{A}{k}[\log_e h_1 - \log_e h_2]$$

or
$$T = \frac{A}{k}\log_e\left(\frac{h_1}{h_2}\right)$$

$$\frac{k}{A} = \frac{2.303}{T}\log_{10}\left(\frac{h_1}{h_2}\right) \tag{14.24}$$

From data recorded for T, h_1, h_2, (k/A), value of well can be determined.

Now for a constant depression head of H,

$$Q \propto H$$

∴
$$Q = kH \tag{14.25}$$

or
$$Q = \left(\frac{k}{A}\right)AH$$

(k/A) is known from Eq. (14.24).

Equation (14.25) gives the yield of well, where (k/A) value is constant for a well and different for different types of soil. For coarse sand $(k/A) = 1$, fine sand $= 0.8$, clay $= 0.25$.

In Eq. (14.25), if $H = 1$ m, and writing the equation as:

$$\frac{Q}{A} = \left(\frac{k}{A}\right)H$$

Then substituting $H = 1$ m,

$$\frac{Q}{A} = \left(\frac{k}{A}\right) \tag{14.26}$$

Equation (14.26) gives the yield or discharge per unit area under unit head. Hence Eq. (14.26) gives the specific yield of the well.

Constant level pumping test: The name of the method implies that the well maintains a constant level at a certain constant discharge or yield from the well. It is a very simple method where Darcy's equation is used for average velocity to the well at constant depression.

To find the yield, water level in the well is depressed first to a considerable depth to create a hydraulic gradient between the well and the aquifer for flow to well. Next step is to regulate the pumping such that a steady state of flow to well is established, i.e., under certain constant pumping rate level in the well remains constant. Amount of water pumped out from the well becomes equal to the amount of flow from the aquifer to the well under a constant depression head equal to H.

Average velocity to the well under this equilibrium condition may be written as:

$$V = Ki$$

where i is the hydraulic gradient in the well and can be expressed as (H/L), L being some characteristic length of aquifer (normally radius of influence under a depression head of H).

If A is the area of flow to the well below the level of H, then constant discharge that the well can yield is:

$$Q = AV$$

or

$$Q = AK\left(\frac{H}{L}\right)$$

or

$$Q = \left(\frac{K}{L}\right)AH$$

or

$$Q = CAH$$

Here C is a percolation intensity coefficient which is constant for a particular formation around the well.

EXAMPLE 14.1

Discharge from a 40 cm diameter tube well fully penetrating an unconfined aquifer of 50-m thickness under depression head of 4 m is 2500 lits/min. Find the discharge of the well under a drawdown of 6 m. Radius of influence may be taken as 320 m.

Solution: $h_e = 50$ m, $h_w = (50 - 4) = 46$ m, $r_e = 320$ m,

$r_w = 40/2 = 20$ cm $= 0.2$ m, $Q_4 = 2500$ lits/min

$= 2.5$ m^3/min

Discharge equation in unconfined aquifer when drawdown is 4 m,

$$Q_4 = \frac{\pi K\left[h_e^2 - h_w^2\right]}{\log_e\left(\frac{r_e}{r_w}\right)} = \frac{\pi K\left(50^2 - 46^2\right)}{\log_e\left(\frac{320}{0.2}\right)}$$

Similarly, if Q_6 is the discharge under a drawdown of 6 m,

$$Q_6 = \frac{\pi K(50^2 - 44^2)}{\log_e\left(\frac{320}{0.2}\right)}$$

$$\therefore \quad \frac{Q_4}{Q_6} = \frac{50^2 - 46^2}{50^2 - 44^2}$$

or

$$Q_6 = Q_4\left(\frac{50^2 - 44^2}{50^2 - 46^2}\right)$$

or $$Q_6 = 2.5 \times \frac{564}{384} = 3.6718$$

∴ $$Q = 0.0612 \text{ m}^3 \text{ sec}$$

EXAMPLE 14.2

Design a tube well from the following data:

(i) Yield required = 0.2 m³/sec, (ii) Thickness of conifer aquifer = 40 m, (iii) Circle of influence = 300 m, (iv) Permeability coefficient = 80 m/day and (v) Drawdown = 6 m.

Solution:

Given Q = 0.2 m³/sec, b = 40 m, k = 80 m/day = 0.000925 m/sec, r_e = 300 m, h_e = 40 m drawdown = 6 m and h_w = (40 – 6) = 34 m

Equation of discharge in confined aquifer is given by

$$Q = \frac{2\pi Kb(h_e - h_w)}{\log_e\left(\frac{r_e}{r_w}\right)} \qquad \text{(i)}$$

Substitute the given values in Eq. (i), we get

$$0.2 = \frac{2\pi \times 0.000925925(40 - 34) \times 40}{\log_e\left(\frac{300}{r_w}\right)}$$

$$\log_e\left(\frac{300}{r_w}\right) = 609812$$

∴ $$\frac{300}{r_w} = 1076.2$$

$$r_w = 0.2787 \text{ m} = 27.87 \text{ cm}$$

Take 30 cm radius for design, i.e., diameter = 60 cm

EXAMPLE 14.3

An artesian tube well has a diameter of 20 cm. With the aquifer thickness of 30 m and permeability of 40 m/day, find the yield under a drawdown of 4 m. Radius of influence is 275 m.

Solution:

An artesian well means a well in confined aquifer. Hence.

$$Q = \frac{2\pi kb(h_e - h_w)}{\log_e\left(\frac{r_e}{r_w}\right)}$$

Here $$K = 40 \text{ m/day} = \left(\frac{40}{24 \times 60 \times 60}\right) \text{m/sec}$$

$$b = 30 \text{ m}$$
$$r_e = 275 \text{ m}$$
$$h_w = (30 - 4) = 26 \text{ m (assuming initial WT = 30 m)}$$
$$h_e = 30 \text{ m} \quad \text{(assumed)}$$
$$r_w = 20 \text{ cm} = 0.2 \text{ m}$$

$\therefore$ $$Q = \frac{2\pi \times 30 \times (40/24 \times 60 \times 60)(30 - 26)}{\log_e\left(\frac{275}{0.2}\right)}$$

or $$Q = 0.0483 \text{ m}^3\text{/sec}$$

EXAMPLE 14.4

Determine the diameter of the well in coarse sand for a yield of 0.005 cumec when working under a depression head of 2.5 m.

Solution:

Yield = 0.005 m^3/sec = (0.005 × 60 × 60) = 18 m^3/hour

For coarse sand, k/A = 1 m^3/hr/m

$\therefore$ $$Q = \left(\frac{k}{A}\right) AH$$

$\therefore$ $$18 = 1 \times A \times 2.5$$

$\therefore$ $$A = 7.2 \text{ m}^2$$

or $$\frac{\pi}{4} D^2 = 7.2$$

$\therefore$ $$D = 3.027 \text{ m}$$

EXAMPLE 14.5

In a recuperation test, water level in the well was depressed to the extent of 3.5 m by pumping and recuperated 2 m in an hour. If the diameter of the well is 3 m, and average working head is 2.5 m, calculate average yield of the well.

Solution:

Using the following equation,

$$Q = \frac{k}{A} AH$$

where $$\frac{k}{A} = \frac{2.303}{T} \log_{10}\left(\frac{h_1}{h_2}\right)$$

Given $T = 1$ hr, $h_1 = 3.5$ m, $h_2 = (3.5 - 2) = 1.5$ m

$$A = \pi/4(3)^2 = \left(\frac{9\pi}{4}\right)\text{m}^2, H = 2.5\text{ m}$$

$\therefore$ $$\frac{k}{A} = \frac{2.303}{1} \log_{10}\left(\frac{3.5}{1.5}\right) = 0.84745$$

Then $$Q = \left(\frac{k}{A}\right) AH$$

or $$Q = 0.84745 \left(\frac{9\pi}{4}\right) \times 2.5$$

$\therefore$ $$Q = 14.975687 \text{ m}^3\text{hr}$$
$$Q = 0.00416 \text{ m}^3/\text{sec}$$

14.12 CONCLUSION

Discussion in this chapter has been started with historical background of groundwater, its occurrence, Darcy's law and its application in both steady and unsteady flow to a well. Unsteady ground water flow equations both in cartesian and cylindrical co-ordinates are derived. Solutions of unsteady equation to a well by different investigators like Theis, Cooper and Jacob and Chow are presented. The use of observation wells and the effect of interference of wells are also analyzed.

EXERCISES

14.1 Design a tube well for the following data:

(a) Yield = 0.06 cumec
(b) Aquifer thickness = 25 m
(c) Radius of influence = 300 m
(d) Permeability = 50 m/day
(e) Drawdown = 4 m

14.2 A tube well penetrates in full into a confined aquifer of thickness 24 m. The radius of the well is 18 cm, and drawdown is 5 m. Radius of influence is 300 m, coefficient of transmissibility is 120×10^{-4} m^2/sec. Find Q and coefficient of permeability.

(***Ans***: $Q = 0.05065$ m^3/sec, $K = 5 \times 10^{-4}$ m/sec)

14.3 During a recuperation test, water level is depressed by pumping to 3.5 m and recuperated 2.5 m in 75 minutes. Determine the yield from the well of diameter 5.5 m under depression head of 3.5 m. Find also the diameter to yield a discharge of 14 litres/sec under a depression head of 3 m. (***Ans***: 0.02315 m^3/sec, 4.7815 m)

14.4 A tube well in unconfined aquifer has yield of 0.1 cumec, thickness 30 m, radius of influence 300 m, and drawdown of 5 m. If the coefficient of permeability is 6.944×10^{-10}m/sec, determine the diameter of the well.

(***Ans*****:** 58 cm)

14.5 The water level in a well was depressed by pumping upto 3 m and level was raised by 1.5 m in 50 minutes after pumping. Determine the yield from the well if diameter is 2.5 m and depression head is 2.5 m. Also find diameter of the well to yield 14 litres/sec under depression head of 3 m.

(***Ans*****:** 2.8359 litres/sec, D = 5.62 m)

14.6 A tube well fully penetrates an unconfined aquifer. Calculate Q if diameter, drawdown length of strainer coefficient of permeability, radius of influence are 16 cm, 5 m, 12 m, 0.05 cm/sec and 300 m respectively.

(***Ans*****:** 0.027587 m^3/sec)

SUGGESTED FURTHER READINGS

Bear, J., *Hydraulics of Groundwater*, McGraw-Hill, New York, 1979.

Bouwer, H., *Ground Water Hydrology*, McGraw-Hill, New York, 1978.

Chow, V.T. (Ed.), *Hand Book of Applied Hydrology*, McGraw-Hill, New York, 1964.

Davis, S.N. and Dewiest, P.J.M., *Hydrogeology*, John Wiley and Sons, New York, 1966.

Fancher, G., Henry Darcy–Engineer and Benefactor of Mankind, *Jour. Petr. Tech.*, vol. 8, pp. 12–14, October 1956.

Halek, V. and Svec, J., *Groundwater Hydraulics*, Elsievier, Amsterdam, 1979.

Marino, M.A. and Luthin, J.N., *Seepage and Groundwater*, Elsievier, Amsterdam, 1982.

Pumer, R.R. and Drinkler, P.A., Resistance to Laminar Flow Through Porous Media, *Jour. Hydro. Div.*, ASCE, Civil Engrs, vol. 92, No. Hy 5, pp. 155–163, 1966.

Smith, W.O. and Sayre, A.N., Turbulence in Groundwater Flow, V.S.G.S. Prof. Paper 402 – E, p. 9, 1964.

Todd D.K. and Bear, J., Seepage Through Layered Anisotropic Porous Media, *Jour. Hydro. Div.*, ASCE, vol. 57, No, Hy 3, 1961.

Walton, W.C., *Ground Water Resource Evaluation*, McGraw-Hill, Kogakusha, Tokyo, 1970.

Ward, J., Turbulent Flow in Porous Media, *Jour. Hy. Div*, ASCE, vol. 90, No. Hy 5, pp. 1–12, 1964.

Chapter 15
Hydrology of Basin Management

15.1 INTRODUCTION

The terms *basin, watershed, catchments* and *drainage area* are used synonymously. Large watersheds are typically called **river basins**. A basin is a natural topographical and hydrological entity that collects and converge all precipitation falling on it to a common outlet. River basins are composed of sub-basins each of which may be further subdivided into more watersheds. Most of the rivers of the basins eventually flow to ocean in a common outlet as shown in Figure 15.1. In Figure 15.1, the basin is a combination of many more sub-basins. Two of such small sub-basins are shown by dotted lines. The basin is composed of such twelve sub-basins and finally drained out its all outflow by a common outlet to ocean or inland sea. Some closed basins of inland seas and salt lakes that have no outlets to oceans are also found in different parts of the world. Those basins include Caspian Sea in Russia and Iran, the Great Salt Lake in Utah and Salton Sea in California, the Aral Sea in Kazakhstan and Uzbekistan, Dead Sea in Israel and Jordan and Lake Balkhash in Russia. Various rivers flow to these seas and lakes.

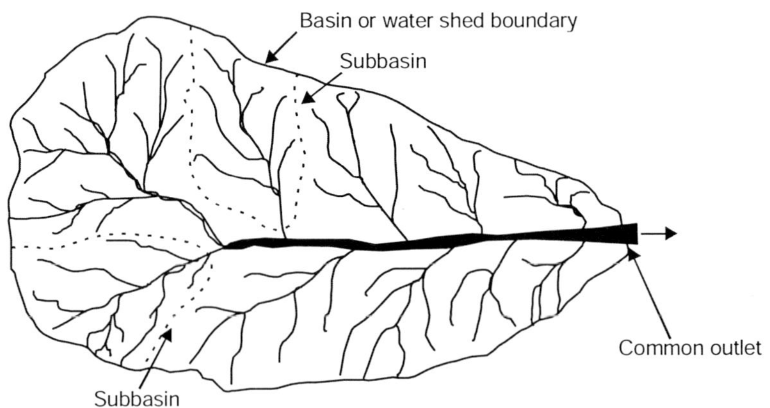

Figure 15.1 Definition sketch of a basin or watershed.

Various hydrologic components of basin like precipitation, infiltration, evapotranspiration, runoff, hydrograph, ground water, etc. have been discussed in Chapters 3, 4, 5, 6, 7 and 14. Control of flood mitigation, river training against erosion and meandering are also already made. Thus, few river basin management studies have already presented. Some of the basin managements works like dam reservoir, reservoir sedimentation and control, basin erosion and control of sediment, environmental policies and practices for protecting ecological systems and water quality, management and maintenance of man made structures for irrigation, water supply, hydropower, fish and wild life, navigation and recreation have not been discussed earlier. Since these management works are intimately related to basin and its water on the surface and subsurface, therefore, discussion in this chapter is termed as *hydrology of basin management.*

15.2 RESERVOIR SYSTEMS

Reservoir systems are very essential to river management due to the following reasons:

1. If a hydrograph of river basin is plotted for few years' data, tremendous variations of stream flow are commonly observed.
2. Seasonal fluctuations of discharge are observed.
3. Sometimes severe droughts for a long time occur.
4. Sometimes extreme flood comes to the river basin.

Due to high and low flow variations, uniform and efficient water management of the basin is not possible without reservoir storage. A reservoir with dam can safely store excess flood water and release it during low flow period. Conservation of storage by reservoir system is meant to store water for uniform demand of water for hydropower, irrigation, water supply, navigation, soil conservation, land reclamation, fishery and wild life, recreation, debris controls, sediment and flood control in river.

Reservoir systems include dams and appurtenant outlet structures, canals, pipes, channel improvements, hydro-electric power plants and transmission system, irrigation head works, pumping plant, navigation locks, fish ladders, spillways, gates, energy dissipaters and various other structures.

15.3 ZONES OR POOLS OF RESERVOIR

Reservoir operation policies involve dividing the storage into various zones or pools shown in Figure 15.2. Surcharge storage is between maximum water level and normal pool level. It is essentially uncontrolled storage and if required may be discharged downstream by opening the crest or spillway gate. The storage between the normal pool level and minimum pool level is the useful storage for multipurpose uses. Storage below minimum pool level is the dead storage. This part also allows the sediment to deposit.

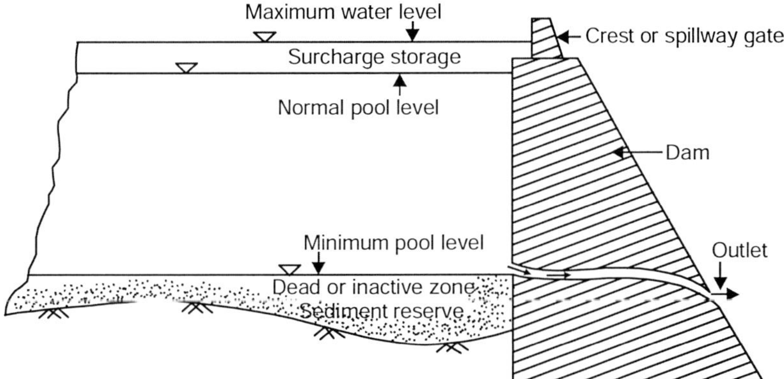

Figure 15.2 Zones or pools of reservoir.

15.4 RESERVOIR SEDIMENTATION

Reservoir sediment is a difficult problem, economical management of which is not yet developed except providing a bit greater dead storage to accommodate the sediment deposit during the life of reservoir. River water carries certain amount of sediment eroded from the basin during heavy rain or due to high velocity of runoff over the ground. The amount of sediment carried by water mainly depends on nature of soil of the basin, topography, vegetation cover of the watershed, intensity of rainfall, soil conservation and watershed management methods adopted in the basin. These sediments at upper layer of the reservoir storage must be controlled or checked from entering the power house where turbines are installed for hydropower generation.

15.5 DENSITY CURRENTS IN RESERVOIR

Density currents or stratified flows occur in reservoir. Water stored in the reservoir is clear with lighter density, whereas inflow to the reservoir is muddy, and thus inflow upstream of reservoir has two different densities. In Figure 15.3, it is shown that heavy turbid water flows along the channel bottom towards the dam under gravity slowly in the form of density currents or stratified flows. This is called **reservoir stratification**. Silting or sedimentation takes place upstream of the reservoir. which can be reduced considerably if the density current is vented by proper location and operation of outlet called **scouring sluices**.

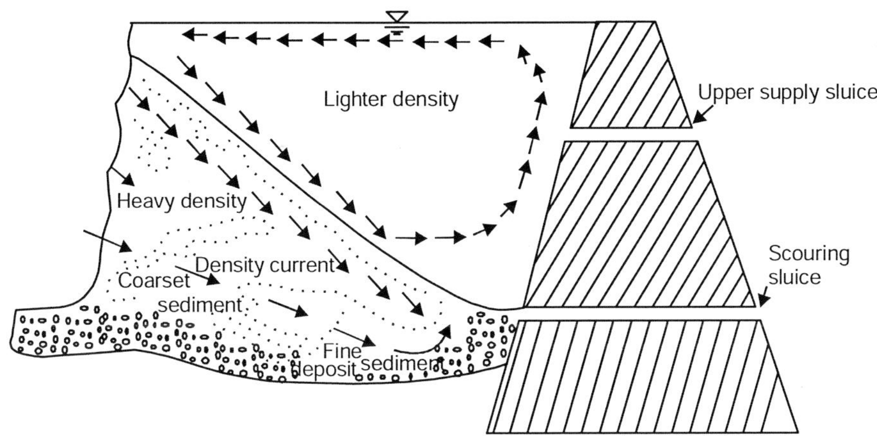

Figure 15.3 Sediment deposit in reservoir and density currents.

15.6 SEDIMENT CONTROL IN RESERVOIR

Sediment in reservoir can be reduced if the following methods in basin and reservoir site are taken:

1. Proper selection of reservoir site with less sediment in flow in the river
2. Control of sediment by constructing series of low dam much upstream of the reservoir
3. Management of basin by vegetal cover to arrest the sediment to flow with runoff
4. Soil conservation and watershed management like strip cropping, control farming, crop rotation, terracing, benching on steep hill slopes, control and management of grazing field, small embankment to prevent erosion and arrest sediment
5. Removal of sediment from reservoir by excavation, draining and flushing by sluices after disturbing the sediment by mechanical method
6. Opening dam sluices to remove sediment coming in during flood time
7. Constructing the dam in stages
8. Afforestation and control of deforestation in the basin

Sedimentation and reservoir function:

1. Sedimentation reduces the useful life of reservoir if not properly managed.
2. Outlet scouring sluices may be blocked by sediment.
3. Aggradations upstream of dam due to sediment deposition.
4. Degradation below reservoir by outflow from scouring sluices.

15.7 TRAP EFFICIENCY OF RESERVOIR

Trap efficiency is the ratio of sediment trapped in reservoir to the sediment transported by water. Experimental analysis shows trap efficiency in the range of 95% to 100% (USBR,[1] 1987).

Trap efficiency depends on

1. Reservoir capacity and inflow ratio.
2. Larger is the silt size, higher is the trap efficiency.
3. The trap efficiency decreases on correct location of scouring sluices.
4. If the shape is diverging, velocity is less near the dam hence trap efficiency is more.
5. With increase in age, basin becomes stable, sediment transportation deceases.

15.8 SOIL EROSION AND CONTROL

Soil erosion in the basin is caused mainly by two natural agents, namely wind and raindrop. In a recent analysis of annual soil erosion rates in India by Narayana et al (1983), it was estimated that about 5334 million tons (16.35 tons/ha) of soil is detached annually due to agriculture and associated activities. Rivers, streams and rivulets carry about 2052 million tons, of which 1572 million tons (29% of total eroded soil) are carried away by various rivers into the sea every year and 480 million tons are being deposited in various reservoirs. Thus, soil erosion and sediment deposition in reservoir appear to be serious problem, control and management of which are very much essential.

15.9 FACTORS AFFECTING SOIL EROSION

Rainfall: If rainfall intensity is more, antecedent moisture condition of soil is high, severe erosion takes place accompanied by high rate of runoff.

Topography of the basin: Topographical factors are degree of ground slope and curvature of slope. If the topography is flat, erosion is negligible. In steep slope, runoff velocity is more, hence erosion is more. In convex slope, erosion and removal of soil takes place. In concave slope, deposition of eroded soil takes place.

Soil surface cover: Vegetal cover on the soil plays an important role to decrease erosion. Storage capacity of soil surface and infiltration increases resulting less runoff and less velocity due to resistance offered by vegetation. Hence erosion becomes small.

Soil properties: A low intensity rainfall on light textured sandy soil may not produce any runoff and hence no erosion. Binding force of such soil is low, detachability is high. If the runoff velocity is not high, soil particles can not be carried by runoff. If the soil is fine textured clay, detachability is difficult unless the high intense rain strikes the bare soil. But once detached, particles can easily be transported.

[1] USBR, *Design of Small Dam*, 3rd ed., U.S. Govt. Printing Office, Denver Co., 1987.

Biological factors: It includes felling of tresses, faulty cultivation practices, overgrazing by cattle, etc. Without such activities, soil remains in a stable condition and hence no erosion takes place.

Incorporated residue: Crop residue every year builds and maintains soil's organic matter which controls soil loss.

15.10 DIFFERENT TYPES AND CAUSES OF EROSION

15.10.1 Types of Erosion

There are two major types of erosion:

Geologic erosion: Geologic erosion is normal process, representing erosion of land in natural environment. It is caused by the effect of rainfall, runoff, wind, topography, atmospheric temperature and gravitational process. It is slow and continuous, but constructive process that has resulted in wearing away mountains, disintegration of rock, building up of flood and coastal plans, and formation of most fertile valleys. Thus, this erosion takes place due to natural geologic action on earth's crust.

Accelerated erosion: Accelerated erosion is due to man made activities, which have brought about changes in natural cover and soil condition. The man made activities are land preparation of raising crops, land use patterns for building houses, infrastructures, industries, hill cutting, deforestation, lack of soil conservation and watershed management. It causes severe deterioration of top surface of the land that disturbs the equilibrium of soil-plant-environment relationship.

15.10.2 Causes of Erosion

The agents of soil erosion and different types of soil erosion are shown in Figure 15.4. The three main agents of erosion are: (i) Wind (ii) Water and (iii) Gravity (force). Soil erosions caused by these agents are explained as follows:

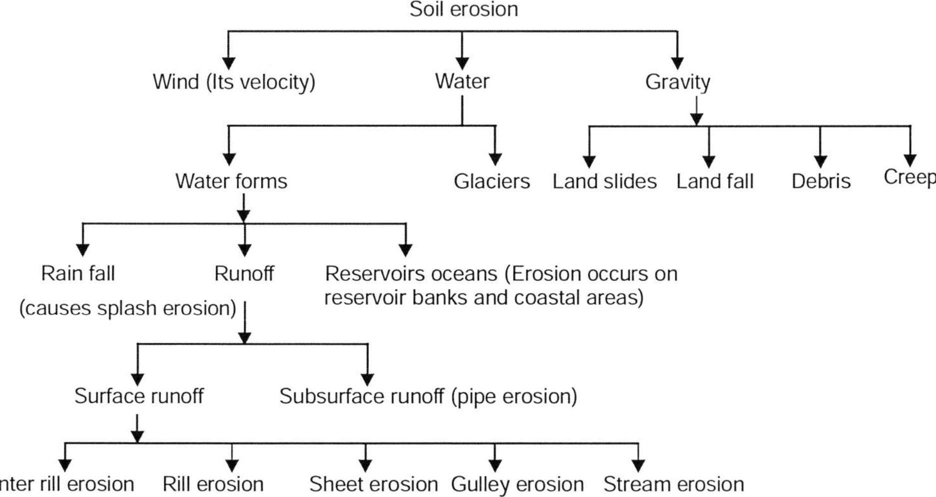

Figure 15.4 Agents of soil erosion and types of soil erosion.

Wind erosion: It is primarily responsible for creation and maintenance of desert areas. Wind erosion is a serious problem in arid and semiarid regions. Finer soil particles from top soil along with organic matter and nutrients are easily detached and removed by wind velocity. Various factors like wind velocity, rainfall, temperature, humidity, soil texture, structure, organic matter content, water holding capacity of soil, bulk density, vegetative cover and height, density and distribution affect the wind erosion. The threshold velocity of wind required to initiate the movement of soil particles is from 3.6 m/sec to 4.2 m/sec. Eddies and cross currents formed by the wind are responsible for lifting and transportation of the soil particles. With reduction of velocity deposition of transported soil particles takes place.

Various empirical equations and models have been proposed by different investigators. For example, Chepil and Woodruff[2] (1963) gave an equation of average velocity of wind erosion. Skidmore and Woodruff[3] (1968) derived the wind erosion force vector equation. Pasak[4] (1967) gave an empirical equation of erodibility of soil by wind. Hsu[5] (1973) gave an empirical formula for determining the weight of soil moved per unit area. Finkel and Noveh[6] (1986) related threshold velocity with the amount of soil moved per unit area. Gillette et al[7] (1972) proposed an empirical model to estimate soil erosion by wind.

Water erosion: Water erosion is the main cause of soil erosion. The impact of rain drop causes splash erosion. Runoff water causes scouring, scrapping and transport of soil particles leading to sheet, rill and gully erosion. Flood water causes erosion of riverbanks. Water waves cause erosion of bank and sides of reservoirs, lakes and oceans. Subsurface seepage water causes soil boiling, erosion and removal by piping. Glacial erosion causes heavy land slides.

Raindrop erosion. Raindrop results splash soil erosion caused by the impact of falling of raindrops. Raindrops strike the soil surface. Laws[8] (1941) measured the terminal fall velocity of raindrop. He claimed that this velocity under normal condition varies from 7 to 10 m/sec creating intense hydrodynamic force at the point of impact. According to Park et al[9] (1982), number of raindrops falling in a unit area directly affects the intensity of rainfall. He develops an empirical formula relating number of raindrops. N_d and intensity of rainfall in mm/hrs, i.e.,

$$N_d = 154\ I^{0.5} \tag{15.1}$$

Again size of raindrops control the terminal velocity of fall V_T of raindrops. Wang[10] (1972) and Park et al (1982) gave

---

[2] Chepil, W.S. and Woodruff, N.P., The Physics of Wind Erosion and Control, Adv. Agronomy, 1963.

[3] Skidmore, E.L. and Woodruff, W.P., Wind Erosion in the United States and their use in predicting soil loss, U.S.D.A., ARS Agric. Handbook, 1968.

[4] Pasak, V., Factors underlying the wind erosion of soil, Vedecke Prace VUM (Czech), 1967.

[5] Hsu, S.A., Computing Aeolian Sand Transport from Shear Velocity Measurement, *J. Geol*, 1973.

[6] Finkel, G.R. and Noveh, *Semi Arid Soil and Water Conservation*, CRC Press Inc. Bocce Raton.

[7] Gillett, D.A. Bliford, I.H. Jr. And Fenster, C.R., Measurement of aero soil size distributions and vertical fluxes of aero soil on land subject to wind erosion, *J. APPL. Meteor.*, 1972.

[8] Laws, J.O., Measurement of Fall Velocity of Water Drops and Rain Drops, Trans. *Am. Geophysical-Union*, 1941.

[9] Park, S.W., Mitchell, J.K. and Bubenger, G.D., Splash Erosion Modeling Physical Analysis, Trans, ASAE, 1982.

[10] Wang, J.Y., Methods of Agrometeorology, *Agri. Meteor.*, W.M.O., Geneva, 1972.

$$V_T = \left(\frac{\rho_w - \rho_a}{1.8}\right) gd \qquad (15.2)$$

where ρ_w and ρ_a are densities of raindrops and air respectively, i.e., $\rho_w = (95.7 \times 9.81)$ N/m^3 at 30°C, and $\rho_a = (1.2985 \times 9.81)$ N/m^3 at 30°C and d is the diameter of raindrops in mm, g is in cm/sec^2. Raindrop erosion increases with the duration of rainfall. However, if the soil is covered with vegetation, damage of erosion is considerably less, and runoff is also less.

Sheet erosion. It is the uniform removal of soil in the layer when runoff or overland flow takes place due to rainfall. Initially top layer of soil gets slowly scrummed off. Erosion in small layers continues with time and thus the fertile top layer is removed under the action of runoff.

Inter rill erosion. In reality, the sheet flow is carried out by very small definable channels called the **inter-rill**. Raindrops detach layers of soil particles through splash and the detached particles are then carried through the inter rills by a thin layer of overland lateral flow.

Rill erosion. When soil erosion and deposition through inter rill is prolonged, the inter rills are widened leading to formation of small channels called **rills**. These channels eventually become well defined. As the depth increases, erosion on bed and bank increases and channels go on widening, but depths remain small.

Gulley erosion. Rills are normally smaller in size and depth. They can be destroyed by tillage operation. If in prolonged occurrence, rills become larger in size and cannot be destroyed by tillage operation, they are transformed into gullies. Shape of gullies may be V-shaped or U-shaped. When subsoil is more resistant to erosion, shape of gully becomes V-shaped. On the other hand, U-shaped gullies are formed in area when surface and subsurface soil are weak and susceptible to erosion by flowing water. Large gullies and their networks are called **ravines**.

Stream erosion. At the alluvial plain, erosion of bed or banks of stream takes place by the action of water. Bed slope of stream is relatively flat, stream erodes the banks, eroded sediment are transported and unable to transport, erosion from concave bank, sediment are deposited on convex bank to develop meandering.

Land slide erosion. When continuous rain fall occurs, water weakens the inter particle cohesive force in the soil and rock on steep hill slopes. As a result, the soil yields to gravity force sliding down as landslide. It occurs as sudden movement of the soil mass along with weathered rocks, due to the weight of the moisture in the soil mass and the downward pull of the gravity. The soil mass, which slides down, flows out with the runoff as viscous flow. It is also called **mudflow** or **debris flow**. Land slides occur when the force caused by the weight of the component of the soil exceed the forces of resistance caused by the shear strength of soil material.

15.11 DIFFERENT METHODS OF EROSION CONTROL

Most of the land under any basin is agricultural land, and therefore, proper emphasis should be given on engineering control measures. Management of land and crop controls erosion on agricultural land.

The following are some of the commonly used methods of soil conservation and erosion:

Contour cultivation: Contour cultivation is the process of cultivation along the contour lines laid along the prevailing slope of the land. All farming operations such as ploughing, seeding, sowing, planting and intercultural are done along the contours. Intercultural operations create contour furrows which along with plant stems act as good barriers to the water flowing down the slope. Infiltration can be increased and erosion can be controlled. Smith and Wischmeier[11] (1962) determined that soil loss due to erosion in contour cultivation decreases with slope. Antal[12] (1986) found experimentally that soil reduction in control cultivation is 30%, increase of soil moisture is 40% and reduction of runoff is 13 times.

Contour bund: It consists of constructing narrow based trapezoidal embankment on contour to impound runoff water behind them, so that this impounded water is absorbed gradually into the soil. Contour bunds reduce the length of slope. They also arrest the eroded soil upstream of them. If the bunds are constructed with some slope, they are called **graded bunds.** If the bunds are constructed along the slope at two sides of contour bunds, they are known as **lateral bunds.** Graded bunds are used for safe disposal of excess runoff in areas of heavy rainfall.

Figure 15.5 shown contour bunds with weir to dispose excess water. Gadkary[13] (1954) recommended the length of field strip from 105 m to 52 m depending on slope of ground smaller to higher. Ghumare[14] (1962) recommended base width from 2.67 m to 4.25 m, top width 0.3 cm to 0.60 m and side slope 1.5H: 1V to 2H: 1V depending on types of soil—shallow, medium and medium deep respectively.

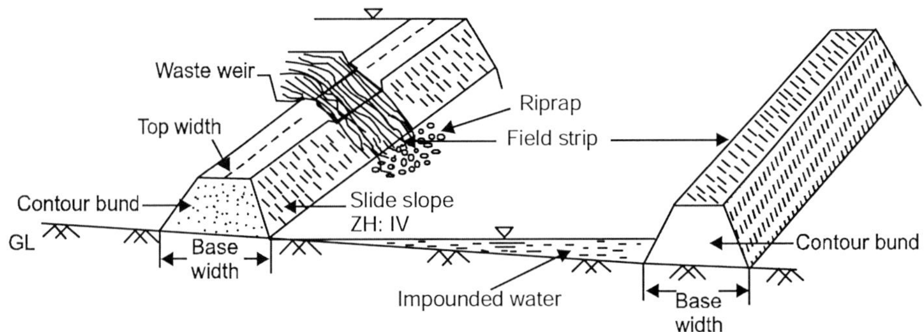

Figure 15.5 Contour bunds with waste weir to dispose excess water.

Bench terracing: In steep hill slope, mere reduction of slope length by contour bunds cannot reduce the scouring action of runoff. In such case, bench terracing is adopted. It is the process

[11] Smith, D.D. and Wischmeier, Rainfall erosion, Adv. Agronomy, 1962.

[12] Antal, J., A New Way of Designing Strip Cropping Patterns to Control Erosion of the Soil (In Slovak), Proc. of VI Nat. Conference – Dev. and prospects of land use planning, 1986.

[13] Gadkary, D.A., Soil Conservation (Engineering Aspects) in Bombay, *J. Soil and Water Conservation*, India 2(3), 1954.

[14] Ghumare, N.K., Studies on Behavior of Contour Bunds, *J. Soil and Water Cons in India*, 1962.

of reducing slopes on hillsides to help in impeding runoff, and thus soil erosion is controlled. In hilly area, bench terracing is extremely necessary. Figure 15.6 shows how slopes are decreased by the construction of bunds.

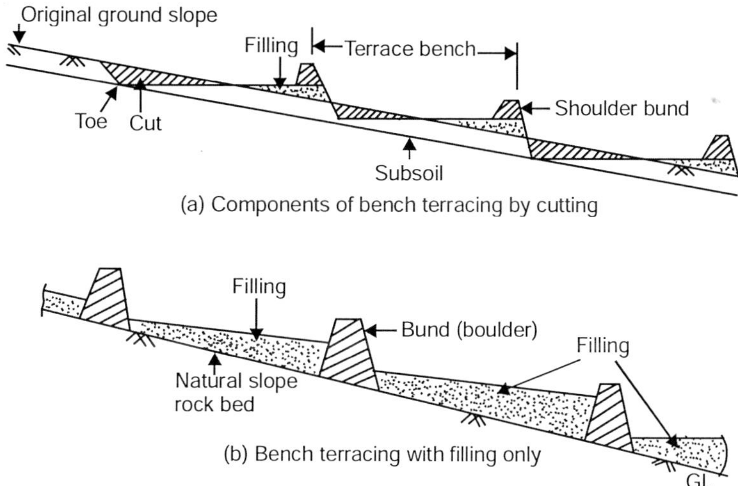

Figure 15.6 Bench terracing on (a) cutting and on (b) filling.

If the subsoil does not contain any rock, it is constructed by cutting [Figure 15.6(a)] and if subsoil is on rock, usually it constructed by only filling [Figure 15.6 (b)].

Grassed waterways: Grassed waterways are also called **vegetated drainage channels** or **vegetated waterways**. Such waterways are either naturally formed or constructed as waterways covered with grass. It is an important soil conservation practice since it is used for disposal of excess runoff from croplands safely and thereby protects the land against rill and gullies. The excess water through this waterway is disposed to rivers, stream and reservoirs.

Check dams: Check dams are constructed in slope with an objective of ponding the water and depositing upstream sediments to reduce the slope with time. The ponded waterway is used for irrigation and other purposes whenever necessary. Check dams are of three types: (i) Temporary (ii) Semi-permanent (iii) Permanent. Series of check dams, shown in Figure 15.7, reduce the velocity of runoff leading to storage of sediment.

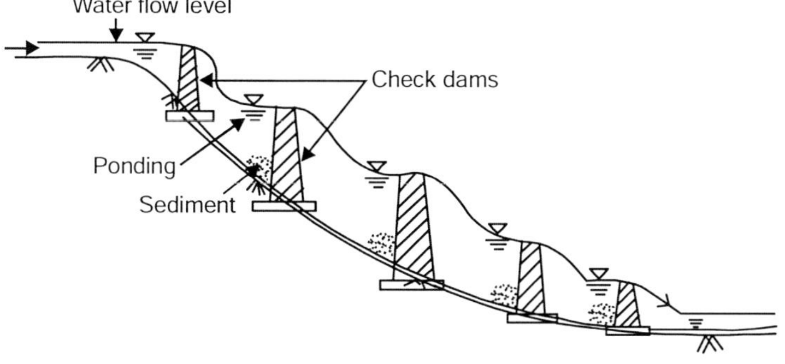

Figure 15.7 Check dam in series.

Temporary check dams are constructed by woven wire dams, brush dams, loses rock dams, wooden plank dams and log dams.

Strip cropping: It is practice of growing strips of crops having poor potential for erosion control, such as root crops, cereals, etc. In strip cropping (Figure 15.8), both close growing and row crops are grown on strips in the same field. These crops prevent erosion, build up soil fertility and improve the soil capability. Such practices include contour farming, proper crop rotation, etc. Three types of strip cropping are in practice:

(a) Contour strip cropping along the contours.

(b) Field strip cropping when land topography is irregular and complex, widths between two contours vary.

(c) Buffer strip consists of planting some grass or legume crops that completely cover the land between contour strips.

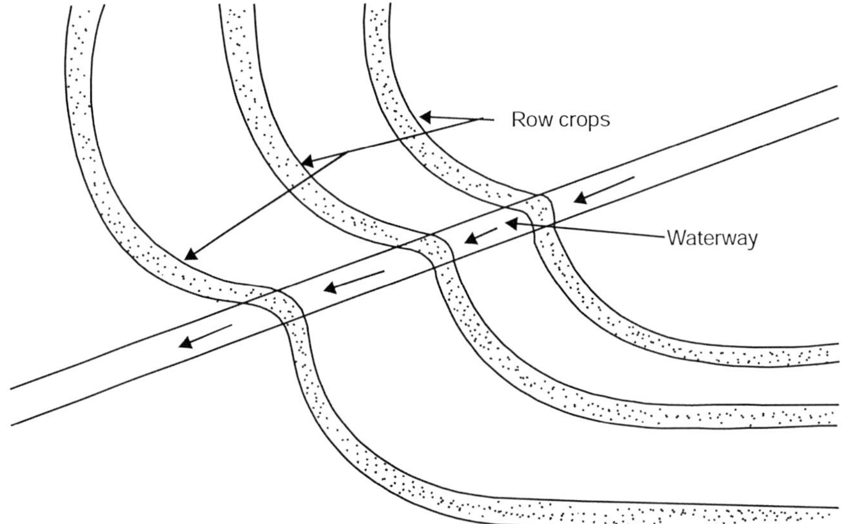

Figure 15.8 Layout of strip cropping pattern.

Mixed cropping: It is a practice of growing more than one mixed crop in the field simultaneously. It results into continuous and better cover of land, good protection from the beating action of rain and almost complete protection against soil erosion.

Mulching: Stubbles, trash, other types of vegetation and polyethylene are some of the most common types of mulches used. Mulches minimise rain splash, reduce evaporation and control weeds. It prevents inter-rill soil erosion.

Land management: Land preparation, including post harvest cultivation and preparatory tillage influences intake rate of water, obstruction to surface and consequently the rate of soil erosion. The concept of minimum tillage can be advantageously applied for soil and water conservation.

Adoption of organic measures and green manures: Fertilizers when applied improve soil physical condition. It improves vegetative growth of crop or crop canopy and thereby reduces soil erosion.

Crop residue: Crop residues, which are generally left in the field after harvesting, can have tremendous influence on soil and water conservation. Crop residue acts as a barrier for soil erosion and water runoff.

Grasses: Inclusion of grasses in crop rotation has been found to reduce soil loss.

Vegetative hedges and buffers: Hedges in and around farm land protect soil from wind and water erosion. Permanent shrubs or grasses are used on gentle slope to reduce the erosion.

Canal lining: Soil erosion in canal bed and sides is controlled by canal lining, i.e., a protective water proofing coating of concrete or masonry or other material is provided.

Hydraulic jump: A condition for creation of hydraulic jump (i.e., subcritical to supercritical) is created. Hydraulic jump dissipates most of the energy of water and hence less energy of flow and less erosion.

Canal fall: Providing hydraulic structures like canal fall slope of the canal is reduced. Hence velocity decreases and erosion is reduced.

Controlling deforestation: Deforestation if controlled, raindrop splash on soil is reduced and detachment of soil and erosion will be less.

Natural hill cutting: Hill cutting obviously increases sediment load of runoff. If hill cutting is really necessary, precautions must be taken so that rainwater cannot carry the bare soil particles along with it. For example, in constructing hill road, hill is required to cut. In some areas, concrete retaining wall with proper engineering design is to be constructed so that soil erosion and landslide cannot occur.

15.12 OTHER WATER MANAGEMENT SECTORS OF BASIN

Water supply: Water from rivers, streams, lakes and aquifers is used to supply for domestic, public, industrial, irrigation and other purposes. To supply water facilities for storing, transporting and distributing are essential. For example, in irrigation, if the scheme is storage type, reservoir for storage, canals with regulators for transportation and field channel for distribution are required. For domestic and industrial uses, additional works of treatment, purification, conveyance and distribution by pipe networks are to be adopted. If the source of water is aquifers (groundwater), construction of wells, pumping, storing, testing the quality and distribution will be necessary. For diversion type scheme in river weir, silt removing devices, fish ladder and regulator are essential. Thus, in water supply system or sector, dams, weirs, reservoirs, canals, pipes, wells, pumps and other appurtenant structures play a dominant role. Users of this water supply should be careful and judicious in using the water. Persons involved in the maintenance and supply must be vigilant in repair, leak detection and protection so that the whole system of supply and use becomes environmental and ecofriendly.

Hydroelectric power development: Flowing water of the basin in rivers and streams has tremendous kinetic energy. Water stored in reservoir by constructing dam at higher altitude can possess plenty of potential energy. These two types of energy are used to produce electricity by the hydroelectric plant. India so far has produced 29,976 MW of hydroelectricity. To produce this hydroelectricity, plant requires components like reservoir, dam, weir, barrage, intake structures, waterway or tunnel or power canal, surge tank, penstock, turbines, electric generator (power house, forebay, pondage, tailrace channel, transformers, transmission lines etc.). Hydropower plant may be storage plant, pumped storage plant, run-of river plant, base load plant, peak load plant, divided plant, concentrated plant, high or medium or low head plant.

Water energy is converted to mechanical energy by the turbines, which are directly coupled to electric generator to convert this mechanical energy to electrical energy.

Navigation: Navigation in river is another major in stream use of each and every basin. Waterways had been the important communication and avenue for commerce since ancient time. The braided river in the basin during low flow creates problems for navigation due to formation sandbars, obstruction, low navigation depths etc. therefore, extensive river improvement works become necessary to remove obstruction, increase the depth of flow. If the canal is navigable, canal lock is also essential.

Heavy machinery, construction materials like iron, steel and other heavy commodities can easily be transported by navigation.

If the depth of flow in the river or canal becomes low even after channel important works, dams or barriers are constructed in stages to create sufficient depth.

Navigation, however, has the disadvantage of being much slower than surface transport. It is also limited to cities located on river banks. For cities away from the bank, other surface transport has to be used to complete the transport. Yet, navigation in the rivers of the basin serves the important mode of transportation.

Waste water management: Waste water produced from residential areas, business, medical and industrial centre is to be treated for removal of environmental pollution and ecological imbalance. This water is collected, pollutants are removed and treated effluent is discharged to the stream system. In India, in every state, usually a pollution control board (PCB) is there to look after the proper treatment of waste water by respective organizations.

Groundwater management: Groundwater is polluted from disposal of waste water, seepage from septic tank, soak hit, polluted swamps and some unwanted chemical present below the ground surface. In contrast to surface water pollution, ground water pollution is difficult to detect and control. With the increase of demand of ground water, attempts have been made to examine, prevent, reduce and eliminate groundwater pollution. The principal source and causes of groundwater pollution and possible pollutants found in groundwater samples are given by Todd[15] (1980). The common sources of ground water pollution are sewer leakage, liquid wastes, solid wastes, biomedical wastes, tank and pipeline leakage of chemicals, mining activities, oil field brines, irrigation return flows, animal wastes, fertilizers and soil amendments, spills and surface discharge and stock piles, septic tank and cesspools, saline water intrusion, interchange through wells and surface water.

[15] Todd, D.K., *Groundwater Hydrology*, 2nd ed., John Wiley & Sons, Singapore, 1980.

The common methods of attenuation of pollution are filtration, chemical processes, microbiological decomposition and dilution.

Urban storm water management: Urban storm water disposal has become a major problem in most of the cities of India. Urban drainage system includes streets, curbs, gutters, ditches, drainage channel, culverts, and bridges, storm sewer networks including pipes, drains, manholes and other appurtenant structures. Drainage improvement is a real need to prevent ponding of water in streets, homes creating great problem of communication for some hour to days even. Infiltration of storm water in urban areas is practically nil due to building, roads, pavement, parking areas, etc., and therefore, all water from storm becomes the runoff. The absence of sufficient drainage with its proper slope towards outlet and absence of regular maintenance of drainage networks aggravates the ponding of storm water on roads and residential campuses. If the urban area is situated near the hill whose slopes towards the city and if the hill is open without any forest or vegetative cover, situation of storm water ponding in urban area becomes serious with sediment transport which may clog the drainage system.

Proper management of urban growth with adequate planning and management of storm water disposal management to the outlet stream is necessary.

15.13 CONCLUSION

Water resources available in basin if properly managed, bring about tremendous development of the society. Properly managed water resource helps our hydraulic turbines to generate hydroelectricity, can nourish croplands and forests, control devastation of flood and erosion, helps floating our ships in shallow water by increasing depth, can preserve and increase wildlife and water growing lives like fishes, provides drinking water, can convert a dry land into beautiful residential and flourished crops land, helps in beautifying the surroundings and environment for recreation, control pollution and mosquito growths. In fact water is an amazing fluid that can lead to overall prosperity of a nation.

EXERCISES

15.1 What is reservoir system? Describe the different zones of reservoir. Explain reservoir sedimentation and density currents in reservoir.

15.2 How do you control sedimentation in reservoir? Describe trap efficiency and the factors on which this trap efficiency depend.

15.3 Explain different types and causes of soil erosion.

15.4 What are the different methods of control of soil erosion? Describe them in brief with sketches where necessary.

15.5 What are the other water management sectors of a basin? Describe each of them.

SUGGESTED FURTHER READINGS

Anderson, J.R. and Dornbush, J.N., Influence of Sanitary Landfill on Groundwater Quality Water Works ASSOC., vol. 59, pp. 457–470, 1967.

CBIP, New Delhi, Sedimentation Studies in Reservoirs, vol. I, Tech Rep. No. 20, River Valley Project,.Reht, 1977.

Central Board of Irrigation & Power (CBIP), New Delhi, Life of Reservoir, Tech Report, No. 19, March 1977.

Das, G., *Hydrology and Soil Conservation Engineering*, Prentice-Hall of India, New Delhi, 2001.

Goel, P.K., *Water Pollution, Causes, Effects and Control*, New Age International *Publishers*, New Delhi, 1997.

Gosuwami, M.D., *Watershed Management,* Ritwik & Gargee Publications, Guwahati, Assam, 2004.

Murty, J.V.S., *Watershed Management in India*, New Age International, Delhi, 1994.

Murty, V.V.N., *Land and Water Management Engineering*, Kalyani Publishers, New Delhi, 1985.

Rajora, R., *Integrated Watched Management*, Rawat Publications, Jaipur and New Delhi, 2002.

Rao, R., Soil Conservation in India, ICAR, Ed., 1974.

Ratan Lal, *Soil Erosion in the Trophies*, McGraw-Hill, New York, 1999.

Senturk, F., Hydraulics of Dams and Reservoirs, Water Resource Publications, Highlands Ranch Co., 1994.

Singh, K.P. and Ali, D.A., New Method for Estimating Future Reservoir Capacity, *Water Resource. Bull.*, vol. 25, No. 2, April 1989.

Todd, D.K. and McNulty, D.E.O., Polluted Groundwater, Water Information Centre, Port Washington, New York, 1976.

U.S. Army Corps of Engineers, Management of Water Control System, Engineers Manual 11103600, Washington, DC, November 1987.

U.S. Army Corps of Engineers, Sedimentation Investigations of River and Reservoirs, Engineers Manual, 1110-2-4000, Washington, DC, December 1989.

Williams, J.R. and Berndt, H.D., Sediment Yield Prediction Based on Watershed Hydrology, Trans., ASAE, 20(6), pp. 1100–1104, 1977.

Wurbs, R.A., *Modeling and Analysis of Reservoir System Operations,* Prentice-Hall, Upper Saddle River, New Jersey, 1996.

———, Water Management Models: A Guide to Software Prentice-Hall, Upper Saddle River, New Jersey, 1995.

Appendix

Introduction to Statistics and Probability

Statistics deals with the methods for collection, classification and analysis of numerical data of some events for drawing valid conclusion and making reasonable decision. It has meaningful application in most of the hydrologic processes such as rainfall, runoff or floods in a basin, evaporation, etc. and in other fields of engineering and science. In this chapter, basic concept of statistics and probability required for analysis of flood and rainfall in the field of hydrology will be discussed.

A.1 FREQUENCY DISTRIBUTION

To present an analysis of frequency distribution, collection of data which constitutes the starting point of any statistical investigation, is required first. The next step is the classification of data as the collected or measured data are not in an easily assimilative form. As such, classification is necessary for making intelligent inferences. To understand this frequency distribution, examples relating to maximum flood flows measured in m^3/sec (cumecs) in a river for 64 years are given as follows:

Year	*Flood in cumecs*	*Year*	*Flood in cumecs*	*Year*	*Flood in cumecs*	*Year*	*Flood in cumecs*
1942	179	1958	165	1974	167	1990	162
1943	188	1959	175	1975	173	1991	167
1944	175	1960	187	1976	181	1992	197
1945	160	1961	174	1977	172	1993	178
1946	193	1962	162	1978	163	1994	185
1947	171	1963	195	1979	176	1995	176
1948	159	1964	178	1980	175	1996	165
1949	185	1965	163	1981	185	1997	171
1950	184	1966	178	1982	180	1998	178
1951	175	1967	182	1983	173	1999	189
1952	182	1968	175	1984	157	2000	161
1953	168	1969	191	1985	188	2001	175
1954	190	1970	177	1986	178	2002	195
1954	162	1971	169	1987	162	2003	160
1955	188	1972	174	1988	176	2004	179
1957	176	1973	168	1989	153	2006	183

The given or measured flood data can conveniently be grouped, and is shown in Table A.1:

Table A.1 Grouping of flood data to represent frequency table.

Class (flood in cumecs)	*Frequency*	*Cumulative frequency*
150–154	1	1
155–159	2	3
160–164	9	12
165–169	7	19
170–174	8	27
175–179	17	44
180–184	6	50
185–189	8	58
190–194	3	61
195–199	3	64

Total = 64

It seems from Table 1 that there is only one flood of magnitude between 150 to 154 cumecs occurs in the river out of 64 years. Similarly, 2 floods of 155 to 159 cumecs, 9 floods of magnitude 160 to 164 cumecs occur and so on. Thus, 64 floods have been put into 10 groups called the **classes**. The width of class is called **class interval,** and number in that is called **frequency**. Table A.1 showing the class of data (floods), frequency and cumulative frequency is called **frequency table**. Thus, a set of raw data summarized by distributing it into a number of classes along with their frequencies is known as a **frequency distribution**.

Cumulative frequency shows the number of floods occurring 27 times less than (say) 175 cumecs out of 64 years, less than 61 times of 195 cumecs and so on.

The condensation of data in the form of frequency distribution is very useful as it converts a long series of data into a compact form. In practice, it is required to compare two or more series of data. Frequency distribution compels to seek for certain constants which could concisely give an insight into the important characteristics of the series. The chief constants which summarize the fundamental characteristics of the series are:

1. Measures of central tendency
2. Measures of dispersion
3. Measures of skewness

A.1.1 Measures of Central Tendency

In the above frequency distribution of floods shown in Table A.1 for 64 years shows the clustering of floods about some central value. Maximum clustering has been seen around the class values of 175 to 179 cumecs (i.e. 17 values). Finding this central value or the average value is of prime importance, as it gives the most representative value of the whole group. Different methods give different averages which are known as the **measures of central tendency**. Commonly used measures of central value are: (a) mean (b) median (c) mode. To understand the three measures of central tendency, following series of data of rainfall and corresponding runoff are taken in a catchment.

Mean: If x_1, x_2, x_3, ..., x_n, are set of n values of a variate (say, flood), then arithmetic mean is given by

$$\bar{X} = \frac{x_1 + x_2 + x_3 + \cdots + x_n}{n} = \sum_{i=1}^{n} x_i$$

If we write Q as flood,

$$\bar{Q} = \frac{13.8 + 28.2 + 38.2 + \cdots + 32.20 + 13.80}{11} \qquad \text{(Here } n = 11\text{)}$$

$$\bar{Q} = \frac{486}{11} = 44.182 \text{ m}^3/\text{sec}$$

If we write rainfall (precipitation) as P,

$$\bar{P} = \frac{2 + 4 + 5 + 57 + \cdots + 4.45 + 2}{n}$$

or

$$\bar{P} = \frac{70}{11} = 6.3636 \text{ mm/hr}$$

Median: If the values of the variables are arranged in the ascending order of magnitude, the median is the middle term (if the number is odd). When the number is even, then mean of the two middle terms is the median.

In Table A.2 (rainfall vs flood), n = 11 which is odd, hence median for flood Q is 79.8 m^3/sec and for precipitation P is 11.5 mm/hr when arranged in ascending order.

Table A.2 Rain fall and Runoff data to determine mean, median and mode

Rainfall (mm/hr, P)	2	4	5.5	7	9	11.5	10	8	6.5	4.5	2
Runoff or flood (m^3/sec, Q)	13.80	28.20	38.20	48.60	62.50	79.80	69.40	55.50	45.00	32.20	13.80

Mode: The mode is defined as that value of the variate which occurs most frequently, i.e., value of maximum frequency. In Table A.2, rainfall of 2 mm/hr and flood of 13.80 m^3/sec both occur 2 times which is the maximum time of occurrence.

Geometric mean (GM): If x_1, x_2, x_3, ..., x_n is a set of n observations, then geometric mean is given by

$$\text{GM} = (x_1.\ x_2,\ \ldots,\ x_n)^{1/n} \tag{A.1}$$

From Table A.2 $\quad$ GM (of rainfall) = $(2 \times 4 \times 5.5 \times 7 \times \cdots \times 4.5 \times 2)^{1/11}$

$$\text{GM (Rainfall } P) = 5.534 \text{ mm/hr}$$

$$\text{GM (flood } Q) = (13.80 \times 28.10 \times 38.20 \times \cdots \times 32.20 \times 13.80)^{1/11}$$

$$\text{GM } Q = 38.526 \text{ m}^3/\text{sec}$$

Harmonic mean (HM): If x_1, x_2, x_3, ..., x_n be a set of n observations, then harmonic mean (HM) is defined as the reciprocal of the arithmetic mean of the reciprocal of quantities, thus,

$$\text{HM} = \frac{1}{\frac{1}{n}\left(\frac{1}{x_1} + \frac{1}{x_2} + \cdots + \frac{1}{x_n}\right)} \tag{A.2}$$

Considering the rainfall,

$$\text{HM (Rainfall)} = \frac{1}{\frac{1}{11}\left(\frac{1}{2} + \frac{1}{4} + \frac{1}{55} + \cdots + \frac{1}{4.5} + \frac{1}{2}\right)}$$

$$\text{HM (Rainfall } P) = 4.6339 \text{ mm/hr}$$

Similarly,

$$\text{HM (Flood } Q) = 32.222 \text{ m}^3\text{/sec}$$

A.1.2 Measures of Dispersion

Although measures of central tendency do exhibit one of the important characteristics of distribution, yet they fail to give any idea as to how the individual values differ from central values, i.e., whether they are closely packed around the central values or widely scattered away from it. Figure A.1 shows two distributions, which have same mean and same total frequency, yet they differ in extent to which individual values spread about the average. The magnitude of such variation is called **dispersion.**

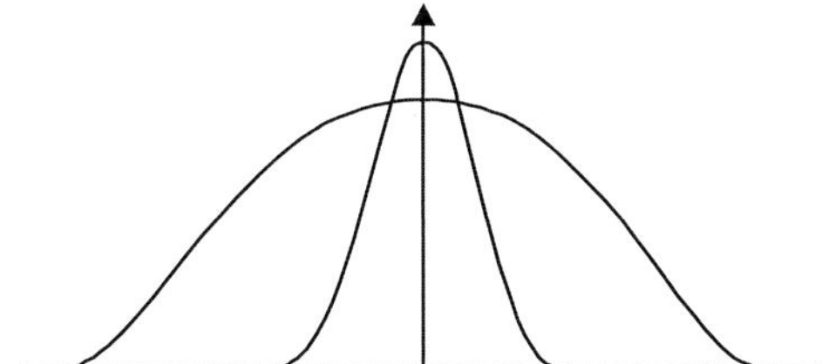

Figure A.1 Same mean; different dispersion.

Mean deviation (MD): It is the absolute deviation of values from their mean.

$$\text{MD} = \frac{\sum_{i=1}^{n} |x_i - \bar{x}|}{n} \tag{A.3}$$

The absolute values of the following MD of rainfall in Table A.2 are considered:

$$\text{MD} = \frac{(2-6.3636)+(4-6.3636)+(5.5-6.3636)+\cdots+(6.5-6.3636)+(4.5-6.3636)+(2-6.3636)}{11}$$

$$= 2.5124 \text{ mm/hr}$$

Similarly MD of flood Q may be determined.

Standard deviation (SD): The most important and powerful measure of dispersion is the standard deviation (SD), generally denoted by σ. It is frequently necessary in the analysis of hydrological data of flood and also other fields of science and engineering. It is the square root of the mean squared deviation of the variates from their mean, and is given by the equation:

$$\sigma = \sqrt{\frac{\sum (x - \bar{x})^2}{n-1}} \tag{A.4}$$

Standard deviation of flood in Table A.2 may be calculated as:

$$\sigma_Q = \sqrt{\frac{(13.8 - 44.272)^2 + (28.2 - 44.272)^2 + \cdots + (45 - 44.272)^2 + (32.2 - 44.272)^2 + (2 - 44.272)^2}{11-1}}$$

$$= \sqrt{\left(\frac{928.54 + 258.31 + 36.87 + 18.73 + 332.26 + 1262.22 + 631.41 + 126.06 + 0.53 + 145.73 + 928.54}{10}\right)}$$

$$= \sqrt{\frac{4747.89}{10}} = 21.789 \text{ m}^3/\text{sec}$$

Variance: The square of the standard deviation is called **variance,** i.e., σ^2

$\therefore$ Variance of flood $Q = \sigma^2 = (21.789)^2 = (4474.76 \text{ m}^3/\text{sec})^2$

Standard deviation for the population σ_p is given by

$$\sigma_p = \sqrt{\frac{\sum (x - \mu)^2}{n}} \tag{A.5}$$

where ì is the mean of population which consists of values of variables (say, flood) from time immemorial at eternity, whereas sample x is the observed values of variates for a finite number of years.

Range (R): The range denotes the difference between the largest and smallest values of the sample, and is given by Hurst (1951) and Klemes (1974) as:

$$R = \sigma \left(\frac{n}{2}\right)^K \quad 0.5 < K < 1 \tag{A.6}$$

$$= 1.25 \sigma_p \sqrt{n} \tag{A.6a}$$

Coefficient of variation (C_v): The standard deviation divided by the mean is called the **coefficient of variation** C_v. It is given by

$$C_v = \frac{\sigma_p}{\mu} \cong \frac{\sigma}{\bar{x}} \tag{A.7}$$

Skewness: Skewness measures the degree of asymmetry or departure from symmetry. If the frequency curve has a longer tail to the right, i.e., the mean is to the right mode [Figure A.2(a)], then distribution is said to have positive skewness. If the curve is elongated to the left [Figure A.2(b)], then it is said to have negative skewness.

Population skewness α is given by

$$\alpha = \frac{\sum (x-\mu)^3}{n} \tag{A.8}$$

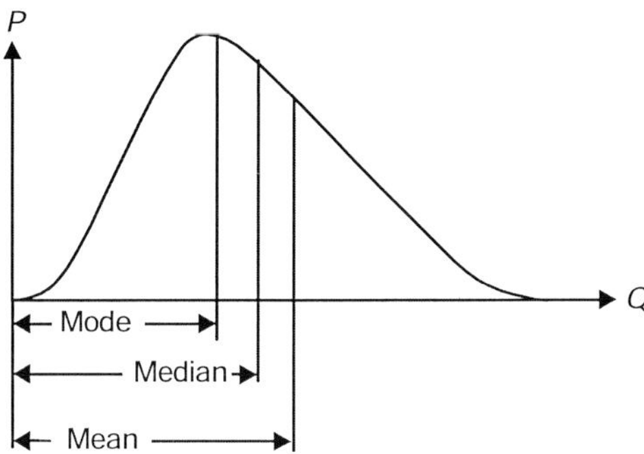

Figure A.2 Positive skewness.

Sample skewness is given by

$$a = \frac{\sum (x-\mu)^3}{n-1} \tag{A.9}$$

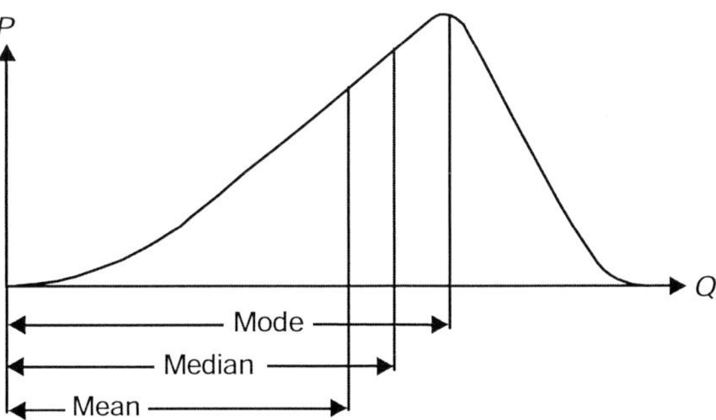

Figure A.3 Negative skewness.

The degree of skewness of the distribution is measured by the coefficient skewness C_s, and is given by

$$C_s = \frac{\alpha}{\sigma_p^3} \cong \frac{a}{\sigma^3} \tag{A.10}$$

Another measure of skewness used in practice is Pearson's skewness S_K which is given as:

$$S_K = \frac{\mu - \text{mod}\, e}{\sigma_p} \cong \frac{\overline{x} - \text{mod}\, e}{\sigma} \tag{A.11}$$

A.2 PROBABILITY

The study of probability provides a mathematical framework for assertions and is essential in every decision-making process. Before defining probability, the following terms are explained:

A set of events is said to be exhaustive if it includes all the possible events. In tossing coin, there are two exhaustive cases either a head or a tail and there is no third possibility.

If one of the events cannot be expected to happen in preference to another, then such events are said to be *equally likely*. In tossing the coin, the chance of coming head or tail is equally likely. Again, when a cubical die is drawn, the turning up the six different faces of the die is exhaustive, mutually exclusive and equally likely.

If there are n exhaustive, mutually exhaustive and equally likely cases of which m are favourable to an event A, then probability P of happening of A is m/n. As there are $(n - m)$ cases in which A will not happen, therefore, chance of A not to happen is:

$$q = \frac{n - m}{n} = 1 - \frac{m}{n} = 1 - P \tag{A.12}$$

$\therefore$

$$p + q = \frac{m}{n} + 1 - \frac{m}{n} = 1 \tag{A.13}$$

In other words, if an event is certain to happen, then its probability is unity. While it is certain not to occur, its probability is zero.

A.2.1 Conditional Probability

Two events are said to be independent, if happening or failure of one does not affect the failure or happening of the other. Otherwise, events are said to be dependent.

For two dependent events A and B, the symbol $P(B/A)$ denotes the probability of B when A has already occurred. It is known as **conditional probability**.

A.2.2 Theorem of Compound Probability

For the probability of an event A happening as a result of trial is $P(A)$ and after A has happened the probability of an event B happening as result of another trial is $P(B/A)$, the probability of both the events A and B happening as a result of two trials is $P(A)$, $P(B/A)$.

A.3 PROBABILITY DISTRIBUTION

If a real variable X be associated with the outcome of a random experiment, since the value which X takes depends on chance, it is called a **random variable** or a **stochastic variable** or simply a **variate**. If in a random experiment, the event corresponding to a number a occurs, then the corresponding random **variable** X is said to assume the value of a, and the probability of the event is denoted by $P(x = a)$. Similarly the probability of the event x assuming any value in the interval $a < X < b$ is denoted by P $(a < X < b)$. The probability of the event $X \leq c$ is written as $P(X \leq c)$.

If a random variable takes a finite set of values, it is called a **discrete variate,** and if it assumes an infinite number of uncountable values, it is called **continuous variate**.

A.3.1 Discrete Probability Distribution

The discrete variate X, which is the outcome of some experiment and probability of X takes the values x_i is p_i, then

$$P(X = x_i) = p_i \text{ or } p(x_i) \tag{A.14}$$

For $i = 1, 2$

1. $$p(x_1) \geq 0; \text{ for all values of } i \tag{A.15}$$
2. $$\sum p(x_i) = 1 \tag{A.16}$$

A.3.2 Distribution Function

The distribution function $F(x)$ of the discrete variate x is defined by

$$F(x) = P(X \leq x) = \sum_{i=1}^{x} p(x_i) \tag{A.17}$$

Here x is an integer. This distribution function is also called **cumulative distribution function**.

A.3.3 Continuous Probability Distribution

When a variate x takes a value in an interval, it gives rise to continuous distribution. For example, distributions defined by the variates like heights are continuous distribution.

If $F(x) = P(X \leq x) = \int_{-\infty}^{x} f(x)\, dx$, then $F(x)$ is defined as the distribution function of continuous variate X. It is the probability that the value of the variate X will be $\leq x$. The distribution function $F(x)$ has the following properties:

1. $F(x) = f(x) \geq 0$ so that $F(x)$ is a non-decreasing function.
2. $F(-\alpha) = 0$
3. $F(\alpha) = 1$
4. $P(a \leq X \leq b) = \int_{a}^{b} (x)\, dx = \int_{-\infty}^{b} f(x)dx + \int_{-\infty}^{a} (x)dx$

   $= F(b) - F(a)$

A.4 BINOMIAL DISTRIBUTION OR BERNOULLI DISTRIBUTION

It was discovered by a Swiss mathematician Jacobs Bernoulli and was published posthumously in 1713. It is concerned with trials of a repetitive nature in which the occurrence or non-occurrence, success or failure, acceptance or rejection, yes or no of a particular event is of interest.

When a coin is tossed, the probability of getting head or tail is 50% or 1/2. If the coin is tossed 3 times, the probability of getting one head and two tails can be combined as H-T-T, T-H-T, T-T-H. The probability of each one of these being $1/2 \times 1/2 \times 1/2 = (1/2)^3$ and their total probability is $3(1/2)^3$. Similarly, if the trial is repeated n times, and if p is probability of success and q of failure, then the probability of r successes and $(n - r)$ failures is given by $p^r q^{n-r}$. But these r successes and $n - r$ failures can occur in any of the ${}^nC_r$ ways in each of which the probability is same.

Thus, the probability of successes is ${}^nC_r p^r q^{n-r}$, where r takes any integral value from 0 to n. The probability of 0, 1, 2, 3, ..., r, ..., n success are therefore, given by

$$q^n,\ {}^nC_1 pq^{n-1},\ {}^nC_2\, p^2 q^{n-2},\ {}^nC_3\, p^3 q^{n-3},\ \ldots,\ {}^nC_r\, p^r q^{n-r},\ \ldots,\ p^n$$

The probability of number of successes so obtained is called the **binominal distribution** for the simple case that the probabilities are the successive terms of the binominal theorem $(q + p)^n$.

$$\therefore \quad \text{Sum of the probabilities} = q^n + {}^nC_1\, pq^{n-1} + {}^nC_2\, p^2 q^{n-2} + \ldots, + p^n = 1 \qquad \text{(A.18)}$$

A.5 POISSON DISTRIBUTION

It was French mathematician S.D. Poisson who introduced it in 1837. This distribution is the limiting case of the binomial distribution by making n very large and p very small as it is the distribution related to probability of events which are extremely rare.

Binominal distribution probability is:

$$P(r) = {}^nC_r p^r q^{n-r}$$

$$= \frac{n(n-1)(n-2)(n-3)\cdots(n-r+1)}{r!} p^r q^{n-r}$$

$$= \frac{np(np-p)(np-2p)\cdots(np-\overline{r-1}p(1-p)^{n-r}}{r!}$$

When $n \to \alpha$, $P \to 0$

and

$$np = m \qquad \text{(A.19)}$$

Now,

$$p(r) = \frac{m^r}{r!} \underset{x \to \alpha}{Lt} \frac{\left(1 + \frac{m}{n}\right)^r}{\left(1 - \frac{m}{n}\right)^r}$$

$$\therefore \qquad P(r) = \frac{m^r}{r!} e^{-m}$$

which gives Poisson distribution with probability function so that the probability of 0, 1, 2, ..., r ..., successes in Poisson distribution are given by

$$e^{-m}, me^{-m}, \frac{m^2 e^{-m}}{2!}, \ldots, \frac{m^r - e^{-m}}{r!}$$

The sum of which is equal to unity.

A.6 NORMAL DISTRIBUTION

Karl Pearson in 1924 reported Abraham De Moivre had discovered this distribution as early as 1733. The continuous distribution having the density

$$f(x) = \frac{1}{\sigma\sqrt{2\pi}} e^{-(x-\mu)^2/2\sigma^2} \tag{A.20}$$

Equation (A.20) gives the normal distribution. Any quantity whose variation depends on random causes is distributed according to the normal law. The curves of normal distribution for different σ values (standard deviation) are shown in Figure A.4. It is bell-shaped and is symmetrical about the mean. Hence its mean, median and mode are the same. The maximum or peak ordinate is $\frac{1}{\sigma\sqrt{2\pi}}$ when $x = \mu$. This distribution is very important in almost all branches of science and engineering for statistical inference.

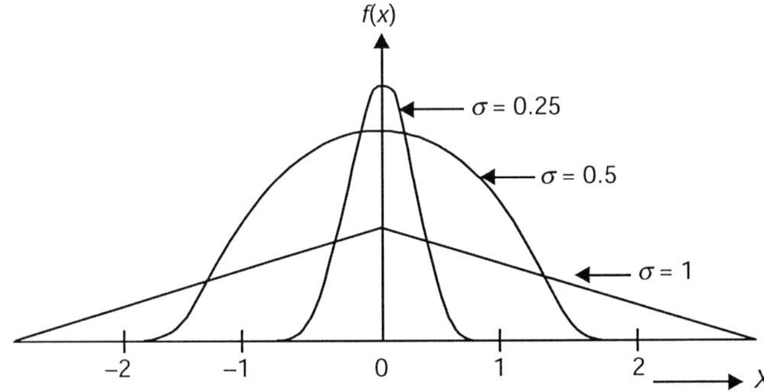

Figure A.4 Normal distribution.

A.7 GAMMA DISTRIBUTION

The density function of this distribution is given by

$$f(x) = \frac{x^a e^{-x/b}}{a!b^{a+1}} \tag{A.21}$$

For $0 \angle x \angle \infty$

$f(x) = 0,$ otherwise

where a, b are two parameters which affect the distribution. Here $\mu = b(a + 1)$ and $\sigma_p^2 = b^2(a + 1)$. Figure A.5 shows the gamma distribution.

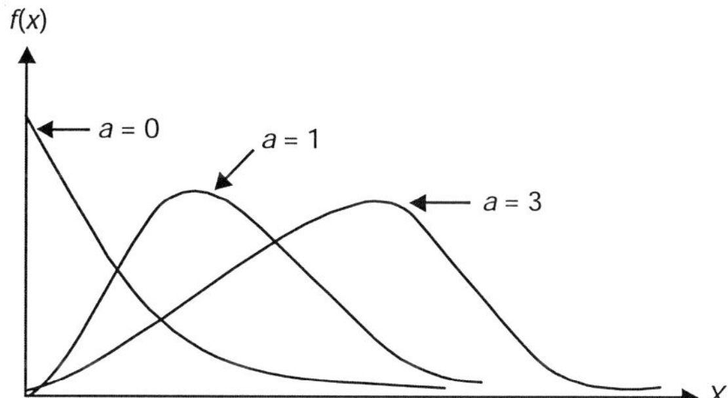

Figure A.5 Gamma distribution.

A.8 OTHER DISTRIBUTIONS

Following are distributions commonly used in hydrology for flood frequency analysis:

1. Gumbel's extreme value distribution
2. Log-Pearson Type III distribution
3. Ven Te Chow method
4. Stochastic method
5. Powell method
6. Fuller's formula

These methods or distributions are already discussed in detail in Chapter 9 with numerical examples of flood frequency analysis.

A.9 CORRELATION AND REGRESSION ANALYSIS

In practice, almost in all the fields of study, there may be some problems involving the use of two or more variables. In hydrology, we can expect some dependence between rainfall and corresponding runoff over a basin, between evaporation and corresponding wind velocity, etc. The degree of dependence between variables under consideration is measured by correlation analysis. The measure of correlation or dependence is called **correlation coefficient** r. The value of correlation coefficient lies between –1 to +1. When the value is very small or tends to zero, the variables are independent, i.e., one does not depend on the other at all.

The correlation coefficient r of the variables x and y is given by

$$r = \frac{\frac{1}{n}\sum (x_i - \bar{x})(y_i - \bar{y})}{\sigma_x \sigma_y} \tag{A.22}$$

If x is plotted against y, extend of correlation between the variables can be qualitatively ascertained. Figures A.6(a) and A.6(b) have been qualitatively ascertained.

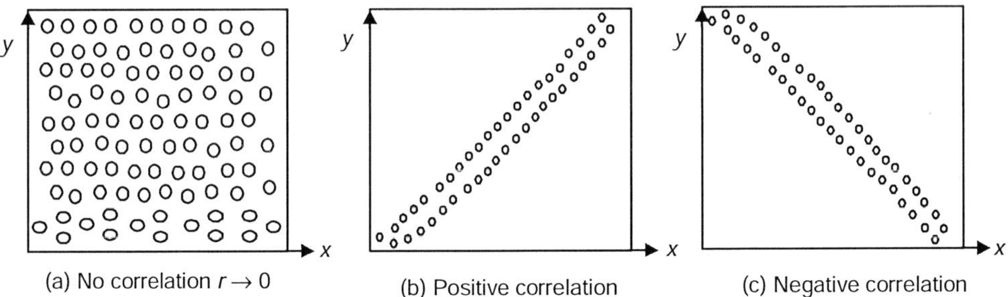

Figure A.6 Scatter diagram of two variables showing the qualitative assessment of r.

In practice, if r lies between –0.5 to + 0.5, it is said to be poor correlation.

A.10 REGRESSION ANALYSIS

When two variables are found to be correlated having good correlation coefficient (like 0.6, 0.7, 0.8, 0.9 or –0.6, –0.7, –0.8, –0.9, etc.) one may be interested in estimating or predicting the values of one when other is given. For example, if rainfall is given, corresponding runoff on catchment may be computed or predicted. Such analysis between variables is called **regression analysis**. This reveals the average relationship between the variables.

A.10.1 Linear Regression

When the analysis is done between two variables, it is called **linear regression**. A regression line is fitted between the two variables. It is desired to fit a regression line for y in terms of independent variables x.

Let $(x_1, y_1), (x_2, y_2), (x_3, y_3), \ldots, (x_n, y_n)$ be n pairs of concurrent obstructions of x and y. The following linear relationship between x and y is assumed:

$$Y = a + bx \tag{A.23}$$

where a, b are constants to be evaluated from regression analysis. Number of lines can be obtained depending on the values of a, b, usually method of least squares is used to select the line that fits the data best.

The principle of least square states that best line for fitting a series of observations is one for which the sum of the squares of departures is minimum. A departure is the difference between the observed values and the obtained from the selected line. Since x is independent, the departures of y are used. The least square line $y = a + bx$ may be obtained by solving for a, b as follows:

Let the line AB is the best fitted line by least square method.

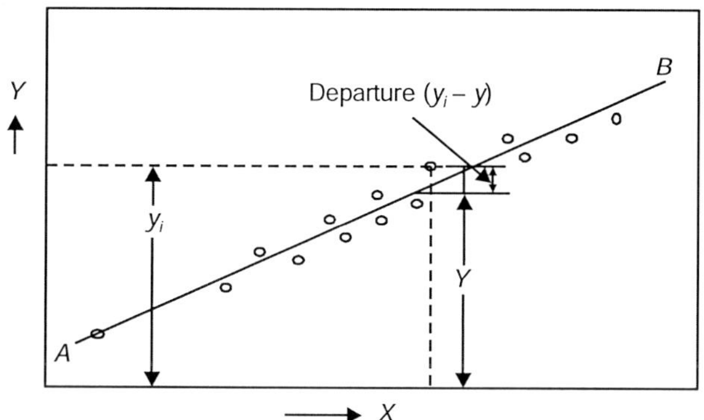

Figure A.7 Regression line AB.

Consider an observed value y_i as shown in Figure A.7. The corresponding value in the best line is y. The deviation or departure of this observed value y_i from the best fitted line is $(y_i - y)$.

But
$$y = a + bx$$

$\therefore$
$$\text{Deviation or departure} = (y_i - a - bx)$$

Obviously the best regression line is the one which gives these deviations as small as possible. The criterion that is used for best values of a and b is that the sum of the squared deviation for all the observations is minimum.

$$\text{Sum of the squared deviations } s = \Sigma(y_i - a - bx)^2$$

Now, differentiating S with respect to a, b and equating to zero for minimum, following normal equations are obtained:

$$\frac{\delta s}{\delta a} = 2\sum_{i=1}^{n} (y_i - a - bx_i) = 0 \qquad \text{(A.24)}$$

$$\frac{\delta s}{\delta b} = 2\sum_{i=1}^{n} (y_i - a - bx_i)(-x_i) = 0$$

$$\frac{\delta s}{\delta b} = 2\sum_{i=1}^{n} (-x_i y_i + ax_i - bx_i^2) = 0 \qquad \text{(A.25)}$$

From Eq. (A.24),
$$\sum_{i=1}^{n} y_i - \sum_{i=1}^{n} a - b\sum_{i=1}^{n} x_i = 0 \qquad \text{(A.26)}$$

or
$$\sum y = na + b\sum x$$

From Eq. (A.25), $$\sum_{i=1}^{n}(x_i y_i + ax_i + bx_i^2) = 0 \tag{A.27}$$

$$\text{or } \sum xy = a\sum x + b\sum x^2$$

From Eq. (A.26), $$a = \frac{\sum y}{n} - \frac{b\sum x}{n} \tag{A.28}$$

$$\therefore \quad a = \bar{y} - b\bar{x}$$

Substitute this value of a given by Eq. (A.28) in original Eq. (A.23), i.e.,

$$y = a + bx$$

$$y = (\bar{y} - b\bar{x}) + bx \tag{A.29}$$

$$\text{or} \quad (y - \bar{y}) = b(x - \bar{x})$$

Equation (A.29) shows that the regression line passes through the point $(\bar{x}, \bar{y})$. Solution for b from Eqs. (A.23) and (A.25) or from (A.26) and (A.27) gives

$$b = r\frac{\sigma_y}{\sigma_x} \tag{A.30}$$

where r is the correlation coefficient given by Eq. (A.22) and σ_y and σ_x are standard deviations of x and y given by Eq. (A.4).

$$b = \frac{\frac{1}{n}\sum_{i=1}^{n}(x_i - \bar{x})(y_i - \bar{y})}{\sigma_x . \sigma_y} \cdot \frac{\sigma_y}{\sigma_x}$$

$$\therefore \quad b = \frac{\frac{1}{n}\sum_{i=1}^{n}(x_i - \bar{x})(y_i - \bar{y})}{\sigma_x{}^2} \tag{A.31}$$

After calculating b, a is obtained from Eq. (A.28).

A.10.2 Multiple Regressions

The least square regression can be extended to the problems where there are more than one independent variables. For example, runoff y may be dependent on rainfall x_1, infiltration x_2, evaporation x_3, transpiration x_4, etc. We may then assume the relationship as:

$$y = a_0 + a_1x_1 + a_2x_2 + a_3x_3 + \cdots + a_m x_m \tag{A.32}$$

Similarly, like the previous case of linear regression sum of the squared deviation is given by

$$s = \sum_{i=1}^{n}(y_i - a_0 - a_1x_1 - a_2x_2 - a_3x_3 - \cdots - a_mx_m)^2 \tag{A.33}$$

Equating the partial derivatives of this sum (when differentiated to regression constants to zero), following simultaneous equations are obtained:

$$\left.\begin{aligned} \Sigma y_0 &= \Sigma a_0 + a_1\Sigma x_1 + a_2\Sigma x_2 + \cdots + a_m\Sigma x_m \\ \Sigma y x_{1i} &= a_0\Sigma x_{1i} + a_1\Sigma x_{1i}^2 + a_2\Sigma x_{2i}X_{1i} + \cdots + a_m\Sigma x_{mi}x_{1i} \\ \Sigma y x_{mi} &= a_0\Sigma x_{mi} + a_1\Sigma x_{1i}x_{mi} + a_2\Sigma x_{2i}x_{mi} + \cdots + a_m\Sigma x_{mi}^2 \end{aligned}\right\} \quad \text{(A.34)}$$

Solving m simultaneous equation given by Eq. (A.34), values of a_0, a_1, a_2, ..., a_m are obtained. We may take the case of three variables like groundwater table (GWT). y depends on three variables. Then multiple regression line with three variables may be written as:

$$y = a + bx_1 + cx_2 \quad \text{(A.35)}$$

If there are n number of data, summation s of squared deviations is:

$$S = \sum_{i=1}^{n} (y - a - bx_1 - cx_2)^2 \quad \text{(A.36)}$$

Differentiating s with respect to a, b and c, and equating to zero for minimum, the following three normal equations are obtained:

$$\sum_1^n y = an + b\Sigma x_1 + c\Sigma x_2 \quad \text{(A.37)}$$

$$\sum_1^n yx_1 = a\Sigma x_1 + b\Sigma x_1^2 + c\Sigma x_1 x_2 \quad \text{(A.38)}$$

$$\sum_1^n yx_2 = a\Sigma x_2 + b\Sigma x_1 x_2 + c\Sigma x_1^2 \quad \text{(A.39)}$$

Now, from the n set of data Σy, Σx_1, Σx_2, Σyx_1, Σyx_2, $\Sigma x_1 x_2$, Σx_1^2, Σx_2^2 may be computed. Substituting these values in Eqs. (A.37), (A.38) and (A.39), a, b and c can be solved and multiple regression equation $y = a + bx_1 + cx_2$ is obtained. Multiple correlation coefficient of GWT with respect to precipitation x_1 and pumping x_2 is given by

$$r_{y.x_1 y_2} = \sqrt{\frac{r_{yx1}^2 + r^2 yx_2 + 2ryx_1 \cdot r^2 yx_{1}.x_2}{1 - r_{x1x2}}} \quad \text{(A.40)}$$

where r_{yx1} = Linear correlation coefficient between y and x_1

r_{yx2} = Linear correlation coefficient between y and x_2

r_{y1x2} = Linear correlation coefficient between y_1 and x_2

Such type of statistical problem involves lot of computation. It is always better to compute by writing a computer program in any of the programming languages to save the time and possibility of error.

ILLUSTRATION:

Data of rainfall x and flood y in Table A.2 are given in Columns 1 and 2 of Table A.3. From these data, required calculations to compute r and linear regression equation's constants a and b are given in subsequent columns of the same table.

Table A.3

1	2	3	4	5	6	7	8	9
Rainfall x (mm)	Flood y m^3/sec	x^2	xy	$x-\bar{x}$	$(x-\bar{x})^2$	$y-\bar{y}$	$(y-\bar{y})^2$	$(x-\bar{x})(y-\bar{y})$
2	13.8	4	27.6	–4.36	19.0	28.38	805.4	123.73
4	28.2	16	112.8	–2.36	5.57	–13.98	195.44	32.99
5.5	38.2	30.25	210.1	0.86	0.74	3.98	15.84	3.42
7	48.6	49	340.2	0.64	0.41	6.42	41.21	4.1
9	62.5	81	571.5	2.64	6.97	21.32	454.54	56.28
11.5	79.8	132.25	917.7	5.14	26.42	37.62	1415.26	193.36
10	69.4	100	694	3.64	13.25	27.22	740.93	99.08
8	55.5	64	444	1.64	2.69	13.32	177.42	21.84
6.5	45.0	42.25	292.5	0.14	0.02	2.82	7.95	0.4
4.5	32.2	20.25	144.9	1.86	3.46	9.98	99.6	18.50
2	13.8	4	27.6	4.36	19.0	28.38	805.4	123.73
70	486	543.25	3782.9	–	97.53	–	4447.89	677.49

$\bar{x} = \dfrac{\sum x}{\eta} = \dfrac{70}{11} = 6.36$

(already calculated)

$\bar{y} = \dfrac{\sum y}{n} = \dfrac{486}{11} = 42.18$

(3) $\Sigma x^2 = 543.25$

(4) $\Sigma xy = 3782.9$

(6) $\Sigma(x-\bar{x})^2 = 97.53$

(8) $\Sigma(y-\bar{y})^2 = 4747.89$

(9) $\Sigma(x-\bar{x})(y-\bar{y}) = 677.49$

Now,

$$\sigma_x = \sqrt{\frac{\Sigma(x_1 - \bar{x})}{n-1}} = \sqrt{\frac{97.49}{11-1}} = 3.1229 \text{ m}^3/\text{sec}$$

Similarly,

$$\sigma_y = \sqrt{\frac{\Sigma(y-\bar{y})^2}{n-1}} = \sqrt{\frac{4747.89}{11-1}} = 21.789 \text{ m}^3/\text{sec}$$

and

$$r = \frac{\frac{1}{n}\sum (x_i - \bar{x})(y - \bar{y})}{\sigma_x \sigma_y}$$

or

$$r = \frac{\frac{1}{11}(677.49)}{3.1229 \times 21.789} = 0.905$$

It is correlation coefficient of rainfall and flood.

If regression equation is:

$$y = a + bx$$

Then
$$b = r\frac{\sigma_y}{\sigma_x}$$

$$= 0.905 \times \frac{21.789}{3.1229} = 6.341$$

and
$$a = y - bx$$

or
$$a = 42.18 - 6.314 \times 6.36 = 2.022 \approx 2$$

Thus, regression equation is

$$y = 2 + 6.314x.$$

Index